Theory and Decision Library C

Game Theory, Social Choice, Decision Theory, and Optimization

Volume 51

The series covers formal developments in game theory, social choice, and decision theory, paying particular attention to the role of optimization in these disciplines. Applications to economics, politics, and other social sciences are also welcome.

All titles in this series are peer-reviewed.

For further information on the series and to submit a proposal for consideration, please contact Johannes Glaeser (Senior Editor Economics and Political Science) Johannes.glaeser@springer.com.

Elena Parilina · Puduru Viswanadha Reddy ·
Georges Zaccour

Theory and Applications of Dynamic Games

A Course on Noncooperative and Cooperative Games Played over Event Trees

Springer

Elena Parilina [iD]
Department of Mathematical Game Theory
and Statistical Decisions
Saint Petersburg State University
Saint Petersburg, Russia

Puduru Viswanadha Reddy [iD]
Department of Electrical Engineering
Indian Institute of Technology Madras
Chennai, India

Georges Zaccour [iD]
Department of Decision Sciences
HEC Montréal
Montreal, QC, Canada

ISSN 0924-6126   ISSN 2194-3044 (electronic)
Theory and Decision Library C
ISBN 978-3-031-16457-6   ISBN 978-3-031-16455-2 (eBook)
https://doi.org/10.1007/978-3-031-16455-2

This Springer imprint is published by the registered company Springer Nature Switzerland AG
The registered company address is: Gewerbestrasse 11, 6330 Cham, Switzerland

Preface

Many problems in economics, engineering, and management science share the following three features:

Strategic interaction: Few agents (individuals, firms, countries) compete or cooperate among themselves, and their payoffs are interdependent, that is, the action of any agent affects the outcomes of all.

Repeated interactions: The agents interact repeatedly over time, and the problem involves accumulation processes, e.g., production capacity, brand reputation, and fish stock. One implication of this is that any actions taken today are relevant to current and future payoffs.

Stochastic environment: Some of the parameter values are uncertain.

Dynamic games theory provides paradigms to analyze repeated strategic interactions in both discrete time (multistage games) and continuous time (differential games). The uncertainty in the parameter values, which is fully expected in any intertemporal problem, can be modeled by various stochastic processes, with each having its pros and cons. For multistage games, we believe that an event tree is a natural approach to representing this uncertainty for two main reasons. First, it mimics how decision makers deal with uncertain parameter values. Indeed, they most likely consider only a few anchor points, e.g., that the next period's inflation rate could be low, medium, or high. Similarly, if the price of oil influences certain choices to be made, e.g., to invest or not in a thermal power plant, the decision maker will consider n sufficiently distinct prices, and rarely a continuum of values. Second, event trees lead to computable and implementable models, and are easily prone to sensitivity analyses, which helps in assessing the robustness of the results with respect to the parameter values.

This book deals with multistage games with uncertain parameters where the stochastic process is described by an event tree. We refer to this class of games as *dynamic games played over event trees* (DGPETs).

An example: For a brief sample of a DGPET model, consider a region served by a few electricity producers (players) who compete or cooperate over time in one or more market segments, e.g., residential, commercial, and industrial markets. In a long-term model, each firm must plan its energy production schedule and investments in different production technologies (solar, thermal, hydro, etc.), while facing some uncertainties in demand or supply, or both. For instance, future demands depend on the not perfectly predictable state of the economy, and the cost of operating a thermal power plant varies greatly with the uncertain price of fuel. In the parlance of dynamic games, the quantities committed by a firm to each market segment and the investments in the different technologies at each period are the player's control variables. The installed mix of production capacities and (if relevant) the capacity of interconnections to export markets are the state variables.

The uncertainties are represented by an event tree. Starting from a root node, the next period's demand can be, e.g., low, medium, or high, depending on the rate of economic growth. Here, the root node has three possible successors, with each having a probability of realization. Each of these successors also has a number of infants (not necessarily three), and so on, until the end of the planning horizon. If the players compete (cooperate), then the equilibrium (optimal) quantities and investments at each node must be determined. In a DGPET, the decisions are indexed over the nodes of the event tree, meaning that they depend on the history of the stochastic process. (In an event tree, there is a unique path between the root node and any given node.) The event tree is given, that is, it is independent of the players' actions. In our example, a certain organization (consulting firm, research center) provides the forecast, in the form of an event tree, of the evolution of the rate of growth over the planning horizon. Each path connecting the root node to a terminal node represents a scenario describing one possible realization of this rate over time.

Contents of the Book

The textbook is self-sufficient and does not require any prior knowledge of game theory. Chapters 1 and 2 introduce some of the main concepts of noncooperative and cooperative games. Chapter 3 is focused on multistage games in a deterministic setup. These chapters do not pretend to provide a full coverage of these classes of games but they give the reader all that is needed to deal with DGPETs. Chapter 4 discusses the problem of the sustainability of cooperation in deterministic dynamic games in discrete time, which sets the stage for handling this type of problem in a DGPET setup.

Chapter 5 deals with the theory of noncooperative DGPETs, while Chaps. 6 and 7 take an in-depth look at cooperative DGPETs. Chapter 8 deals with cooperative equilibria in DGPETs.

Each chapter includes a series of exercises with varying levels of difficulty, and a list of additional readings for readers interested in going beyond the presented material.

Saint Petersburg, Russia   Elena Parilina
Chennai, India   Puduru Viswanadha Reddy
Montreal, Canada   Georges Zaccour

Acknowledgements

This book uses some results from joint research carried out with many different co-authors. A special thank to Alain Haurie with whom the third author initiated the theory and applications of DGPETs. We would like to thank our colleagues and friends Michèle Breton, Ekaterina Gromova, Steffen Jørgensen, Elnaz Kanani Kuchesfehani, Jacek B. Krawczyk, Guiomar Martin-Herrán, Vladimir Mazalov, Leon Petrosyan, Pierre-Olivier Pineau, Hasina Rasata, Artem Sedakov, and Sihem Taboubi.

We also would like to thank Josée Lafrenière who patiently proofread all the chapters.

Elena Parilina is grateful for the support received from Saint Petersburg State University.

Puduru Viswanadha Reddy is grateful for the support received from the Department of Electrical Engineering, IIT Madras. He also acknowledges the generous financial support from the Center for Continuing Education, IIT Madras, through the grant CCE/BWS/03/PVR/EE/19-20.

Georges Zaccour acknowledges the generous financial support of HEC Montréal.

Contents

Part I
Static Games

Chapter 1
Noncooperative Games

In this chapter, we review the basic concepts of noncooperative game theory. In Sect. 1.1, we deal with finite games in strategic form and, in Sect. 1.2, with continuous games in strategic form. In Sect. 1.3, games in extensive form are introduced, and the Stackelberg equilibrium is defined in Sect. 1.4. The chapter ends with a series of exercises.

1.1 Finite Games in Strategic Form

By definition, a noncooperative game in normal or strategic form Γ involves the following elements:

1. A set of players $M = \{1, \ldots, m\}$;
2. A finite set of pure strategies S_i for each player $i \in M$, where a strategy profile is a vector $s = (s_1, s_2, \ldots, s_m) \in S_1 \times S_2 \times \cdots \times S_m = S$;
3. A payoff function $u_i : S_1 \times S_2 \times \cdots \times S_m \to \mathbb{R}$ for each player $i \in M$.

We shall refer to a game in strategic form (with finite or continuous strategy sets) by $\Gamma = (M, \{S_i\}_{i \in M}, \{u_i\}_{i \in M})$. Whereas in an optimization problem, the outcome of a decision maker depends only on her actions, in a game, the payoff of one player depends on the strategies selected by all players. We illustrate the above components with the famous *prisoner's dilemma* game.

Example 1.1 (*prisoner's dilemma*) Two criminals are arrested and imprisoned in separate cells, with no possibility of communicating with each other. The prosecutor lacks sufficient evidence to convict them on a principal charge, but she has enough to convict both on a lesser charge. She simultaneously offers each prisoner the following bargain: "If you confess to the principal charge and the other suspect denies, you walk free and the other gets 15 years. If both of you deny, then each gets one year

E. Parilina et al., *Theory and Applications of Dynamic Games*, Theory and Decision Library C 51, https://doi.org/10.1007/978-3-031-16455-2_1

in prison for the minor offense. If both of you confess to the principal charge, then each gets eight years."

In this example, the players are the two prisoners, i.e., $M = \{1, 2\}$. The set of strategies is the same for both players, that is, $S_1 = S_2 = \{Confess, Deny\}$. Each strategy profile assigns an outcome u_i to Player $i = 1, 2$. The data for this game are summarized in the table below, where the first (resp., second) entry in each cell is the gain of Player 1 (resp., Player 2):

Prisoner 2

Confess   Deny

Confess $(-8, -8)$ $(0, -15)$

Prisoner 1

Deny $(-15, 0)$ $(-1, -1)$

Mixed strategies: A *mixed strategy* of Player i is a probability distribution over the set of pure strategies, i.e., $\xi_i = (\xi_i^1, \ldots, \xi_i^{|S_i|})$, $\xi_i \geq 0$, $\sum_{j=1}^{|S_i|} \xi_i^j = 1$. The set of mixed strategies is denoted by $\Delta(S_i)$.

Definition 1.1 A mixed extension of the noncooperative game Γ in strategic form is defined by $G = (M, \{\Delta(S_i)\}_{i \in M}, \{\hat{u}_i\}_{i \in M})$, where $\Delta(S_i)$ is the set of mixed strategies of Player i, and $\hat{u}_i$ is the payoff function of Player i from $\Delta(S_1) \times \cdots \times \Delta(S_m)$ to $\mathbb{R}$, which associates to any mixed-strategy profile $(\xi_1, \ldots, \xi_m)$ the payoff

$$\hat{u}_i(\xi_1, \ldots, \xi_m) = \sum_{(s_1, \ldots, s_m) \in S} u_i(s_1, \ldots, s_m) \xi_1^{s_1} \times \cdots \times \xi_m^{s_m},$$

where ξ_i^j is the probability of Player i choosing strategy j.

Zero-sum game: A two-player game in strategic form is called a *zero-sum game* if, in any strategy profile, the sum of the players' payoffs is equal to zero, that is, if one player's gain is equal to the other's loss. In such a case, the players' payoffs can be represented by a matrix with one entry in each cell giving one of the players' payoffs. For this reason, the game is called a matrix game. The solution concept for this class of games is a *saddle point*.[1]

Bimatrix game: A two-player non-zero-sum finite game in strategic form is called a *bimatrix game* because the payoff functions of the players are defined by two matrices. In the prisoner's dilemma example, these two matrices were combined into one table, with each cell having two entries. We could have written them as

$$A_1 = \begin{pmatrix} -8 & 0 \\ -15 & -1 \end{pmatrix}, \quad A_2 = \begin{pmatrix} -8 & -15 \\ 0 & -1 \end{pmatrix},$$

---

[1] At a saddle point, a function of several variables, e.g., a player's payoff function, has partial derivatives equal to zero, but at that point the function has neither a maximum nor a minimum.

where A_i is the payoff matrix of Player $i = 1, 2$. This idea can be extended to m-person finite strategic-form games.

1.1.1 Domination and Nash Equilibrium

In an optimization problem, the decision maker seeks a maximizer (or minimizer) to an objective function, subject to some constraints. In a game, each player also optimizes her payoff, but the result depends on the strategies chosen by the other players. This implies that in a game setting several outcomes are possible depending upon the strategic interaction of the players.

Definition 1.2 A strategy s_i' dominates the strategy s_i'' for Player i if

$$u_i(s_i', s_{-i}) \geq u_i(s_i'', s_{-i}), \quad \forall s_{-i} \in S_{-i},$$

where $S_{-i} = \prod_{j \in M, j \neq i} S_j$.

Definition 1.3 A strategy s_i' is a dominant strategy for Player i if, for all $\hat{s}_i \in S_i$ and $\hat{s}_i \neq s_i'$, it holds that

$$u_i(s_i', s_{-i}) \geq u_i(\hat{s}_i, s_{-i}), \quad \forall s_{-i} \in S_{-i}.$$

In the prisoner's dilemma example, the strategy *Confess* is a dominant strategy for both players.

Definition 1.4 Strategy profile $s \in S$ Pareto dominates a strategy profile $s' \in S$ if

1. no players receive a worse payoff with s than with s', that is, $u_i(s) \geq u_i(s')$ for all $i \in M$;
2. at least one player receives a better payoff with s than with s', that is, $u_i(s) > u_i(s')$ for at least one player $i \in M$.

Definition 1.5 Strategy profile $s \in S$ is Pareto optimal, or strictly Pareto efficient, if there is no strategy $s' \in S$ that Pareto dominates s.

In the prisoner's dilemma example, the strategy profile (*Deny, Deny*) Pareto dominates every other strategy profile. In fact, the strategy profile (*Deny, Deny*) is Pareto optimal.

The main solution concept of a noncooperative game is the Nash equilibrium (Nash 1950). Let $s_{-i} = (s_1, \ldots, s_{i-1}, s_{i+1}, \ldots, s_m)$, that is, a strategy profile of all players but Player i.

Definition 1.6 The strategy profile $s^* = (s_1^*, \ldots, s_m^*)$ is a Nash equilibrium in game Γ if

$$u_i(s_i^*, s_{-i}^*) \geq u_i(s_i, s_{-i}^*), \quad \forall s_i \in S_i, \forall i \in M.$$

In other words, a Nash equilibrium has the property that no player $i \in M$ can improve her payoff by deviating unilaterally from her equilibrium strategy while all the other players stick to their equilibrium strategies. A deviation to s_i by Player i is profitable if $u_i(s_i, s^*_{-i}) > u_i(s^*_i, s^*_{-i})$. However, in a Nash equilibrium, no player can have a profitable deviation.

The Nash equilibrium can also be defined in terms of best-reply strategies.

Definition 1.7 Player i's strategy $s'_i \in S_i$ is a best reply to s_{-i} if

$$u_i(s'_i, s_{-i}) = \max_{s_i \in S_i} u_i(s_i, s_{-i}).$$

Definition 1.8 The strategy profile s^* is a Nash equilibrium if, for any player i, strategy s^*_i is the best reply to s^*_{-i}.

In the prisoner's dilemma example, the unique Nash equilibrium is the pair of strategies (*Confess, Confess*). Indeed, if Player 2 chooses *Confess*, then the best reply of Player 1 is to choose *Confess*. If Player 2 chooses *Deny*, then Player 1's best reply is to choose *Confess*. Therefore, whatever Player 2 chooses, Player 1 is better off confessing. A similar reasoning applies to Player 2. In fact, *Confess* is a dominant strategy for both players.

A dominant strategy does not always exist. But, if a Nash equilibrium exists and player i has a dominant strategy s'_i, then s'_i is necessarily part of this equilibrium. A question that naturally follows is when does a Nash equilibrium exist?

Theorem 1.1 (Nash 1950, 1951) *Any finite game in strategic form Γ has a Nash equilibrium in the mixed strategies.*

The above theorem, whose proof is based on Kakutani's fixed-point theorem (Kakutani 1941), gives a hint as to the mathematical rationale for considering the (somehow counterintuitive) concept of mixed strategies. Why would a rational player randomize over the set of pure strategies? Suppose that Player i has $|S_i|$ pure strategies. Then, the set of mixed strategies is a simplex of $|S_i|$ dimensions, which is a compact and convex set.[2] To compute her best reply, Player i must solve an optimization problem in which the other players' strategies are fixed at s^*_{-i}. The question of the existence of an equilibrium then becomes this: when does a best response for Player i exist? The answer comes from a direct application of the Weierstrass theorem on the existence of a maximizer.

Theorem 1.2 (Weierstrass) *If f is a real-valued continuous function on a nonempty compact domain S, then there exists an $x \in S$ such that $f(x) \geq f(y)$ for all y in S.*

Example 1.2 (*matching pennies*) Consider a two-player game having the same set of strategies $S = \{H, T\}$. The payoffs are given in the following table:

---

[2] For the computation of mixed-strategy saddle points in matrix games and mixed-strategy Nash equilibria in bimatrix games, we refer the reader to Haurie et al. (2012).

Player 2

H T

$$H \ (1, -1) \ (-1, 1)$$

Player 1

$$T \ (-1, 1) \ (1, -1)$$

This zero-sum game does not have an equilibrium in pure strategies, but it has one in mixed strategies, given by

$$\left(\xi_1^*, \xi_2^*\right) = \left((1/2, 1/2), (1/2, 1/2)\right),$$

that is, each player assigns a probability of one half to each pure strategy.

Example 1.3 (*hawk-dove game*) Two players each have two pure strategies: "F" (Fight) and "Y" (Yield). Strategy F gives payoff $v > 0$ if the other player uses strategy Y and gets zero payoff. Either player has the same probability of winning the fight, i.e., in the event of a fight, either player gets the payoff of $(v - c)/2$, where $c > 0$ is the cost of losing a fight. When both players yield, each gets a payoff of $v/2$. We assume that the cost of the fight exceeds the gain of a victory, i.e., $v < c$. Therefore, the resulting bimatrix game is defined as follows:

Player 2

F Y

$$F \ \left(\tfrac{v-c}{2}, \tfrac{v-c}{2}\right) \ (v, 0)$$

Player 1

$$Y \ \ (0, v) \ \ \ \left(\tfrac{v}{2}, \tfrac{v}{2}\right)$$

In this game, there are two equilibria in pure strategies: (F, Y) and (Y, F). The strategy profile (F, F) (as well as the strategy profile (Y, Y)) is not a Nash equilibrium because there is a profitable deviation for Player 1 from F to Y (from Y to F). In hawk-dove game, there additionally exists a Nash equilibrium in mixed strategies, that is, $\left(\xi_1^*, \xi_2^*\right)$, where $\xi_1^* = \xi_2^* = (v/c, 1 - v/c)$, with equal negative payoffs $\left(-v(v - c)^2/(2c^2), -v(v - c)^2/(2c^2)\right)$.

Example 1.4 (*coordination game*) Two players have the same set of strategies $\{s_1, s_2\}$. If both players choose strategy s_1, each player will receive a payoff 2. If both players choose strategy s_2, each player will receive a payoff 1. In the event that they choose different strategies, they both get zero payoff. The following table represents the data for this bimatrix game:

Player 2

s_1 s_2

$$s_1 \ (2, 2) \ (0, 0)$$

Player 1

$$s_2 \ (0, 0) \ (1, 1)$$

It is easy to verify that this game has two Nash equilibria in pure strategies, i.e., (s_1, s_1) and (s_2, s_2), giving payoffs $(2, 2)$ and $(1, 1)$ to the players, respectively. There also exists a Nash equilibrium in mixed strategies, $\left(\xi_1^*, \xi_2^*\right)$, where $\xi_1^* = \xi_2^* = (1/3, 2/3)$, which gives $2/3$ to each player.

Examples 1.1–1.4 allow the following observations:

1. A game may have a unique equilibrium (prisoner's dilemma), no equilibrium in pure strategies (matching pennies), or multiple equilibria (in hawk-dove game and coordination game).
2. A Nash equilibrium is generally not Pareto optimal. In the prisoner's dilemma example, both players would be better off denying the principal charge, but there is no way to reach this outcome as an equilibrium of a noncooperative game. In this sense, a Nash equilibrium is typically not collectively optimal.
3. In the hawk-dove game, we have three Nash equilibria: two in pure strategies and one in mixed strategies, which is Pareto dominated by the two others. In the coordination game, there is one equilibrium (s_1, s_1), which dominates the two others; see Definition 1.2. The existence of multiple equilibria raises the fundamental question of which one will actually be played? The short answer is that the model of noncooperative game in strategic form does not include any mechanism for predicting which equilibrium will ultimately be selected by the players.

1.2 Continuous Games in Strategic Form

Up to now, we considered a setup in which each player has a finite number of pure strategies. In this section, we assume that the strategy set is continuous, that is, it contains an infinite number of strategies.

A *continuous game* Γ is defined by

1. The set of players $M = \{1, \ldots, m\}$;
2. A set of pure strategies S_i of Player $i \in M$, where S_i is a compact and convex subset of $\mathbb{R}^n$, and the set of strategy profiles is the Cartesian product of the sets of strategies $S = \prod_{i \in M} S_i$;
3. A utility function $u_i : S \to \mathbb{R}$ of Player $i \in M$, which is continuous.

Typical examples of a continuous game are the Cournot duopoly, where Player i chooses her output q_i from a feasible set $Q_i = [0, \infty)$, and auctions, in which a participant can bid any amount to acquire an auctioned item. In these examples, there is no conceptual reason preventing the selection of any real number. However, one might argue that prices and quantities are not infinitely divisible in practice, that is, that "reality" is discrete, and a continuum of actions is a mathematical abstraction. One reason for considering continuous strategy sets is that they are easier mathematical objects to work with than finite sets with large numbers of actions. Dasgupta and Maskin (1986) suggest discretizing the continuous strategy sets if the game does not have an equilibrium. Naturally, the (equilibrium) result of the discretized game is sensitive to the chosen grid.

Example 1.5 (*Cournot competition*) Two firms compete in quantity in a market. They noncooperatively and simultaneously choose their output levels q_i from feasible sets $Q_i = [0, \infty)$, $i = 1, 2$. The firms sell their products at the market price $p(q)$, where $q = q_1 + q_2$. Let firm i's production cost be given by $c_i(q_i)$ and its payoff (profit) by

$$u_i(q_1, q_2) = q_i p(q) - c_i(q_i).$$

The feasible sets Q_i and the payoff functions π_i define a strategic-form game with continuous strategy sets. The best-reply strategy of firm i is a function $r_i : Q_j \to Q_i$, $i, j = 1, 2, i \neq j$, which gives firm i's output as a function of its rival's output.

If functions u_i, $i = 1, 2$, are differentiable and strictly concave and the boundary conditions are satisfied, we can find the best-reply strategies using the first-order conditions:

$$p(q_1 + r_2(q_1)) + \frac{dp(q_1 + r_2(q_1))}{dq_1} r_2(q_1) - \frac{dc_2(r_2(q_1))}{dq_1} = 0, \qquad (1.1)$$

$$p(q_2 + r_1(q_2)) + \frac{dp(q_2 + r_1(q_2))}{dq_2} r_1(q_2) - \frac{dc_1(r_1(q_2))}{dq_2} = 0. \qquad (1.2)$$

An intersection (if it exists) of the two best-reply functions r_1 and r_2 is a Nash equilibrium of the Cournot game, that is, a case where neither firm can increase its profit by unilaterally changing its output, given the rival's output.

Suppose that the inverse-demand function is linear and given by $p(q) = \max\{0, 1 - q\}$ and the cost function is also linear, i.e., $c_i(q_i) = cq_i$, $i = 1, 2$, $c \in [0, 1]$. Then, from (1.1) and (1.2), we obtain the following best-reply strategies:

$$r_1(q_2) = \frac{1 - q_2 - c}{2}, \qquad (1.3)$$

$$r_2(q_1) = \frac{1 - q_1 - c}{2}. \qquad (1.4)$$

The Nash equilibrium (q_1^*, q_2^*) is unique and satisfies the conditions $q_1^* = r_1(q_2^*)$ and $q_2^* = r_2(q_1^*)$, from which it follows that $q_1^* = q_2^* = (1 - c)/3$.

Remark 1.1 The best-reply functions characterize the type of strategic interaction between the players:

- If $r_i(q_j)$ is increasing in q_j, then the two strategies are complements (strategic complementarity).
- If $r_i(q_j)$ is decreasing in q_j, then the two strategies are substitutes (strategic substitutability).
- If $r_i(q_j)$ is independent of q_j, then there are no strategic interactions between the players' decisions.

We see clearly from (1.3)–(1.4) that the strategies are substitutes in the Cournot model.

1.2.1 Mixed Strategies

When the number of strategies is infinite, a player's mixed strategy cannot be described as the collection of probabilities of each individual strategy. Therefore, we need to refer to the probability space formalism defining the probability measure of any measurable subset of the set of strategies S_i of Player i. We define the collection of Borel subsets $\mathcal{B}_i$ of set S_i and define the set of probability measures $\Delta(S_i)$ over set S_i. The probability distribution $\sigma_i \in \Delta(S_i)$ assigns a nonnegative number $\sigma_i(D) \in [0, 1]$ for any $D \in \mathcal{B}_i$ and satisfies two properties:

1. $\sigma_i(S_i) = 1$;
2. For any countable collection of sets $\{D_k\}_{k=1}^{\infty}$ such that $D_k \in \mathcal{B}_i$ for any k, and $D_k \cap D_l = \varnothing$ for any $k \neq l$, the following equality

$$\sigma_i\left(\bigcup_{k=1}^{\infty} D_k\right) = \sum_{k=1}^{\infty} \sigma_i(D_k)$$

holds.

Player i's payoff when the players use mixed strategies is defined as follows:

$$u_i(\sigma) = \int_{s_m \in S_m} \cdots \int_{s_1 \in S_1} u_i(s) d\sigma_1(s_1) \times \cdots \times d\sigma_m(s_m),$$

for any $\sigma \in \prod_{j \in M} \Delta(S_j)$, where $\sigma = (\sigma_1, \ldots, \sigma_m)$ and $s = (s_1, \ldots, s_m)$.

Definition 1.9 A Nash equilibrium in a continuous game Γ is a profile of strategies $\tilde{\sigma} = (\tilde{\sigma}_1, \ldots, \tilde{\sigma}_m)$ such that, for any player $i \in M$ and any strategy $\sigma_i \in \Delta(S_i)$, the inequality

$$u_i(\tilde{\sigma}_i, \tilde{\sigma}_{-i}) \geq u_i(\sigma_i, \tilde{\sigma}_{-i})$$

holds.

In the next section, we provide the sufficient conditions for the existence of a Nash equilibrium in pure or mixed strategies.

1.2.2 Existence of a Nash Equilibrium

The existence proof of a Nash equilibrium rely on the existence of a fixed point of a set-valued mapping. Kakutani (1941) gives the conditions for the existence of such a fixed point.

Theorem 1.3 (Kakutani 1941) *Let* $\Phi : S \rightrightarrows S$ *be a set-valued upper semicontinuous mapping, where S is a nonempty compact (closed and bounded) subset of $\mathbb{R}^m$, and Φ is such that for any $x \in S$, $\Phi(x)$ is a nonempty convex subset of set S. Then, there exists a fixed point for mapping Φ, i.e., there exists $x^* \in S$ such that $x^* \in \Phi(x^*)$.*

In the above theorem, Kakutani (1941) uses the property of an upper semicontinuous function. A definition follows.

Definition 1.10 Let Φ be a set-valued mapping from X to Y, where X and Y are metric spaces, and $\Phi(X) \subset Y$. The mapping Φ is upper semicontinuous if, for any sequence, $\{x_k\}_{k=1,2,\ldots}$ converges in X toward x_0 and any limit y_0 of the sequence $\{y_k\}_{k=1,2,\ldots}$ in Y, $y_k \in \Phi(x_k)$ for any k, is such that $y_0 \in \Phi(x_0)$.

Before stating the existence theorem of a Nash equilibrium in a continuous pure-strategy game, we recall the definition of a quasi-concave function.

Definition 1.11 A function f defined on X is quasi-concave, if

$$f(x) \geq \min\{f(x_1), f(x_2)\}, \qquad \forall x \in (x_1, x_2).$$

Theorem 1.4 (Debreu 1952; Glicksberg 1952; Fan 1952) *Consider a strategic-form game in which the players' strategy sets S_i are nonempty compact convex subsets of a Euclidean space. If the payoff functions u_i are continuous in s and quasi-concave in s_i, then there exists at least one pure-strategy Nash equilibrium.*

Proof Consider any player i. Define the best-response function of Player i, denoted by b_i, which is a set-valued mapping from $S_{-i} = \prod_{j \neq i, j \in M} S_j$ to S_i, that is,

$$b_i(s_{-i}) = \{s_i' : u_i(s_i', s_{-i}) \geq u_i(s_i, s_{-i}) \text{ for all } s_i \in S_i\},$$

where $s_{-i} \in S_{-i}$. Define for any $s \in S$, the vector $B(s) = (b_1(s_{-1}), \ldots, b_m(s_{-m}))$ of best-response functions. The vector function $B : S \rightrightarrows S$ is a set-valued mapping. To apply Kakutani's fixed-point theorem to the best-response function $B : S \rightrightarrows S$, we need the following four properties:

1. S is a nonempty compact convex set. This is the case because S is the Cartesian product of strategy sets S_i, $i \in M$, which are nonempty compact convex subsets of a Euclidean space.
2. Vector function $B(s)$ is nonempty because, by definition,

$$b_i(s_{-i}) = \arg\max_{s_i \in S_i} u_i(s_i, s_{-i}),$$

where S_i is nonempty and compact, and payoff function u_i is continuous in s by assumption. Then by Weierstrass's Theorem 1.2, $B(s)$ is nonempty.
3. $B(s)$ is a convex-valued mapping. This follows from the quasi-concavity of $u_i(s_i, s_{-i})$ in s_i. Suppose this is not true; then, there exists a player $i \in M$ and some $s_{-i} \in S_{-i}$ such that $b_i(s_{-i}) \in \arg\max_{s_i \in S_i} u_i(s_i, s_{-i})$ is not convex. This means that there exists s_i' and s_i'' belonging to S_i such that $s_i', s_i'' \in b_i(s_{-i})$, and $\lambda s_i' + (1 - \lambda)s_i'' \notin b_i(s_{-i})$, $\lambda \in [0, 1]$, which implies

$$\lambda u_i(s_i', s_{-i}) + (1 - \lambda)u_i(s_i'', s_{-i}) > u_i(\lambda s_i' + (1 - \lambda)s_i'', s_{-i}),$$

which violates the quasi-concavity of u_i in s_i, that is,

$$u_i(\lambda s_i' + (1 - \lambda)s_i'', s_{-i}) \geq \min\{u_i(s_i', s_{-i}), u_i(s_i'', s_{-i})\}.$$

Therefore, $B(s)$ is a convex set.
4. $B(s)$ is an upper semicontinuous mapping. This follows from Berge's maximum theorem (see pp. 115–117, Berge 1997). If u_i is continuous (i.e., both upper and lower semicontinuous) at s, then $b_i(s_{-i})$ is an upper semicontinuous mapping.

The result then follows from Kakutani's Theorem 1.3.   □

Remark 1.2 In Theorem 1.4, the condition that the payoff functions u_i are quasi-concave in s_i can be substituted by the stronger condition that the payoff functions u_i are concave in s_i because the quasi-concavity of a function follows from its concavity.

We notice that Theorem 1.1 (Nash's) is a special case of Theorem 1.4. Indeed, the set of mixed strategies in Theorem 1.1 is a simplex, which is a compact convex subset of a Euclidean space. The payoffs are polynomial, and therefore, quasi-concave in each player's own mixed strategy. The quasi-concavity of the payoff functions requires some strong conditions on the second derivatives, which may be hard to satisfy in some contexts. (In Example 1.5, quasi-concavity is straightforwardly satisfied when the demand and cost are linear functions.) Note that these conditions are sufficient, but not necessary, meaning that an equilibrium may still exist even if they are not satisfied.

The above theorem provides sufficient conditions for the existence of a pure-strategy Nash equilibrium in continuous games. We know from Nash (1950) that finite games always admit an equilibrium in mixed strategies. The following theorem by Glicksberg (1952) shows that a continuous game also admits an equilibrium in mixed strategies.

Theorem 1.5 (Glicksberg 1952) *Any continuous game* Γ *admits a mixed-strategy Nash equilibrium.*

We note that the payoff functions need not be quasi-concave for the existence of an equilibrium in mixed strategies.

1.2.3 Concave Games

In this subsection we study an important class of continuous games that admit an equilibrium in pure strategies.

1.2.3.1 Problem Formulation

In optimization theory, it is well known that the strict concavity of the objective function to be maximized and the compactness and convexity of the constraint set make it possible to prove the existence and uniqueness of a solution. The generalization of this result to games is due to Rosen (1965).

Formally, a *concave m*-person game is defined by the following:

1. A set of players $M = \{1, \ldots, m\}$;
2. A set of strategies S_i of Player $i \in M$, which is a compact and convex subset of $\mathbb{R}^{l_i}$;
3. A payoff function $u_i : S_1 \times \cdots \times S_m \to \mathbb{R}$ of Player $i \in M$, which is continuous in any s_j and concave in s_i;
4. A *coupled-constraint* set, which is a proper subset S of the set of strategy profiles $S_1 \times \cdots \times S_m$. Any strategy profile chosen by the players must belong to S, which is assumed to be a convex set.

Definition 1.12 A *coupled-constraint equilibrium* is an m-tuple $\tilde{s} = (\tilde{s}_1, \ldots, \tilde{s}_m) \in S$, such that, for each player $i \in M$, the inequality

$$u_i(\tilde{s}_i, \tilde{s}_{-i}) \geq u_i(s_i, \tilde{s}_{-i}) \tag{1.5}$$

must hold for any $s_i \in S_i$ and $(s_i, \tilde{s}_{-i}) \in S$.

Remark 1.3 In some works (see, e.g., Facchinei et al. 2007; Harker 1991), a coupled-constraint equilibrium is called a generalized Nash equilibrium (GNE). We refer to Facchinei et al. (2007) for a comprehensive survey of the GNE and numerical solutions to this type of equilibrium.

Example 1.6 Consider an oligopolistic industry producing a homogeneous product. Each firm maximizes its profit, given by the difference between its revenue and its production and environmental damage costs. As there is a monotone relationship between emissions and production, Player i's revenues can be expressed as a function of emissions $e_i \geq 0$, $i \in M$. Denote by $f_i(e_i)$ the net revenue function, that is, the gross revenue minus production cost, and let $e = \sum_{j \in M} e_j$.

Players suffer from pollution, and each player's damage cost depends on all players' emissions and is denoted by $d_i(e)$, $i \in M$. Player i's payoff function is given by

$$u_i(e_1, \ldots, e_m) = f_i(e_i) - d_i(e).$$

A regulator imposes the following constraint on the industry:

$$e \leq E,$$

that is, the total emissions cannot exceed a given upper bound E.

The set of strategy profiles is $\mathbb{R}_+^m$, and the coupled-constraint set is defined by

$$\left\{ (e_1, \ldots, e_m) \in \mathbb{R}_+^m \text{ such that } e \leq E \right\}.$$

Clearly, the strategy sets of the players are convex and compact. Further, if $f_i(e_i)$ is a concave increasing function and $d_i(e)$ is a convex increasing function, then we have a coupled-constraint concave m-person game.

1.2.3.2 Existence of a Coupled-Constraint Equilibrium

In a coupled-constraint equilibrium, no player can increase her payoff by individual deviation while keeping the profile of strategies in set $\mathcal{S}$. In this section, we recall the results regarding the existence of a coupled-constraint equilibrium in any concave game.

Rosen (1965) formulates the equilibrium conditions of Definition 1.12 using point-to-set mapping $\theta : \mathcal{S} \times \mathcal{S} \times \mathbb{R}_+^m \to \mathbb{R}$ defined as

$$\theta(s, t, r) = \sum_{i=1}^m r_i u_i(s_1, \ldots, s_{i-1}, t_i, s_{i+1}, \ldots, s_m), \qquad (1.6)$$

where $s, t \in \mathcal{S}$ are strategy profiles satisfying the coupled constraints, and $r = (r_1, \ldots, r_m)$, $r_i > 0$, $i \in M$, is a vector of weights assigned to players' payoffs. In (1.6), the players from the set $M \setminus \{i\}$ choose their strategies according to profile $s \in \mathcal{S}$, and Player i chooses strategy t_i from profile t.

The strategy profiles s and t satisfy the coupled constraints, i.e., $s, t \in \mathcal{S}$, the profile $(s_1, \ldots, s_{i-1}, t_i, s_{i+1}, \ldots, s_m)$ is assumed to belong to a set of all strategy profiles $S_1 \times \cdots \times S_m$. The function $\theta(s, t, r)$ defined by (1.6) is continuous in s and concave in t for any fixed s.

Lemma 1.1 *Let the strategy profile $\tilde{s} \in \mathcal{S}$ be such that*

$$\theta(\tilde{s}, \tilde{s}, r) = \max_{s \in \mathcal{S}} \theta(\tilde{s}, s, r), \qquad (1.7)$$

then, $\tilde{s}$ is a coupled-constraint equilibrium.

Proof Suppose that $\tilde{s}$ satisfies (1.7) but is not a coupled-constraint equilibrium, that is, it does not satisfy (1.5). Then, there exists at least one player $j \in M$ such that the strategy profile $\hat{s} = (\tilde{s}_1, \ldots, \tilde{s}_{j-1}, s_j, \tilde{s}_{j+1}, \ldots, \tilde{s}_m) \in \mathcal{S}$ and

$$u_j(\hat{s}) > u_j(\tilde{s}).$$

Then, $\theta(\tilde{s}, \hat{s}, r) > \theta(\tilde{s}, \tilde{s}, r)$, which contradicts (1.7). $\qquad\square$

Lemma 1.1 shows that to prove the existence of an equilibrium, we need to prove the existence of a fixed point for an appropriately defined reaction function or best-reply strategy ($\tilde{s}$ is the best reply to $\tilde{s}$). To state a coupled-constraint equilibrium existence theorem, we use Theorem 1.3 (Kakutani 1941) on the existence of a fixed point to a point-to-set upper semicontinuous mapping.

Theorem 1.6 (Rosen 1965) *A coupled-constraint equilibrium exists for every concave m-person game.*

Proof Consider the point-to-set mapping

$$\Gamma(s, r) = \left\{ t \mid \theta(s, t, r) = \max_{q \in S} \theta(s, q, r) \right\},$$

which is called a coupled-reaction mapping, defined for any vector r of positive weights. The fixed point of this mapping is a vector $\tilde{s} \in S$ such that $\tilde{s} \in \Gamma(\tilde{s}, r)$. If such a fixed point exists, then it is a coupled-constraint equilibrium by Lemma 1.1.

Because of the continuity of $\theta(s, q, r)$ and the concavity in q for fixed s, it follows that $\Gamma(s, r)$ is an upper semicontinuous mapping that maps each vector of the convex compact set S into a closed convex subset of S for any fixed vector r. Then, by Theorem 1.3, there exists a vector $\tilde{s} \in S$ such that $\tilde{s} \in \Gamma(\tilde{s}, r)$ or

$$\theta(\tilde{s}, \tilde{s}, r) = \max_{q \in S} \theta(\tilde{s}, q, r), \tag{1.8}$$

for any given vector r.

Then, by Lemma 1.1, we easily prove that $\tilde{s}$ is a coupled-constraint equilibrium in a concave game. $\qquad\square$

Theorem 1.6 does not provide a computational method for finding an equilibrium. A constructive link between mathematical programming and concave games was studied by Haurie et al. (2012).

In an optimization problem, the uniqueness of the optimum follows from the strict concavity of the objective function to be maximized. In a game setting, however, as illustrated by the following example, strict concavity of the payoff functions does not imply uniqueness of the equilibrium.

Example 1.7 Consider a two-player game, where $S_i = [0, 1]$ for $i = 1, 2$, and the payoff functions are given by $u_1(s_1, s_2) = s_1 s_2 - \frac{1}{2} s_1^2$ and $u_2(s_1, s_2) = s_1 s_2 - \frac{1}{2} s_2^2$. Note that $u_i(s_1, s_2)$ is strictly concave in s_i. It is easy to verify that the best-response correspondences (which are unique-valued) are given by $b_1(s_2) = s_2$ and $b_2(s_1) = s_1$. Plotting the best-response curves shows that any pure-strategy profile $(s_1, s_2) = (x, x)$, $x \in [0, 1]$ is a pure-strategy Nash equilibrium.

Rosen (1965) shows that a continuous game admits a unique pure-strategy Nash equilibrium under a stringent requirement on the payoff functions, also referred to as *diagonal strict concavity*. In order to discuss the uniqueness of an equilibrium, we provide a more explicit description of the strategy sets of the players. In particular, we assume that the joint strategy S is given by

$$S := \{s \in \mathbb{R}^l \mid h(s) \geq 0\}, \tag{1.9}$$

where $l = l_1 + \cdots + l_m$ and $h : \mathbb{R}^l \to \mathbb{R}^p$ is a concave function. Since h is concave, it follows that the set S is convex.[3] Denote by $U(s, r)$ the linear combination of players' payoffs, that is,

$$U(s, r) = \sum_{i \in M} r_i u_i(s),$$

and by $g(s, r)$ the *pseudo-gradient* of this function, which is a row vector

$$g(s, r) = \left(r_1 \frac{\partial u_1(s)}{\partial s_1}, \ldots, r_m \frac{\partial u_m(s)}{\partial s_m} \right).$$

Definition 1.13 The function $U(s, r)$ is diagonally strictly concave on S if the following condition

$$g(s^1, r)(s^2 - s^1) + g(s^2, r)(s^1 - s^2) > 0$$

holds for any $s^1, s^2 \in S$.

Theorem 1.7 (Rosen 1965) *Let $U(s, r)$ be diagonally strictly concave for some $r \in \mathbb{R}^m_+$ on the convex set S, and let the assumptions ensuring the existence of Karush-Kuhn-Tucker multipliers hold; then there exists a unique normalized equilibrium for the weighted scheme r.*

Proof Assume that for some $r \in \mathbb{R}^m_+$ there exist two equilibria, s^1 and s^2. Then, we have for $q = 1, 2$, $k = 1, \ldots, p$:

$$h_k(s^q) \geq 0, \tag{1.10}$$

and there exist multipliers $\lambda^1 \geq 0$ and $\lambda^2 \geq 0$ such that

$$(\lambda^q)' h(s^q) = 0, \quad q = 1, 2,$$

for which the following equations are true for any player $i \in M$:

---

[3] For example, the set $S = [0, 1]$ can be represented as $S = \{s \in \mathbb{R} \mid h(s) \geq 0\}$, where $h(s) = s(1 - s)$. Notice, $h : \mathbb{R} \to \mathbb{R}$ is a concave function.

$$r_i \frac{\partial u_i(s^q)}{\partial s_i} + (\lambda^q)' \frac{\partial h(s^q)}{\partial s_i} = 0, \tag{1.11}$$

for any $q = 1, 2$. We multiply (1.11) with $q = 1$ by $(s^2 - s^1)'$, with $q = 2$ by $(s^1 - s^2)'$ and sum up over the set of players. We group the terms of the sum to obtain the expression $\beta + \gamma = 0$, where γ contains the terms with multipliers λ_k^1 and λ_k^2. Due to concavity of the constraint functions h_k and, nonnegativity condition (1.10), we obtain

$$\begin{aligned}
\gamma &= \sum_{i \in M} \sum_{k=1}^{p} \left\{ \lambda_k^1 \frac{\partial h_k(s^1)}{\partial s_i} (s^2 - s^1) + \lambda_k^2 \frac{\partial h_k(s^2)}{\partial s_i} (s^1 - s^2) \right\} \\
&\geq (\lambda^1)'(h(s^2) - h(s^1)) + (\lambda^2)'(h(s^1) - h(s^2)) \\
&= (\lambda^1)'h(s^2) + (\lambda^2)'h(s^1) \geq 0,
\end{aligned}$$

and

$$\beta = \sum_{i \in M} r_i \left[\frac{\partial u_i(s^1)}{\partial s_i}(s^2 - s^1) + \frac{\partial u_i(s^2)}{\partial s_i}(s^1 - s^2) \right].$$

Since $U(s, r)$ is diagonally strictly concave, we have $\beta > 0$, which contradicts $\beta + \gamma = 0$ and proves the theorem. $\qquad\Box$

Remark 1.4 A sufficient condition of $U(s, r)$ being diagonally strictly concave is that the symmetric matrix $[G(s, r) + G'(s, r)]$ be negative definite for $s \in S$, where $G(s, r)$ is the Jacobian of $g(s, r)$ with respect to s.

Example 1.8 We revisit Example 1.6 with two firms. The strategy spaces of the players are coupled due to the constraints imposed by the regulator, that is, $S := \{(e_1, e_2) \in \mathbb{R}^2 \mid e_1 \geq 0, \ e_2 \geq 0, \ e_1 + e_2 \leq E\}$. Clearly, the strategy sets of the players are convex and compact. We assume that Player i's revenue function $f_i(e_i)$ is twice differentiable and a concave increasing function. Further, we assume that Player i's damage costs vary according to $d_i(e)$, which is twice differentiable and a convex increasing function. So, the duopoly game is a concave game and, from Theorem 1.6, there exists a coupled-constraint equilibrium denoted by (e_1^R, e_2^R). The KKT conditions associated with the implicit optimization problem (1.8) are given by

$$r_1 \left(\frac{df_1}{de_1}(e_1^R) - \frac{dd_1}{de}(e^R) \right) = \lambda^R, \tag{1.12}$$

$$r_2 \left(\frac{df_2}{de_2}(e_2^R) - \frac{dd_2}{de}(e^R) \right) = \lambda^R, \tag{1.13}$$

$$r_1 \geq 0, \ r_2 \geq 0, \ \lambda^R \geq 0, \ e_1^R + e_2^R \leq E, \ \lambda^R(E - e_1^R - e_2^R) = 0. \tag{1.14}$$

Here, λ^R denotes the (common) multiplier associated with the coupled constraint $e_1 + e_2 \leq E$, with $\lambda^R > 0$ whenever $e_1^R + e_2^R = E$. As there is a common constraint, a question that arises in this context is does there exist a vector of weights (r_1, r_2) that leads to the cooperative outcome? We answer this question by showing that if one defines the r_i as proportions, then one can choose them as a means to reach the cooperative solution, at least in region $e_1 + e_2 = E$.

The cooperative outcome involves players jointly maximizing their payoffs resulting in Pareto solutions. As the players' payoffs are concave, the Pareto solutions can be obtained by solving the following weighted sum optimization problem.

$$\max_{e_1, e_2} \alpha_1 (f_1(e_1) - d_1(e)) + \alpha_2 (f_2(e_2) - d_2(e)), \ \alpha_1, \alpha_2 \in (0, 1), \ \alpha_1 + \alpha_2 = 1,$$

$$\text{subject to } e_i \geq 0, \ e_1 + e_2 \leq E. \tag{1.15}$$

The KKT conditions associated with the above maximization problem are given by

$$\alpha_1 \left(\frac{df_1}{de_1}(e_1^C) - \frac{dd_1}{de}(e^C) \right) - \alpha_2 \frac{dd_2}{de}(e^C) = \lambda^C, \tag{1.16}$$

$$\alpha_2 \left(\frac{df_2}{de_2}(e_2^C) - \frac{dd_2}{de}(e^C) \right) - \alpha_1 \frac{dd_1}{de}(e^C) = \lambda^C, \tag{1.17}$$

$$\lambda^C \geq 0, \ e_1^C + e_2^C \leq E, \ \lambda^C (E - e_1^C - e_2^C) = 0. \tag{1.18}$$

The Hessian associated with the joint payoff is given by

$$H(e_1^C, e_2^C) = \begin{bmatrix} \zeta_1 - \gamma & -\gamma \\ -\gamma & \zeta_2 - \gamma \end{bmatrix},$$

where $\zeta_1 = \alpha_1 \frac{d^2 f_1}{de_1^2}$, $\zeta_2 = \alpha_2 \frac{d^2 f_2}{de_2^2}$, and $\gamma = (\alpha_1 \frac{d^2 d_1}{de^2} + \alpha_2 \frac{d^2 d_2}{de^2})$. As the function f_i is concave increasing, and d_i is convex decreasing, we have $\zeta_1 < 0$, $\zeta_2 < 0$ and $\gamma > 0$. Then, for any $x := (x_1, x_2) \neq (0, 0)$ we have $x^T H x = -\gamma(x_1^2 + x_2^2) + \zeta_1 x_1^2 + \zeta_2 x_2^2 < 0$, implying $H(e_1, e_2)$ is negative definite. So, the weighted sum payoff is strictly concave, implying that the KKT conditions are sufficient and (e_1^C, e_2^C) is the unique cooperative solution.

We consider the region where the constraint $e_1 + e_2 = E$ is satisfied, and see if the cooperative solution can be implemented by the coupled-constraint equilibrium. Using $e_1^R = e_1^C$ and $e_2^R = e_2^C$ in (1.12)–(1.14), and from (1.16)–(1.18) we get

$$\lambda^R = \frac{r_i}{\alpha_i} \left(\lambda^C + \alpha_j \frac{dd_j}{de}(e^C) \right), \ i \neq j, \ i, j = 1, 2. \tag{1.19}$$

Now, consider the noncooperative game where players' payoff functions are given by

$$v_i(e_1, e_2) = f_i(e_i) - d_i(e) + \frac{\lambda^R}{r_i}(E - e_1 - e_2).$$

Here, the term $\frac{1}{r_i}\lambda^R(E - e_1 - e_2)$ can be interpreted as a penalty or tax imposed by the regulator on Player i for violating the constraint $e_1 + e_2 \leq E$. It can be easily verified that the payoff functions (v_1, v_2) are strictly diagonally concave so there exists a unique Nash equilibrium, denoted by (e_1^*, e_2^*), for this game, which is obtained by solving the following KKT conditions:

$$\frac{df_i}{de_i}(e_i^*) - \frac{dd_i}{de}(e^*) = \frac{\lambda^R}{r_i}, \quad i = 1, 2.$$

Using (1.19) we get

$$\frac{df_i}{de_i}(e_i^*) - \frac{dd_i}{de}(e^*) = \frac{1}{\alpha_i}\left(\lambda^C + \alpha_j \frac{dd_j}{de}(e^C)\right)$$

$$\Rightarrow \alpha_i\left(\frac{df_i}{de_i}(e_i^*) - \frac{dd_i}{de}(e^*)\right) - \alpha_j \frac{dd_j}{de}(e^C) = \lambda^C, \quad i, j = 1, 2, \ i \neq j.$$

From the uniqueness of the cooperative solution and the Nash equilibrium, we immediately have that the solution of the above equation is $e_i^* = e_i^C, i = 1, 2$.

1.3 Games in Extensive Form

1.3.1 Game Tree

A game in extensive form is defined on a *tree graph,* that is, a connected acyclic undirected graph (X, E), where X is a set of nodes with a root node $x_0 \in X$, and E is the set of edges, such that, for every node x, there is a unique path from x_0 to x. The edge in a tree graph connects two nodes: an origin node, or parent, and its descendant. The root node is a single node in a tree without a *parent*, and nodes without descendants are called *leaves*.

In an extensive-form game, nodes indicate the positions of the game, each being reached by a succession of moves that define the history of play. The edges correspond to the possible actions of the player who has the move in a given position.

Example 1.9 Consider a three-player game in which players move sequentially. The graph tree is represented in Fig. 1.1. The tree represents the sequence of actions that influence the players' outcomes. First, Player 1 moves, choosing her action from the set $\{U, D\}$. Next, Player 2 chooses her action. Then, Player 3 chooses an action after Player 2's choice of actions a and b. Finally, the game terminates at one of the leaves, and the players get their payoffs.

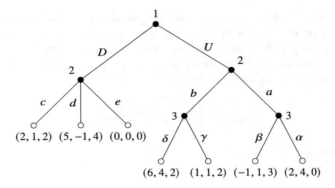

Fig. 1.1 A game tree

1.3.2 Information Sets

A game in extensive form is described by the following:

1. A set of players, which may include Nature as a dummy player that plays randomly, i.e., its moves are not chosen in view of achieving a certain objective;
2. A set of *positions,* which corresponds to the set of nodes;
3. A payoff function for each player, defined over the set of leaves.

The unique move sequence between the root node and any other node is called *history.* At each node, one particular player has the move, i.e., she selects one possible action from an admissible set represented by the edges emanating from the node. This selection defines the transition to a new node. The game stops when a terminal node (or leaf) is reached and the players receive their payoffs as defined on the leaves.

The information that a player possesses at each node is referred to as the *information structure* of the game. In general, the player moving at the node may not know exactly at which node of a game tree she is located. She has the following information: *she knows that the current position of the game is within a given subset of nodes, however, she does not know which particular node it is.* These subsets of nodes are called *information sets* and are represented by dotted lines. The notion of an information set was introduced in von Neumann and Morgenstern (1944).

In Fig. 1.2, the set of nodes in which Player 3 moves are linked by a dotted line. These nodes form the information set of Player 3. We note that the number of edges emanating from each node of the information set is the same. Moreover, the set of actions available to the player at any node of the information set must coincide. So, the player does not distinguish between the game positions represented by the connected nodes, but she knows the set of actions $\{\alpha, \beta\}$ originating from any (connected) node.

Example 1.10 A two-player game in extensive form, with Nature (chance) having moves, is represented in Fig. 1.3. The positions of Nature are marked by n. For any node in which Nature moves, the probability distribution over the set of emanating

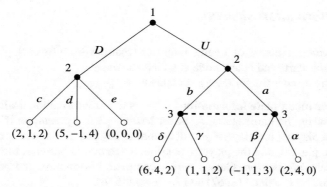

Fig. 1.2 Representation of an information set

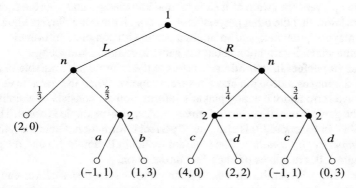

Fig. 1.3 A game in extensive form in which Nature intervenes

nodes should be defined. Nature chooses an action with a given probability distribution. We notice that Player 2 has two information sets: the first one contains a unique node, and the second contains two nodes, which means that Player 2 does not know Nature's move when she chooses an action from the set $\{c, d\}$. At the terminal nodes, the game stops and the players obtain their payoffs.

The representation of a game in extensive form using a tree graph structure is inspired by games like chess, poker, bridge, etc., which can be accurately described by this framework. In this context, Nature's moves correspond to, e.g., a card draw or a dice toss. The extensive form provides a very detailed description of a game. If the player knows at which node the game is now, she may remember the history of play, i.e., the sequence of moves made by all players that has led to this node. However, the extensive form is rather impractical for analyzing even simple games because the size of the tree increases very fast with the number of steps. Describing a complex game like bridge or chess using the extensive form is not feasible. Still, the extensive form is useful to conceptualize the structure of the game, even when the actions are described by continuous variables.

1.3.3 Information Structure

The *information structure* of a game indicates what is known by each player at the time the game starts and at any node at which she moves.

We briefly discuss the following structures:

Complete or incomplete information: refers to the information available to the players when the game starts. A player has *complete information* if she knows the set of players, the players' sets of actions at any position, the information sets of all the players, and the players' possible outcomes. Otherwise, the player has *incomplete information*. Games with incomplete information and beliefs were introduced by Harsanyi (1967) and Aumann (1976).

Common knowledge: The information structure in a game is *common knowledge* if all the players are aware of it (it is mutual knowledge) and, moreover, if all the players know that the other players know, and if all the other players know that all other players know that, and so on. In a game with complete information, before the game starts, information about the game is common knowledge.

Perfect or imperfect information: refers to the information available to a player when she makes a decision about a specific move. When the game is defined in extensive form and when any player's information set consists of a single node, then the game is one of *perfect information*. A typical example is chess. The game described in Example 1.9 is also one of perfect information. Otherwise, the game is of *imperfect information*. The Cournot model in Example 1.5 and the game in Example 1.10 are games of imperfect information.

Perfect recall: When the information structure is such that a player can always remember all the past moves she has made, along with all the information sets she has attained, then the game is one of *perfect recall*. Otherwise, the game is of *imperfect recall*. The notion of perfect recall was introduced by Kuhn (1953).

Commitment: is when a player takes an action she will be bound by. This binding and the action itself are known to other players. By making a commitment, a player can persuade other players to take actions that are favorable to her. *Threats* are a particular class of commitment. To be effective, a commitment must be *credible*, that is, it must be in the player's best interest to implement the binding action if faced with such a choice. The basic concepts of commitments were introduced by Schelling in his seminal book (Schelling 1960).

Binding agreements: are restrictions on the possible actions that can be decided on by two or more players, under a contract that forces the implementation of an agreement. A binding agreement usually requires an outside authority that can monitor the agreement at no cost and impose sanctions on violators that are severe enough to prevent cheating.

Fig. 1.4 A game in
extensive form

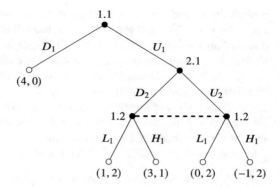

1.3.4 *Equilibrium in Behavior Strategies*

In normal-form games, randomization over the set of strategies leads to mixed strategies. In extensive-form games, a similar idea can be achieved if, from every information set, the player chooses one of her possible actions randomly. This model of
randomization leads to the concept of behavior strategies, where players randomly
select their moves at each information set. Kuhn (1953) showed that, in any game
with perfect recall, a player's behavior strategy is equivalent to a mixed strategy,
and vice versa. Also, he showed that such a game has an equilibrium in behavior
strategies.

To explain the connection between behavior and mixed strategies, we consider the
two-player game represented in Fig. 1.4. In this game, Player 1 has two information
sets denoted by $\mathcal{I}_{1.1}$, containing a single (root) node, and $\mathcal{I}_{1.2}$, containing two nodes.
The first subscript refers to the player, and the second to the a player's number of
information sets. The set of actions of Player 1 in $\mathcal{I}_{1.1}$ is $\{U_1, D_1\}$, and $\{H_1, L_1\}$ in
$\mathcal{I}_{1.2}$. Player 2 has one information set, denoted by $\mathcal{I}_{2.1}$, with the set of actions being
$\{U_2, D_2\}$.

The set of pure strategies of Player 1 is

$$S_1 = \{U_1 H_1, U_1 L_1, D_1 H_1, D_1 L_1\},$$

and the set of pure strategies of Player 2 is

$$S_2 = \{U_2, D_2\}.$$

Players' mixed strategies are defined as probability distributions over the sets of
pure strategies. We denote the set of mixed strategies of Player 1 and Player 2 by Σ_1
and Σ_2, respectively.

Now, we give various interpretations of the realization of mixed strategies. If a
player has only one move (a unique information set), such as Player 2 in Fig. 1.4, then
there is a single way to implement a mixed strategy, i.e., to choose action U_2 with

probability γ, and action D_2 with probability $1 - \gamma$. Player 1 has two information sets in the game. The first way to implement a mixed strategy is to define a probability distribution over the set $\{U_1 H_1, U_1 L_1, D_1 H_1, D_1 L_1\}$; e.g., $(0.25; 0.00; 0.50; 0.25)$ is a mixed strategy of Player 1. Another way is to choose randomly between actions U_1 and D_1 (with probabilities α and $1 - \alpha$, respectively) when at information set $\mathcal{I}_{1.1}$, and then to choose randomly between H_1 and L_1 (with probabilities β and $1 - \beta$, respectively) when at information set $\mathcal{I}_{1.2}$. In other words, instead of randomizing over the set of pure strategies that determines the player's actions at each information set, the player randomly chooses her action once she is at a particular information set. Such a strategy is called a behavior strategy.

Definition 1.14 A behavior strategy for a player in an extensive-form game is a function mapping of each of her information sets to a probability distribution over the set of possible actions at that information set.

A behavior strategy b_i of Player i is a vector of probability distributions defined for each information set of Player i. Denote the set of behavior strategies of Player i by $\mathcal{B}_i$.

The following question naturally arises: can players obtain higher payoffs using mixed or behavior strategies? Kuhn (1953) finds the conditions under which it makes no difference which of these strategy concepts is used. First, we demonstrate the equivalence of the sets of mixed and behavior strategies in the game shown in Fig. 1.4. Then, we introduce the theoretical results in Kuhn (1953).

From Fig. 1.4, the equivalence of these sets for Player 2 is clear because she has a single information set. Player 1's behavior strategy is described by two probability distributions: $(\alpha, 1 - \alpha)$ and $(\beta, 1 - \beta)$, or by two probabilities $\alpha \in [0, 1]$, $\beta \in [0, 1]$. The set $\mathcal{B}_1$ is equivalent to the set

$$\{(\alpha, \beta) : \alpha \in [0, 1], \beta \in [0, 1]\}.$$

The set of mixed strategies Σ_1 is defined as

$$\{p = (p_1, p_2, p_3, p_4) : p_j \in [0, 1], \sum_{j=1}^{4} p_j = 1\},$$

where p_1, p_2, p_3, p_4 are the probabilities of choosing strategies $U_1 H_1, U_1 L_1, D_1 H_1, D_1 L_1$, respectively.

So, $\mathcal{B}_1 \in \mathbb{R}^2$, and $\Sigma_1 \in \mathbb{R}^4$, which may suggest that Σ_1 is a larger set than $\mathcal{B}_1$. But in this example, any behavior strategy is equivalent to a mixed strategy. Specifically, any behavior strategy $\big((\alpha, 1 - \alpha), (\beta, 1 - \beta)\big)$ is equivalent to a mixed strategy $\big(\alpha\beta, \alpha(1 - \beta), (1 - \alpha)\beta, (1 - \alpha)(1 - \beta)\big)$. Therefore, the probabilities defining the mixed strategy p can be obtained from the behavior strategy $\big((\alpha, 1 - \alpha), (\beta, 1 - \beta)\big)$ in the following way:

$$p_1 = \alpha\beta, \quad p_2 = \alpha(1 - \beta), \quad p_3 = (1 - \alpha)\beta, \quad p_4 = (1 - \alpha)(1 - \beta).$$

Now, we demonstrate how to define the behavior strategy, i.e., the probabilities (α, β), if we have a mixed strategy $p = (p_1, p_2, p_3, p_4)$. To do this, we need to find probabilities p_i, $i = 1, \ldots, 4$ s.t. $\sum_{j=1}^{4} p_j = 1$, which are the solutions of the following system:

$$\begin{cases} \alpha\beta = p_1, \\ \alpha - \alpha\beta = p_2, \\ \beta - \alpha\beta = p_3, \\ (1 - \alpha)(1 - \beta) = p_4. \end{cases}$$

If the first three equations are true, then the last one is obviously satisfied, and we may omit it. From the first three equations, we obtain the system

$$\begin{cases} \alpha\beta = p_1, \\ \alpha - \beta = p_2 - p_3, \end{cases}$$

which has a unique solution:

$$\alpha = \frac{p_2 - p_3 + \left((p_2 - p_3)^2 + 4p_1\right)^{0.5}}{2},$$

$$\beta = \frac{-(p_2 - p_3) + \left((p_2 - p_3)^2 + 4p_1\right)^{0.5}}{2}.$$

The above calculations illustrate the theoretical result given in the following theorem.

Theorem 1.8 (Kuhn 1953) *In every game in extensive form, if Player i has perfect recall, then for every mixed strategy of Player i, there exists an equivalent behavior strategy.*

Theorem 1.9 (Kuhn 1953) *If in the game in extensive form, all the players have perfect recall, then there exists a Nash equilibrium in behavior strategies.*

Proof If the players have perfect recall, then there is an equivalence between the player's set of mixed strategies and her set of behavior strategies by Theorem 1.8. Therefore, the existence of a Nash equilibrium in behavior strategies follows from the existence of a Nash equilibrium in mixed strategies. □

1.3.5 Subgame Perfect Equilibrium

Selten (1975) proposed the concept of a subgame perfect equilibrium, which is a refinement of the Nash equilibrium for games in extensive form. The idea behind this refinement is to rule out noncredible threats.[4] Selten (1975) proved the existence of a subgame perfect equilibrium in games with perfect information.

In an extensive-form game Γ, consider the subgame starting at node x, which is a root of the subtree, and denote it by Γ_x. This subgame is derived from the initial game by deleting all nodes and edges that do not follow x.

Definition 1.15 A subgame perfect equilibrium of an extensive-form game Γ is an equilibrium in behavior strategies, such that, for any subgame Γ_x of game Γ, beginning at any possible subroot x, the restriction of these behavior strategies to this subgame is also an equilibrium in the behavior strategies for the subgame Γ_x.

Theorem 1.10 *Every finite game in extensive form with perfect information has a subgame perfect equilibrium in pure strategies.*

Proof First, we construct the profile in behavior strategies using a specific algorithm, and then prove that this profile is subgame perfect.

Let the length of the maximal path in a tree be called the length of the game and be denoted by T. Let the length of the game be $T + 1$. Denote the set of positions in the game by $X = X_0 \cup \cdots \cup X_T$, where X_t is the set of positions, which can be obtained from the root node x_0 by $T - t$ moves.

Denote the subset of the set of nodes where Player i moves as $\mathcal{I}_i, i \in M$. Therefore, $X = \mathcal{I}_1 \cup \cdots \cup \mathcal{I}_m \cup \mathcal{I}_{m+1}$, where $\mathcal{I}_{m+1}$ is the set of terminal positions, at which the payoff functions are defined.

First, consider the set of terminal positions X_0 at which no player moves and the payoffs to the players are defined. For any position $x_0 \in X_0$, let the players' payoffs be $h_i(x_0), i \in M$.

Step 1. Transit from the set X_0 to a position $x_1 \in X_1$, and x_1 is not a terminal position. Let this position belong to an information set of Player $i(x_1)$. The algorithm prescribes that Player $i(x_1)$ choose an action x from the set of possible actions at position x_1, i.e., $x \in Z(x_1)$ maximizing the payoff

$$h_{i(x_1)}(\bar{x}_0) = \max_{x \in Z(x_1)} h_{i(x_1)}(x), \tag{1.20}$$

where $h_i(x)$ is a payoff to Player i in a terminal position x.

The maximum in (1.20) may be nonunique. Denote the set of actions or nodes satisfying (1.20) as

$$\tilde{Z}_{i(x_1)}(x_1) = \left\{ y : h_{i(x_1)}(y) = \max_{x_0 \in Z(x_1)} h_{i(x_1)}(x_0) \right\}.$$

Construct the behavior strategy $\bar{b}_{i(x_1)}$ of Player $i(x_1)$ in position $x_1 \in X_1$ to choose an alternative $y \in \tilde{Z}_{i(x_1)}(x_1)$ with probability $p_{x_1}(y) \in [0, 1]$ s.t. $\sum_{y \in \tilde{Z}_{i(x_1)}(x_1)} p_{x_1}(y) =$

---

[4] A noncredible threat in a sequential game refers to those threats a rational player would not actually carry out (as it would not be in her best interest to do so).

1. These probabilities define the behavior strategy $\bar{b}_{i(x_1)}$ of Player $i(x_1)$ in position $x_1 \in X_1$. If the maximum in (1.20) is unique and consists of a unique position x_0, then the behavior strategy $\bar{b}_{i(x_1)}$ prescribes choosing this position with probability $p_{x_1}(x_0) = 1$. The payoff to Player $j \in M$ is

$$\sum_{y \in \tilde{Z}_{i(x_1)}(x_1)} p_{x_1}(y) h_j(y).$$

If position $x_1 \in X_1$ is a terminal one, then the payoffs to the players are defined by the functions $h_j(x_1)$, $j \in M$.

If the set $\tilde{Z}_{i(x_1)}(x_1)$ contains more than one node, then the path is not uniquely defined by the behavior strategy, and thus we have a subtree $G(x_1)$ of the tree.

Making the constructions in a similar way, we may define such a subtree for any node $x_1 \in X_1$. Therefore, we may define the analog of Bellman function

$$H_i^1 : X_1 \to \mathbb{R}^1$$

for any $i \in M$. In notation H_i^1, the superscript "1" corresponds to index "1" in the notation of set X_1.

We may interpret H_i^1 as the expected payoff of Player j in the subgame Γ_{x_1}, in position $x_1 \in X_1$ when the behavior strategies are defined as above, that is,

$$H_j^1(x_1) = \begin{cases} h_j(x_1), & \text{if } x_1 \in I_{m+1}, \\ \sum_{y \in \tilde{Z}_{i(x_1)}} p_{x_1}(y) h_j(y), & \text{if } x_1 \notin I_{m+1}, \end{cases}$$

where $\sum_{y \in \tilde{Z}_{i(x_1)}(x_1)} p_{x_1}(y) = 1$.

The behavior strategy $\bar{b}_i$ of Player i who moves in position $x_1 \in X_1 \cap I_i$ is defined as follows:

$$\bar{b}_i(x_1) = \begin{cases} \bar{x}_0, & \bar{x}_0 = \arg \max_{y \in Z_i(x_1)} h_i(y), & \text{if } |\tilde{Z}_i(x_1)| = 1, \\ p_{x_1} = (p_{x_1}(y) : y \in \tilde{Z}_i(x_1)), & \text{if } |\tilde{Z}_i(x_1)| > 1, \end{cases}$$

where p_{x_1} is a probability distribution over the set of nodes $\tilde{Z}_i(x_1)$.

Now, assume that functions $H_i^l(x_l)$, where $x_l \in X_l, l < t$, and behavior strategies b_i are defined in a similar way in all positions x_l. Therefore, for any position $x_{t-1} \in X_{t-1}$, functions H_j^{t-1} are defined in the way explained above, that is,

$$H_j^{t-1} : X_{t-1} \to \mathbb{R}^1, \quad j \in M.$$

Step t. Define the set

$$\tilde{Z}_{i(x_t)}(x_t) = \left\{ x_{t-1} \in X_{t-1} : x_{t-1} = \arg \max_{y \in Z(x_t)} H_{i(x_t)}^{t-1}(y) \right\}.$$

The behavior strategy $\bar{b}_i(x_t)$ of Player i in position $x_t \in X_t \cap I_i$, at which Player i moves, is defined as follows:

$$\bar{b}_i(x_t) = \begin{cases} \bar{x}_{t-1}, & \bar{x}_{t-1} = \arg \max_{y \in Z_i(x_t)} H_i^{t-1}(y), & \text{if } |\tilde{Z}_i(x_t)| = 1, \\ p_{x_t} = (p_{x_t}(y) : y \in \tilde{Z}_i(x_t)), & & \text{if } |\tilde{Z}_i(x_t)| > 1, \end{cases}$$

where p_{x_t} is a probability distribution over set $\tilde{Z}_i(x_t)$. The payoff to Player $j \in M$ in the subgame starting from position $x_t \in X_t$ equals

$$\sum_{y \in \tilde{Z}_i(x_t)} p_{x_t}(y) H_j^{t-1}(y).$$

Again, if $x_t \in X_t \cap I_{m+1}$, then the players' payoffs are defined by the functions $h_j(x_t)$, $j \in M$.

The functions $H_j^t : X_t \to \mathbb{R}^1$, $j \in M$ are defined for any $x_t \in X_t$ by

$$H_j^t(x_t) = \begin{cases} h_j(x_t), & \text{if } x_t \in I_{m+1}, \\ \sum_{y \in \tilde{Z}_{i(x_t)}(x_t)} p_{x_t}(y) H_j^{t-1}(y), & \text{if } x_t \notin I_{m+1}. \end{cases}$$

Denote by $\bar{b}^{x_t} = (\bar{b}_i^{x_t} : i \in M)$ the behavior strategy profile for the subgame Γ_{x_t} starting from position x_t, defined above at steps 1 to t of the algorithm. Then, we proceed to define the behavior strategies in a similar way until we reach a root node $x_0 \in X_T$ and obtain the behavior strategy profile $\bar{b}^{x_0} = (\bar{b}_i^{x_0} : i \in M)$ for the game with perfect information.

Now, we can easily prove that the behavior strategy profile $\bar{b}(\cdot)$ defined by the algorithm is a subgame perfect equilibrium in the game.

Let

$$E_j(x, \bar{b}^x(\cdot)) = E_j(x, (\bar{b}_1^x(\cdot), \dots, \bar{b}_m^x(\cdot)))$$

be the expected payoff to Player $j \in M$, in the subgame starting from position x when players implement behavior strategies $\bar{b}_1^x, \dots, \bar{b}_m^x$. Then, according to the above algorithm, for any position $x_t \in X_t$, we have

$$E_j(x_t, \bar{b}^{x_t}(\cdot)) = H_j^t(x_t), \quad j \in M.$$

To prove the subgame perfectness of the profile $\bar{b}(\cdot)$ we need to prove the inequality

$$E_j(x_t, \bar{b}^{x_t}(\cdot)) \geq E_j(x_t, (\bar{b}_{-j}^{x_t}(\cdot), b_j^{x_t}(\cdot))), \tag{1.21}$$

for any $x_t \in X_t$, any player $j \in M$ and any behavior strategy $b_j^{x_t}(\cdot) \in \mathcal{B}_j^{x_t}$. The behavior strategy profile $(\bar{b}_{-j}^{x_t}(\cdot), b_j^{x_t}(\cdot))$ is such that all players except player j use their strategies $\bar{b}_i^{x_t}(\cdot)$, $i \in M \setminus \{j\}$, and Player j uses strategy $b_j^{x_t}(\cdot)$. The strategy set $\mathcal{B}_j^{x_t}$ is the restriction of the set $\mathcal{B}_j$ on the subgame Γ_{x_t} starting from position x_t.

The proof of inequality (1.21) can be made by induction along the game paths starting from positions for which the subgames have a length of one. For all such positions, the inequality (1.21) takes the form of an equality because the players do not move in such positions. Now, we consider any node $x_t \in X_t$ such that the length of the game starting from this position is no greater than T. Assume that the theorem is true (inequality (1.21) is satisfied) for all subgames with lengths of less than $T + 1$, and prove it for the game with a length of $T + 1$.

Let Player i moves in position x_T, i.e., $i = i(x_T)$. Taking into account that for any, x_{T-1}, the behavior strategy profile $\bar{b}^{x_{T-1}}$ is subgame perfect in the subgame Γ_{T-1} starting from position x_{T-1}, we have

$$
\begin{aligned}
E_{i(x_T)}(x_T, \bar{b}^{x_T}(\cdot)) &= H_{i(x_T)}^T(x_T) = \max_{x_{T-1} \in Z(x_T)} H_{i(x_T)}^{T-1}(x_{T-1}) \\
&= H_{i(x_T)}^{T-1}(\bar{x}_{T-1}) \geq H_{i(x_T)}^{T-1}(x_{T-1}) = E_{i(x_T)}(x_{T-1}, \bar{b}^{x_{T-1}}(\cdot)) \\
&\geq E_{i(x_T)}(x_{T-1}, (\bar{b}_{-i(x_T)}^{x_{T-1}}(\cdot), b_{i(x_T)}^{x_{T-1}}(\cdot))) \\
&= E_{i(x_T)}(x_T, (\bar{b}_{-i(x_T)}^{x_T}(\cdot), b_{i(x_T)}^{x_T}(\cdot))).
\end{aligned}
$$

If Player j does not move in position x_T, i.e., $j \neq i(x_T)$, then

$$
\begin{aligned}
E_j(x_T, \bar{b}^{x_T}(\cdot)) &= \sum_{y \in \tilde{Z}_{i(x_T)}(x_T)} p_{x_T}(y) H_j^{T-1}(y) \\
&= \sum_{y \in \tilde{Z}_{i(x_T)}(x_T)} p_{x_T}(y) E_j(x_{T-1}, \bar{b}^y(\cdot)) \\
&\geq \sum_{y \in \tilde{Z}_{i(x_T)}(x_T)} p_{x_T}(y) E_j(x_{T-1}, (\bar{b}_{-j}^y(\cdot), b_j^y(\cdot))) \\
&= E_j(x_T, (\bar{b}_{-j}^{x_T}(\cdot), b_j^{x_T}(\cdot))),
\end{aligned}
$$

which finally proves the theorem.   □

The proof of Theorem 1.10 is constructive and provides an algorithm to find subgame perfect equilibria in an extensive-form game with perfect information.

Theorem 1.11 *Every game in extensive form with perfect recall has a subgame perfect equilibrium in mixed strategies.*

Proof Every subgame of an extensive-form game with perfect recall is also an extensive-form game with perfect recall. Therefore, any subgame has a Nash equilibrium in mixed strategies. Then, using the algorithm from the proof of Theorem 1.10,

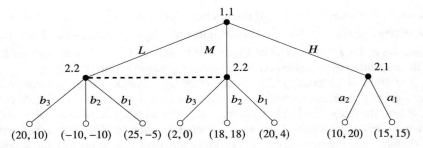

Fig. 1.5 A game in extensive form

we consider all subgames and construct the behavior strategies of the players forming the subgame perfect equilibrium in the game. The existence of the pure-strategy
equilibrium cannot be guaranteed because of imperfect information. By Theorem
1.8, we prove the equivalence between the behavior set and the mixed strategies in
the game and any of its subgames.   □

We illustrate the concept of subgame perfect equilibrium using the two-player
game depicted in Fig. 1.5.

The set of pure strategies in a normal-form equivalent game is $\{H, M, L\}$ for
Player 1 and $\{a_1b_1, a_1b_2, a_1b_3, a_2b_1, a_2b_2, a_2b_3\}$ for Player 2. The normal-form representation of the game is given in Table 1.1.

This game admits two equilibria in pure strategies and three others in mixed strategies, all shown in Table 1.2. These equilibria can be written in behavior strategies,
as given in Table 1.3. Among these Nash equilibria, only Eq. 5 is subgame perfect. If
we consider the subgame starting from node 2.1 where Player 2 moves, the equilib-

Table 1.1   The game in normal form

	a_1b_1	a_1b_2	a_1b_3	a_2b_1	a_2b_2	a_2b_3
H	(15, 15)	(15, 15)	(15, 15)	(10, 20)	(10, 20)	(10, 20)
M	(20, 4)	(18, 18)	(−5, 25)	(20, 4)	(18, 18)	(−5, 25)
L	(25, −5)	(−10, −10)	(20, 10)	(25, −5)	(−10, −10)	(20, 10)

Table 1.2   Nash equilibria in mixed strategies in the normal-form game

	H	M	L	a_1b_1	a_1b_2	a_1b_3	a_2b_1	a_2b_2	a_2b_3
Equilibrium 1	1	0	0	0	0	0	$\frac{44}{173}$	$\frac{65}{173}$	$\frac{64}{173}$
Equilibrium 2	1	0	0	0	0	0	0	$\frac{15}{23}$	$\frac{8}{23}$
Equilibrium 3	1	0	0	0	0	0	0	$\frac{1}{3}$	$\frac{2}{3}$
Equilibrium 4	0	0	1	0	0	1	0	0	0
Equilibrium 5	0	0	1	0	0	0	0	0	1

Table 1.3 Nash equilibria represented in behavior strategies the normal-form game

	H	M	L	a_1	a_2	b_1	b_2	b_3
Equilibrium 1	1	0	0	0	1	$\frac{44}{173}$	$\frac{65}{173}$	$\frac{64}{173}$
Equilibrium 2	1	0	0	0	1	0	$\frac{15}{23}$	$\frac{8}{23}$
Equilibrium 3	1	0	0	0	1	0	$\frac{1}{3}$	$\frac{2}{3}$
Equilibrium 4	0	0	1	1	0	0	0	1
Equilibrium 5	0	0	1	0	1	0	0	1

rium in this subgame is action a_2. Therefore, from Table 1.3, we conclude that Eq. 4 is not subgame perfect because it prescribes that Player 2 choose action a_1. Next, consider the subgame starting from information set 2.2. In this subgame, the Nash equilibrium is action b_3 for Player 2 (it dominates actions b_1 and b_2). Therefore, Eq. 1–3 are not subgame perfect. The unique subgame perfect equilibrium is the strategy profile $(L; a_2 b_3)$, with payoffs 20 and 10 to Player 1 and 2, respectively.

Every game is a subgame, and therefore every subgame perfect equilibrium is a Nash equilibrium. We use backward induction to find the subgame perfect equilibrium. This method is akin to dynamic programming. We start by selecting the optimal action in the last stage t, for every state of the system at stage $t - 1$. We continue by optimizing the action at stage $t - 1$ for every state of the system at stage $t - 2$, and so on.

1.4 Stackelberg Equilibrium Solution

The concept of the Nash equilibrium assumes that the players choose their strategies independently and simultaneously. A duopoly model was proposed in von Stackelberg (1934) with sequential sequential moves by the two players, called leader and follower, respectively. The leader first announces her decision, and next, the follower announces hers, taking into account the leader's announcement.

To introduce the Stackelberg equilibrium concept, we consider a bimatrix game. Player 1 has two pure strategies, $S_1 = \{U, D\}$, and Player 2 has three pure strategies, $S_2 = \{\alpha, \beta, \gamma\}$, with the payoffs given below:

$$
\begin{array}{cccc}
& & \text{Player 2} & \\
& \alpha & \beta & \gamma \\
U & (2, 4) & (5, 1) & (-1, 3) \\
\text{Player 1} & & & \\
D & (2, 1) & (3, 5) & (5, 4)
\end{array}
$$

In this game, (U, α) is the unique Nash equilibrium in pure strategies with payoffs $(2, 4)$. The representation of this game in extensive form is given in Fig. 1.6.

Fig. 1.6 A game in
extensive form

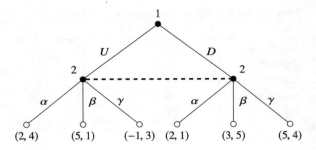

Now assume that the players' roles in the game are not symmetric and that Player 1 is a leader. This means she chooses the strategy and announces it to the follower, Player 2. If the leader chooses strategy U, then the follower's best response is α, that is, $\alpha = R_2(U)$, and the payoffs are $(2, 4)$. If Player 1 chooses D, then Player 2 chooses strategy β, i.e., $\beta = R_2(D)$, and the payoffs are $(3, 5)$. The leader, Player 1, being aware of the follower's reaction functions, chooses strategy D, giving her a payoff of 3 instead of the payoff of 2 she would have received had she chosen strategy U. Therefore, the leader's strategy is a response function to function R_2, i.e., $D = R_1(R_2(\cdot))$. The Stackelberg strategy of Player 1 is $s_1^* = D \in S_1$; for Player 2, it is $s_2^* = \beta \in S_2$. The Stackelberg equilibrium in the game when Player 1 is the leader is (D, β) with payoffs $(u_1^*, u_2^*) = (3, 5)$. The described game when Player 1 is the leader is represented in extensive form in Fig. 1.7. We notice that the Stackelberg equilibrium solution is the realization of the subgame perfect equilibrium in the game. The subgame perfect equilibrium in the described game is $(D, (\alpha, \beta))$, which defines the Nash equilibrium behavior of the players in any subgame. If Player 2 is the leader, then she chooses her strategy first and announces it to the follower, Player 1. If the leader chooses strategy α, then Player 1 is indifferent between U and D, that is, $R_1(\alpha) = \{U, D\}$, and the payoffs are $(2, 4)$ or $(2, 1)$. If Player 2 chooses strategy β, then Player 1 chooses strategy U, i.e., $U = R_1(\beta)$, and the payoffs are $(5, 1)$. If Player 2 chooses strategy γ, then Player 1 chooses strategy D, i.e., $D = R_1(\gamma)$, and the payoffs are $(5, 4)$. The leader, Player 2, being aware of the follower's reaction functions, chooses strategy γ, giving her a payoff of 5. Therefore, the leader's strategy is a reaction function of function R_2, i.e., $\gamma = R_2(R_1(\cdot))$. The Stackelberg strategy

Fig. 1.7 A game in
extensive form

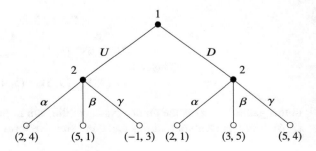

Fig. 1.8 A game in extensive form

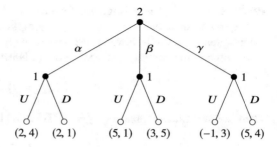

of Player 1 is $s_1^* = D \in S_1$, and it is $s_2^* = \gamma \in S_2$ for Player 2. The Stackelberg equilibrium solution in the game when Player 2 is the leader is (D, γ) with payoffs $(u_1^*, u_2^*) = (5, 4)$. The game when Player 2 is the leader is represented in extensive form in Fig. 1.8. We note that there are two subgame perfect equilibria in this game, namely, $((U, U, D), \gamma)$ and $((D, U, D), \gamma)$.

The example above shows that the role (leader or follower) played by a player influences her payoff.

Definition 1.16 In a two-player finite game where Player 1 acts as the leader, the strategy s_1^* is a Stackelberg equilibrium strategy of Player 1 if

$$u_1^* = \min_{s_2 \in R_2(s_1^*)} u_1(s_1^*, s_2) = \max_{s_1 \in S_1} \min_{s_2 \in R_2(s_1)} u_1(s_1, s_2), \tag{1.22}$$

where $R_2(s_1) := \arg\max_{s_2 \in S_2} u_2(s_1, s_2)$ denotes the best response of the follower to the leader's strategy s_1, and u_1^* is the leader's Stackelberg equilibrium payoff. Then, any strategy $s_2 \in R_2(s_1^*)$ is a Stackelberg equilibrium strategy of the follower (Player 2). The profile (s_1^*, s_2^*) is called a Stackelberg solution of the game with Player 1 as the leader, and the payoffs $(u_1(s_1^*, s_2^*), u_2(s_1^*, s_2^*))$ are the Stackelberg equilibrium outcome.

Clearly, any two-player finite game admits a Stackelberg strategy for the leader, but this strategy does not have to be unique. The best-response mapping $R_2 : S_1 \to S_2$ of the follower can be multi-valued. Then (1.22) states that the leader's Stackelberg strategy is obtained by choosing the follower's strategy in $R_2(s_1)$ that minimizes $u_1(s_1, s_2)$.

How does the leader's payoff in the Stackelberg equilibrium compare to her payoff in any Nash equilibrium? Intuitively, the former should not be less than the latter. Denote by u_1^N any Nash equilibrium payoff of Player 1.

Proposition 1.1 *In any two-player finite game with Player 1 as the leader, if $R_2(s_1)$ is a singleton for any $s_1 \in S_1$, then*

$$u_1^* \geq u_1^N.$$

Proof Since $R_2(s_1)$ is a singleton for each $s_1 \in S_1$, then there exists a mapping $F_2 : S_1 \to S_2$ such that for any $s_2 = R_2(s_1)$ implies $s_2 = F_2(s_1)$, which follows from the uniqueness of the follower's response strategy to any strategy of the leader.

Assume to the contrary that there exists a Nash equilibrium $(\tilde{s}_1, \tilde{s}_2)$ such that

$$u_1(\tilde{s}_1, \tilde{s}_2) > u_1^*. \tag{1.23}$$

Then, we substitute expression $\tilde{s}_2 = F_2(\tilde{s}_1)$ into (1.23) to obtain

$$u_1^* < u_1(\tilde{s}_1, F_2(\tilde{s}_1)),$$

which contradicts the condition that $u_1^* = \max_{s_1 \in S_1} u_1(s_1, F_2(s_1))$. $\square$

The above example assumed a finite game. In the following example, we illustrate the Stackelberg equilibrium solution with a duopoly game à la Cournot, that is, a game with continuous strategy sets.

Example 1.11 We reconsider the Cournot competition model from Example 1.5. Suppose that firm 1 is the leader. Firm 1 announces its output level $q_1 \in Q_1 = [0, \infty)$, and then firm 2 chooses its output $q_2 = r_2(q_1) \in Q_2 = [0, \infty)$. Recall that the profit of firm i is

$$u_i(q_1, q_2) = q_i \, p(q) - cq_i,$$

where $q = q_1 + q_2$, $p(q) = \max\{0, 1 - q\}$ is the market price and $c \in (0, 1)$ is the unit cost.

Maximization of firm 2's profit yields the following response function:

$$q_2 = r_2(q_1) = \frac{1 - q_1 - c}{2}.$$

Substituting for the above response function in the leader's profit, we get

$$u_1(q_1, r_2(q_1)) = q_1 \left(1 - q_1 - \frac{1 - q_1 - c}{2} \right) - cq_1.$$

The first-order condition gives the leader's Stackelberg strategy, that is,

$$q_1^* = \frac{1 - c}{2}.$$

Substituting for q_1^* into the response function $r_2(q_1)$ of the follower, we obtain the following Stackelberg strategy of the follower:

$$q_2^* = \frac{1 - c}{4}.$$

The Stackelberg solution of the duopoly game when firm 1 is the leader is $(q_1^*, q_2^*) = \left(\frac{1-c}{2}, \frac{1-c}{4}\right)$ with the firms' profits being $\left(\frac{(1-c)^2}{8}, \frac{(1-c)^2}{16}\right)$. Recall the Nash equilibrium profits are $\left(\frac{(1-c)^2}{9}, \frac{(1-c)^2}{9}\right)$, so one can easily see that the leader's profit in the Stackelberg solution is larger than in the Nash equilibrium.

Remark 1.5 In some situations, the game naturally involves more than one follower. A typical example is a supply chain made up of one supplier offering its product through n competitive retailers. To solve for a Stackelberg equilibrium, one starts by determining a Nash equilibrium for the noncooperative game played by the retailers, to obtain their n response functions to the leader's announcement. Next, one substitutes for these functions in the supplier's optimization problem and computes its Stackelberg strategy.

1.5 Additional Readings

Game theory, especially in its noncooperative version, has been the subject of more than 100 textbooks and a large series of lecture notes, some available for free on the Internet. Almost all textbooks cover the basic concepts of noncooperative games in some details. As the choice of a textbook depends essentially on two attributes, namely, the desired level of mathematical detail and sophistication, and the application fields of interest (economics, management science, engineering, political sciences, etc.), it is hard to make specific recommendations for additional readings. We limit our suggestions to books that have been around for almost 30 years, are highly cited, and are popular among educators: Fudenberg and Tirole (1991), Myerson (2013), Gibbons (1992), Osborne and Rubinstein (1994), and Owen (2013) (the first edition of Owen's book dates back to 1968).

1.6 Exercises

Exercise 1.1 Find all Nash equilibria of the following bimatrix game:

$$\begin{pmatrix} (1,0) & (-1,1) & (0,0) \\ (2,2) & (-3,10) & (2,9) \end{pmatrix}.$$

Exercise 1.2 Consider an industry with two identical firms producing a homogeneous good at a unit cost of c. Denote by q_i the output of firm $i = 1, 2$. The inverse-demand law is given by

$$p = \max\{1 - q_1 - q_2, 0\},$$

and the profit of firm i by
$$\pi_i(q_1, q_2) = (p - c)q_i.$$

Firms aim to maximize their profits.

1. Compute the Nash equilibrium for the simultaneous game. Is the equilibrium unique?
2. Suppose that both firms merge; find the optimal output.
3. Assume that firm 1 moves before firm 2. Find the Stackelberg equilibrium of this game.
4. Draw the extensive and normal forms of the games under the assumption that the output choice is discrete and that the possible output levels are the Cournot-Nash, monopoly, and Stackelberg equilibria levels.
5. Assume that the firms are not identical and their unit cost can be low or high, i.e., the costs are from the set $\{c_L, c_H\}$. Before the start of the game, let Nature choose the cost level of firm 2 with probability distribution $(0.3, 0.7)$ over the set $\{c_L, c_H\}$. Let firm 1 not know whether the cost of firm 2 is low or high. However, both firms know their own cost level. Find the Nash equilibrium for the Cournot game, assuming that the firms maximize their expected payoffs.

Exercise 1.3   Consider a two-player game in which the strategy sets of the players are the intervals $U_1 = [0, 100]$ and $U_2 = [0, 50]$, respectively. The payoff functions are $\psi_1(u_1, u_2) = 25u_1 + 10u_1u_2 - 4u_1^2$ and $\psi_2(u_1, u_2) = 100u_2 - 10u_1 - u_1u_2 - 4u_2^2$, where $u_1 \in U_1$ and $u_2 \in U_2$.

1. Find the equilibrium for this game.
2. Is the equilibrium unique?

Exercise 1.4   A two-player game in extensive form is defined in Fig. 1.9.

1. Represent the game in normal form.
2. Define the set of mixed and behavior strategies in the game.
3. Show that for every behavior strategy there exists an equivalent mixed strategy.

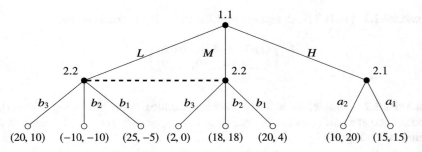

Fig. 1.9   A game in extensive form for Exercise 1.4

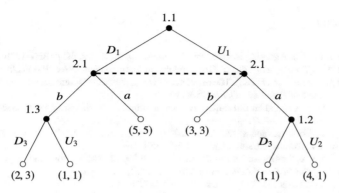

Fig. 1.10 A game in extensive form for Exercise 1.6

Exercise 1.5 Consider a two-player constant-sum game. First, Player 1 chooses a number from the set $\{1, 2, 3\}$ secretly, i.e., does not tell her number to Player 2. Then, Player 2 makes a move guessing which number was selected and choosing a number from the same set $\{1, 2, 3\}$. Next, the numbers are announced and the players get their payoffs. Let Player i's action be denoted by $a_i \in \{1, 2, 3\}$. The payoff of Player 2 is defined by

$$H_2(a_1, a_2) = \begin{cases} 2, & \text{if } a_1 = a_2, \\ 1, & \text{if } |a_1 - a_2| = 1, \\ 0, & \text{if } |a_1 - a_2| = 2, \end{cases}$$

and the payoff of Player 1 is given by

$$H_1(a_1, a_2) = 2 - H_2(a_1, a_2).$$

1. Draw the extensive form of the game.
2. Construct the corresponding strategic form of the game.
3. Find all Nash equilibria. Which ones are subgame perfect Nash equilibria?

Exercise 1.6 A two-player game is represented in Fig. 1.10.

1. Represent the game in normal form.
2. Define the set of mixed and behavior strategies.
3. Find all Nash equilibria in mixed strategies.
4. Find the equivalent representations of the Nash equilibria in behavior strategies.
5. Find all subgame perfect equilibria.

References

Aumann, R. J. (1976). Agreeing to disagree. *The Annals of Statistics*, 4(6):1236–1239.

Berge, C. (1997). *Topological Spaces*. Dover books on mathematics. Dover Publications.

Dasgupta, P. and Maskin, E. (1986). The existence of equilibrium in discontinuous economic games, i: Theory. *Review of Economic Studies*, 53(1):1–26.

Debreu, G. (1952). A social equilibrium existence theorem. *Proceedings of the National Academy of Sciences*, 38(10):886–893.

Facchinei, F., Fischer, A., and Piccialli, V. (2007). On generalized Nash games and variational inequalities. *Operations Research Letters*, 35(2):159–164.

Fan, K. (1952). Fixed-point and minimax theorems in locally convex topological linear spaces. *Proceedings of the National Academy of Sciences*, 38(2):121–126.

Fudenberg, D. and Tirole, J. (1991). *Game Theory*. MIT Press.

Gibbons, R. S. (1992). *Game Theory for Applied Economists*. Princeton University Press.

Glicksberg, I. L. (1952). A further generalization of the Kakutani fixed point theorem with application to Nash equilibrium points. *Proc. Amer. Math. Soc.*, 3:170–174.

Harker, P. T. (1991). Generalized Nash games and quasi-variational inequalities. *European Journal of Operational Research*, 54(1):81–94.

Harsanyi, J. C. (1967). Games with incomplete information played by "bayesian" players, i-iii part i. the basic model. *Management Science*, 14(3):159–182.

Haurie, A., Krawczyk, J. B., and Zaccour, G. (2012). *Games and Dynamic Games*. World Scientific, Singapore.

Kakutani, S. (1941). A generalization of Brouwer's fixed point theorem. *Duke Mathematical Journal*, 8(3):457–459.

Kuhn, H. W. (1953). *11. Extensive Games and the Problem of Information*, pages 193–216. Princeton University Press.

Myerson, R. (2013). *Game Theory: Analysis of Conflict*. Harvard University Press.

Nash, J. (1951). Non-cooperative games. *Annals of Mathematics*, 54(2):286–295.

Nash, J. F. (1950). Equilibrium points in n-person games. *Proceedings of the National Academy of Sciences*, 36(1):48–49.

Osborne, M. and Rubinstein, A. (1994). *A Course in Game Theory*. The MIT Press. MIT Press.

Owen, G. (2013). *Game Theory*. Emerald Group Publishing Limited.

Rosen, J. B. (1965). Existence and uniqueness of equilibrium points for concave n-person games. *Econometrica*, 33(3):520–534.

Schelling, T. (1960). *The Strategy of Conflict: With a New Preface by the Author*. Harvard University Press.

Selten, R. (1975). Reexamination of the perfectness concept for equilibrium points in extensive games. *International Journal of Game Theory*, 4(1):25–55.

von Neumann, J. and Morgenstern, O. (1944). *Theory of Games and Economic Behavior*. Princeton University Press.

von Stackelberg, H. (1934). *Marktform und Gleichgewicht*. Springer.

Chapter 2
Cooperative Games

In this chapter, we review the main ingredients of cooperative games, which will be needed in the following chapters. In Sect. 2.1, we introduce the elements of a cooperative game, and define the characteristic function in Sect. 2.2. Some solutions to a cooperative game are presented in Sect. 2.3 and the Nash bargaining solution in Sect. 2.4. The chapter ends with suggested additional readings and a series of exercises.

2.1 Elements of a Cooperative Game

Recall that, by definition, a noncooperative game involves the following elements:

1. A set of players $M = \{1, 2, \ldots, m\}$;
2. For each player $i \in M$, a set of strategies S_i: we denote by $S := S_1 \times S_2 \times \cdots \times S_m$ the set of joint strategies, and by $s := (s_1, s_2, \ldots, s_m)$ the strategy profile of players; and we let $s_{-i} := (s_1, \ldots, s_{i-1}, s_{i+1}, \ldots, s_m)$ and $(s_i, s_{-i}) := (s_1, \ldots, s_{i-1}, s_i, s_{i+1}, \ldots, s_m)$;
3. For each player $i \in M$, a payoff function u_i that associates a real number to each strategy profile $s \in S$.

In a noncooperative game, the relevant information is found at the individual level. In a cooperative (or coalitional) game, the focus is on what groups of players (coalitions) can achieve when they join forces. A coalition of k players is denoted by K and its complement by $M \setminus K$. We denote by $s_K := (s_i)_{i \in K}$ and $s_{-K} := (s_j)_{j \in M \setminus K}$ the strategies of the players in coalition K and its complement $M \setminus K$, respectively. The set of all strategies of K is denoted by $S_K := \times_{i \in K} S_i$.

In principle, defining a cooperative game only requires that the set of players be specified, along with the value (payoff) that each coalition K can get. The latter value is measured by the characteristic function, that is, a function $v(\cdot)$ defined by

$$v : \mathcal{P}(M) \to \mathbb{R}, \qquad v(\varnothing) = 0, \tag{2.1}$$

E. Parilina et al., *Theory and Applications of Dynamic Games*, Theory and Decision Library C 51, https://doi.org/10.1007/978-3-031-16455-2_2

where $\mathcal{P}(M)$ is the power set of M. (Recall that $\mathcal{P}(M) = 2^M$.) The characteristic function (CF) measures the outcome that any coalition can achieve. In particular, $v(\{i\})$ is the gain that a player can secure by acting alone, while $v(M)$ is the optimal outcome that the grand coalition can realize. As we will see later on, individual behavior and strategies are not completely excluded from the analysis of a cooperative game, especially when the game involves externalities, that is, when a coalition's gain also depends on the strategies of the left-out players.

The set of imputations (or feasible allocations) is denoted by Y and given by

$$Y = \left\{ (y_1, \ldots, y_m) \mid y_i \geq v(\{i\}), \forall i \text{ and } \sum_{i=1}^{m} y_i = v(M) \right\}.$$

Defining the set of imputations involves two conditions. The first condition, individual rationality, states that no player i will accept an imputation that is lower than what she can get by acting alone, i.e., $y_i \geq v(\{i\})$, $\forall i$. The second condition, efficiency, means that the grand coalition's total payoff $v(M)$ must be allocated, i.e., there is no deficit or wastage.

To wrap up, we refer to a cooperative game by (M, v), that is, a set of players $M = \{1, \ldots, m\}$ and a characteristic function $v : \mathcal{P}(M) \to \mathbb{R}$, satisfying $v(\varnothing) = 0$.

To illustrate the above concepts, consider the following example.

Example 2.1 There are four parties represented in a parliament, with $15, 20, 25$, and 40 seats, respectively. A coalition wins (forms the government) if it has at least half the seats. The characteristic function can be defined as follows:

$$v(K) = \begin{cases} 1, & \text{if } \sum_{i \in K} w_i \geq 0.5, \\ 0, & \text{otherwise}, \end{cases}$$

where w_i is the share of seats of party $i \in M = \{1, \ldots, 4\}$. The CF values are

$$v(\{i\}) = 0 \text{ for all } i \in M,$$
$$v(\{1, 2\}) = v(\{1, 3\}) = v(\{2, 3\}) = 0,$$
$$v(\{1, 4\}) = v(\{2, 4\}) = v(\{3, 4\}) = 1,$$
$$v(\{i, j, k\}) = 1 \text{ for all } i, j, k \in M,$$
$$v(M) = 1.$$

The set of imputations is given by

$$Y = \left\{ (y_1, \ldots, y_4) \mid y_i \geq 0, \forall i \text{ and } \sum_{i=1}^{m} y_i = 1 \right\}.$$

Intuitively, cooperation is attractive if it procures a higher payoff to participating players than would noncooperation.

Definition 2.1 The characteristic function is superadditive if

$$v(K \cup L) \geq v(K) + v(L) \text{ for all } K, L \subset M \text{ and } K \cap L = \varnothing.$$

Superadditivity means that when two coalitions join forces, they can achieve at least the same payoff as they would acting separately. If a cooperative game is such that

$$\sum_{i \in M} v(\{i\}) = v(M),$$

then it is said to be *inessential*, that is, there is no incentive to cooperate.

A less demanding property than superadditivity is cohesiveness.

Definition 2.2 The function $v(K)$ is cohesive if

$$v(M) \geq \sum_{j=1}^{J} v(T_j) \text{ for every partition } \{T_1, \dots T_J\} \text{ of } M,$$

where $v(\varnothing) = 0$.

For some cooperative games, the incentives for joining a coalition increase as the coalition grows. Shapley (1971) refers to this phenomenon as the snowballing effect, which is captured by the notion of convex games.

Definition 2.3 The cooperative game (M, v) is convex if the function $v(K)$ satisfies

$$v(K \cup L) + v(K \cap L) \geq v(K) + v(L) \text{ for all } K, L \subseteq M.$$

To simplify the analysis, it is sometimes useful to model the cooperative game in normalized form.

Definition 2.4 A cooperative game with CF v is in 0–1 normalized form if

$$v(\{i\}) = 0, \forall i \in M, \ v(M) = 1.$$

The normalized form makes it possible to assess at a glance the relative strategic strength of any intermediate coalition, with respect to acting alone and to the grand coalition.

Definition 2.5 Two m-person games with CF v and v' defined over the same set of players are strategically equivalent if there exist $q > 0$ and numbers $c_i, i \in M$ such that

$$v'(K) = qv(K) + \sum_{i \in K} c_i \text{ for all } K \subseteq M. \tag{2.2}$$

The strategic equivalence property in (2.2) induces equivalence between imputations, that is,

$$y_i' = q y_i + c_i.$$

Example 2.2 Three neighboring towns want to build a facility to treat contaminated water. An engineering company provides the following costs for all alliance combinations, i.e., each town builds its own facility, two players build a common facility, or all players build one facility:

$$C(\{1\}) = 100, \ C(\{2\}) = 120, \ C(\{3\}) = 140,$$
$$C(\{1, 2\}) = 190, \ C(\{1, 3\}) = 220, \ C(\{2, 3\}) = 240,$$
$$C(\{1, 2, 3\}) = 300.$$

We transform the costs into gains by letting

$$v(K) = \sum_{i \in K} C(\{i\}) - C(K),$$

that is, the gain of a coalition is the difference between the sum of the members' individual costs and the coalition's cost. Consequently, we get

$$v(\{1\}) = 0, \ v(\{2\}) = 0, \ v(\{3\}) = 0,$$
$$v(\{1, 2\}) = 30, \ v(\{1, 3\}) = 20, \ v(\{2, 3\}) = 20,$$
$$v(\{1, 2, 3\}) = 60.$$

Applying the transformations in (2.2), we get

$$q = \frac{1}{60}, \ c_1 = c_2 = c_3 = 0,$$

and

$$v'(\{1\}) = v'(\{2\}) = v'(\{3\}) = 0,$$
$$v'(\{1, 2\}) = 1/2, \ v'(\{1, 3\}) = 1/3, \ v'(\{2, 3\}) = 1/3,$$
$$v'(\{1, 2, 3\}) = 1.$$

It is easy to verify that the CF $v(\cdot)$ is superadditive and convex. Clearly, $v'(\cdot)$ inherits the same properties.

2.2 Defining the Characteristic Function

The definition of a characteristic function in (2.1) does not tell us precisely how to compute its values. In their seminal book, von Neumann and Morgenstern (1944) interpreted the value $v(K)$ as the sum of the gains that coalition K can *guarantee* its members. In cooperative games without externalities, i.e., when the payoff of a coalition K is independent of the actions of the players in $M \backslash K$, the value $v(K)$ is obtained by optimizing the (possibly weighted) sum of the coalition members' payoffs. In this case, we need to solve $2^m - 1$ optimization problems to obtain the values of $v(\cdot)$ for all nonempty coalitions. In games with externalities, the outcome of K depends on the behavior of the players in $M \backslash K$, which leads to different approaches to computing the values of $v(\cdot)$.

α **characteristic function**: The α characteristic function (α CF) was introduced in von Neumann and Morgenstern (1944) and is defined as

$$v^\alpha(K) := \max_{s_K \in S_K} \min_{s_{M \backslash K} \in S_{M \backslash K}} \sum_{i \in K} u_i(s_K, s_{M \backslash K}). \qquad (2.3)$$

In other words, $v^\alpha(K)$ represents the maximum payoff that coalition K *can guarantee for itself* irrespective of the strategies used by the left-out players, that is, the players in $M \backslash K$. In this concept, coalition K moves first, choosing a joint strategy s_K that maximizes its payoff, before the joint strategy $s_{M \backslash K}$ of the players in $M \backslash K$ is chosen.

β **characteristic function**: The β characteristic function is defined as

$$v^\beta(K) := \min_{s_{M \backslash K} \in S_{M \backslash K}} \max_{s_K \in S_K} \sum_{i \in K} u_i(s_K, s_{M \backslash K}), \qquad (2.4)$$

that is, $v^\beta(K)$ gives the maximum payoff that coalition K *cannot be prevented from getting* by the players in $M \backslash K$. In this case, the players in coalition K choose their joint strategy s_K after the left-out players in $M \backslash K$ have chosen their joint strategy $s_{M \backslash K}$. For a discussion and a comparison of the α and β characteristic functions, see Aumann and Peleg (1960).

The α and β CFs are constructed under the (pessimistic) assumption that the left-out players form an anti-coalition, whose sole objective is to minimize the payoff of K. Clearly, coalition structures other than $\{K, M \backslash K\}$ can be considered. Thrall and Lucas (1963) proposed a CF in partition function form, with the more general coalition structure $\{K, \{R_l\}_{l=1}^L (1 \leq l < L)\}$, where $R_i \cap R_j = \emptyset$ for each $1 \leq i, j \leq L$, $i \neq j$, and $\cup_{l=1}^L R_l = M \backslash K$. In this setting, the CF value of coalition K depends on how the set of left-out players $M \backslash K$ is partitioned. These considerations led to the introduction of other CFs.

γ **characteristic function**: In the γ CF, $v^\gamma(K)$ is defined as the partial equilibrium outcome of the noncooperative game between coalition K and the left-out players acting individually. The resulting characteristic function is in partition function form, with the coalition structure given by $\{K, \{j\}_{j \in M \setminus K}\}$, that is, each left-out player takes note of the formation of coalition K and gives a best reply to the other players' strategies. Formally, for any coalition $K \subseteq M$,

$$v^\gamma(K) := \sum_{i \in K} u_i(s_K^\gamma, \{s_j^\gamma\}_{j \in M \setminus K}), \tag{2.5}$$

$$s_K^\gamma := \arg\max_{s_K \in S_K} \sum_{i \in K} u_i(s_K, \{s_j^\gamma\}_{j \in M \setminus K}),$$

$$s_j^\gamma := \arg\max_{s_j \in S_j} u_j(s_K^\gamma, s_j, \{s_i^\gamma\}_{i \in M \setminus \{K \cup j\}}), \text{ for all } j \in M \setminus K.$$

Determining $v^\gamma(K)$ requires finding an equilibrium for a noncooperative game with $m - k + 1$ players. In capturing the strategic interactions of left-out players, we see from (2.5) that the γ CF involves solving for a total of $\sum_{k=1}^{m-1} m C_k^m + m = \sum_{k=1}^{m} m\, C_k^m = m(2^m - 1)$ variables, i.e., $2^m - 2$ equilibrium problems and one optimization problem (the grand coalition's problem). The γ CF was proposed by Chander and Tulkens (1997) for a class of games with multilateral externalities. See also Chander (2007) for a discussion of coalition formation in this context, and Germain et al. (2003) for an application in environmental economics.

δ **characteristic function**: The δ CF makes a different assumption than the γ CF on the behavior of left-out players, and its computation involves a two-step procedure. In the first step, an m-player noncooperative equilibrium is computed. Denote this equilibrium by $(\tilde{s}_1, \ldots, \tilde{s}_m)$. In the second step, each coalition K optimizes its joint payoff, assuming that left-out players stick to their Nash equilibrium actions in the m-player noncooperative game. Formally, for any coalition $K \subseteq M$,

$$v^\delta(K) := \sum_{i \in K} u_i(s_K^\delta, \{\tilde{s}_j\}_{j \in M \setminus K}), \tag{2.6}$$

$$s_K^\delta := \arg\max_{s_K \in S_K} \sum_{i \in K} u_i(s_K, \{\tilde{s}_j\}_{j \in M \setminus K}),$$

$$\tilde{s}_j := \arg\max_{s_j \in S_j} u_j(s_j, \{\tilde{s}_i\}_{i \in M \setminus \{j\}}), \text{ for all } j \in M.$$

In the δ CF, the coalition structure is given by $\{K, \{j\}_{j \in M \setminus K}\}$. The computation of δ CF values involves solving for a total of $m + \sum_{k=2}^{m} k\, C_k^m = \sum_{k=1}^{m} k\, C_k^m = m2^{m-1}$ variables, i.e., one equilibrium problem and $2^m - 2$ optimization problems. As solving an optimization problem is easier than solving an equilibrium one, the δ CF has a clear computational advantage over the γ CF. The limitation of the δ CF is the assumption that the left-out players do not make a best reply to the formation of the coalition, but stick to their Nash equilibrium actions in the m-player noncooperative game. The δ CF was introduced in Petrosyan and Zaccour (2003) and further discussed in Zaccour (2003), Reddy and Zaccour (2016).

Other characteristic functions making different assumptions on the behavior of the left-out players have been proposed in the literature; see, e.g., Gromova and Petrosyan (2017).

A naturally arising question is how the values resulting from the different CFs compare. Without specifying the payoff functions, we have the following result.

Theorem 2.1 *The characteristic function values compare as follows:*

$$v^\alpha(K) \le v^\beta(K) \le v^\delta(K) \text{ for all } K \subset M,$$
$$v^\alpha(K) \le v^\beta(K) \le v^\gamma(K) \text{ for all } K \subset M.$$

Proof First, we define the best response strategy associated with a coalition K as $\bar{s}$

$$\bar{s}_K(s_{-K}) := \arg\max_{s_K \in S_K} \sum_{i \in K} u_i(s_K, s_{-K}), \ s_{-K} \in S_{-K}.$$

Let the strategies $(s_K^\beta, s_{-K}^\beta)$ and $(s_K^\delta, s_{-K}^\delta)$ be such that $v^\beta(K) = \sum_{i \in K} u_i(s_K^\beta, s_{-K}^\beta)$ and $v^\delta(K) = \sum_{i \in K} u_i(s_K^\delta, s_{-K}^\delta)$. From (2.4), we have

$$v^\beta(K) = \sum_{i \in K} u_i(\bar{s}_K(s_{-K}^\beta), s_{-K}^\beta) \le \sum_{i \in K} u_i(\bar{s}_K(s_{-K}), s_{-K}), \text{ for all } s_{-K} \in S_{-K}.$$

Then, from (2.6), it is clear that, for a particular s_{-K}^δ, we have

$$v^\beta(K) = \sum_{i \in K} u_i(\bar{s}_K(s_{-K}^\beta), s_{-K}^\beta) \le \sum_{i \in K} u_i(\bar{s}_K(s_{-K}^\delta), s_{-K}^\delta) = v^\delta(K).$$

In a similar fashion, from (2.5), it is clear that $v^\beta(K) \le v^\gamma(K)$ for all $K \subset M$. The inequality $v^\alpha(K) \le v^\beta(K)$ follows from (2.3), (2.4), and max-min inequality. $\square$

The above theorem reflects the idea that the α and β CFs are too pessimistic. In general, it is not clear how $v^\gamma(K)$ and $v^\delta(K)$ are related for a coalition $K \subset M$.

Example 2.3 Consider a symmetric three-player cooperative game. The utility function of Player i is given by

$$u_i = as_i - \frac{1}{2}bs_i^2 - \frac{1}{2}cs^2, \ i \in M = \{1, 2, 3\},$$

where $s = \sum_{i \in M} s_i$, and a, b, and c are positive parameters. Table 2.1 reports the results for the γ and δ CFs for all feasible coalitions. As expected, the strategies and payoff values only differ for two-player coalitions.

To simplify the interpretation of the results, we transform the game into a strategically equivalent 0–1 normalized game, that is,

$$\hat{v}(\{i\}) = 0, \ i = 1, 2, 3, \text{ and } \hat{v}(M) = 1.$$

Table 2.1 Strategies and payoffs

K	δ CF		γ CF	
	$s_i,\ i \in K$	$u_i,\ i \in K$	$s_i,\ i \in K$	$u_i,\ i \in K$
$\{1\},\{2\},\{3\}$	$\dfrac{a}{3c+b}$	$\dfrac{a^2(b-3c)}{2(b+3c)^2}$	$\dfrac{a}{3c+b}$	$\dfrac{a^2(b-3c)}{2(b+3c)^2}$
$\{i,j\},\{k\}$	$\dfrac{a(c+b)}{(b+4c)(3c+b)}$	$\dfrac{(b^2+bc-3c^2)a^2}{2(b+4c)(b+3c)^2}$	$\dfrac{a(b-c)}{5cb+b^2}$	$\dfrac{(b^2+bc-11c^2)a^2}{2b(b+5c)^2}$
$\{i,j,k\}$	$\dfrac{a}{9c+b}$	$\dfrac{a^2}{2(b+9c)}$	$\dfrac{a}{9c+b}$	$\dfrac{a^2}{2(b+9c)}$

We obtain the following δ and γ CF values for $i, j = 1, 2, 3$, $i \neq j$:

$$\hat{v}^{\delta}(\{i\}) = 0; \quad \hat{v}^{\delta}(\{i,j\}) = \frac{1}{6}\frac{9c+b}{4c+b}; \qquad \hat{v}^{\delta}(M) = 1,$$

$$\hat{v}^{\gamma}(\{i\}) = 0; \quad \hat{v}^{\gamma}(\{i,j\}) = \frac{1}{6}\frac{(9c+b)\left(b^2+2cb-11c^2\right)}{b\left(5c+b\right)^2}; \quad \hat{v}^{\gamma}(M) = 1,$$

and we have

$$\hat{v}^{\delta}(\{i,j\}) - \hat{v}^{\gamma}(\{i,j\}) = \frac{2c\left(99c^3+74c^2b+16cb^2+b^3\right)}{3b\left(4c+b\right)\left(5c+b\right)^2} > 0.$$

Therefore,

$$\hat{v}^{\delta}(K) \geq \hat{v}^{\gamma}(K) \quad \text{for all } K \subset M.$$

2.3 Solutions to a Cooperative Game

The set of imputations and the characteristic function values are the main ingredients for defining a solution to a cooperative game. The set of imputations gives what is individually rational and feasible, whereas the CF values inform us about the strategic strength of each possible coalition. A solution can be a value, that is, one particular imputation, or it can be a subset of the set of imputations. Many solutions have been proposed over time and defined through given properties, e.g., fairness, stability of the allocation(s), etc. The earliest in the history of cooperative games are the *stable set* (von Neumann and Morgenstern 1944), *core* (Gillies 1953), *Shapley value* (Shapley 1953), *bargaining set* (Aumann and Maschler 1965), *kernel* (Davis and Maschler 1965), and the *nucleolus* (Schmeidler 1969). In this section, we only present the core and the Shapley value, which are by far the most used solutions in applications of cooperative games and which are of interest in the following chapters. For a comprehensive coverage of the theory of cooperative games, see the list of additional readings provided at the end of this chapter.

2.3.1 The Core

To define the core, we need to compare imputations. Let $y = (y_1, \ldots, y_m)$ and $y' = (y'_1, \ldots, y'_m)$ be two imputations of the cooperative game (M, v).

Definition 2.6 Imputation y dominates imputation y' through a coalition K if the following two conditions hold:

$$\text{feasibility} : \sum_{i \in K} y_i \leq v(K),$$

$$\text{preferability} : y_i > y'_i, \ \forall i \in K.$$

Definition 2.7 The core is the set of all undominated imputations.

In other words, the core is the set of imputations under which no coalition has a value greater than the sum of its members' payoffs. Therefore, no coalition has incentive to leave the grand coalition to receive a larger payoff. Although Definition 2.7 characterizes the imputations that belong to the core, it does not provide a scheme for finding them. The following theorem does.

Theorem 2.2 (Gillies 1953) *An imputation* $y = (y_1, \ldots, y_m)$ *is in the core if*

$$\sum_{i \in K} y_i \geq v(K), \ \forall K \subseteq M. \tag{2.7}$$

It is readily seen from (2.7) that the core is defined by $2^m - 2$ inequalities and one equality $\left(\sum_{i \in M} y_i = v(M) \right)$.

In Example 2.2, the set of imputations is given by

$$X = \{(x_1, x_2, x_3) \,|\, x_i \geq 0, \forall i, \ x_1 + x_2 + x_3 = 60\}.$$

The remaining inequalities in (2.7) that need to be considered are

$$x_1 + x_2 \geq 30 \Rightarrow x_3 \leq 30,$$
$$x_1 + x_3 \geq 20 \Rightarrow x_2 \leq 40,$$
$$x_2 + x_3 \geq 20 \Rightarrow x_1 \leq 40.$$

Therefore, the core is the set

$$C = \{(x_1, x_2, x_3) \,|\, 0 \leq x_1 \leq 40, \ 0 \leq x_2 \leq 40, \ 0 \leq x_3 \leq 30, \ x_1 + x_2 + x_3 = 60\}.$$

Figure 2.1 gives a geometrical representation of the core.

By assigning the set of payoffs that cannot be improved upon by any coalition, the core yields stable outcomes and eliminates further bargaining between players. A drawback is that the core may be empty.

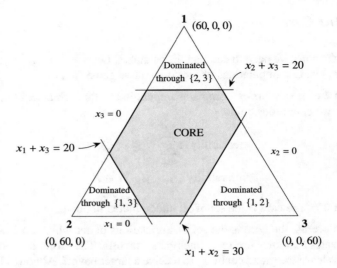

Fig. 2.1 Core is illustrated by the shaded region

Example 2.5 Consider the three-player cooperative game having the following CF values:

$$v(\{1\}) = v(\{2\}) = v(\{3\}) = 0,$$
$$v(\{1, 2\}) = v(\{1, 3\}) = v(\{2, 3\}) = a, \quad v(\{1, 2, 3\}) = 1,$$

where $0 < a < 1$.

We have three cases:

1. If $0 < a < 2/3$, then the core is given by

   $$C = \{(x_1, x_2, x_3) \mid 0 \le x_i \le 1 - a, \text{ for } i = 1, \ldots, 3 \text{ and } x_1 + x_2 + x_3 = 1\}.$$

2. If $a = 2/3$, then the core is the allocation $\left(\frac{1}{3}, \frac{1}{3}, \frac{1}{3}\right)$.
3. If $a > 2/3$, then the core is empty.

The intuition behind having an empty core when $a > 2/3$ is that intermediate coalitions are relatively too strong with respect to the grand coalition and cannot be contended with. A legitimate question is under what conditions the core is nonempty. The Bondareva-Shapley theorem answers this question (see Bondareva 1963; Shapley 1967).

We start by considering the linear programming problem:

$$\sum_{i=1}^{m} x_i \to \inf, \tag{2.8}$$

$$\sum_{i \in K} x_i \geq v(K) \quad \text{for any } K \subset M, \ K \neq \varnothing, M. \tag{2.9}$$

Obviously, the set of x satisfying system (2.9) is nonempty. Therefore, linear programming problem (2.8)–(2.9) has a solution. Let the minimal objective function value be denoted as v^*. The following theorem gives a necessary and sufficient condition for the core to be nonempty (see Naumova and Solovieva 2017).

Proposition 2.1 *The core C of cooperative game (M, v) is nonempty if and only if $v(M) \geq v^*$.*

Proof $\Rightarrow$ Let $C \neq \varnothing$, then any vector $x = (x_1, \ldots, x_m) \in C$ satisfies the system of inequalities (2.9). Obviously, $\sum_{i \in M} x_i \geq v^*$. By definition of the core, $\sum_{i \in M} x_i = v(M)$. Thus, we obtain that $v(M) \geq v^*$.

$\Leftarrow$ Let $v(M) \geq v^*$. Denote the solution of problem (2.8)–(2.9) by $\hat{x} = (\hat{x}_1, \ldots, \hat{x}_m)$. Define vector $x \in \mathbb{R}^m$ as follows:

$$x_i = \hat{x}_i + v(M) - v^*,$$
$$x_j = \hat{x}_j, \ j \in M, j \neq i,$$

where i is any player from the set M. It can be easily verified that such a vector x belongs to the core. Therefore, the core is nonempty. $\qquad\square$

Proposition 2.1 provides an intuition for the Bondareva-Shapley theorem.

Denote by C the set of possible coalitions. Let $C_i = \{K \in C : i \in K\}$ be the subset of those coalitions that have Player i as a member. A vector $(\lambda_K)_{K \in C} : \lambda_K \in [0, 1]$ for all $K \in C$ is called a balanced collection of weights if for all $i \in M$ we have $\sum_{K \in C_i} \lambda_K = 1$. A balanced collection is called minimal balanced if it does not contain a proper balanced subcollection. We refer to Peleg (1965), who proposed an inductive method for finding all the minimal balanced collections over set M for any m.

Theorem 2.3 (Bondareva 1963; Shapley 1967) *A cooperative game has a nonempty core if, and only if, for every minimal balanced collection of weights, $\sum_{K \in C} \lambda_K v(K) \leq v(M)$.*

We illustrate the Bondareva-Shapley theorem with an example.

Example 2.6 Consider the data in Example 2.4. To determine all minimal balanced collections of weights, we state the following linear programming problem (2.8)–(2.9) and solve its dual problem. The linear programming problem is given by

$$x_1 + x_2 + x_3 \rightarrow \min$$
$$x_1 \geqslant 0,$$
$$x_2 \geqslant 0,$$
$$x_3 \geqslant 0,$$
$$x_1 + x_2 \geqslant 30,$$
$$x_1 + x_3 \geqslant 20,$$
$$x_2 + x_3 \geqslant 20,$$

and the dual problem by

$$30\lambda_4 + 20\lambda_5 + 20\lambda_6 \rightarrow \max$$
$$\lambda_1 + \lambda_4 + \lambda_5 = 1,$$
$$\lambda_2 + \lambda_4 + \lambda_6 = 1,$$
$$\lambda_3 + \lambda_5 + \lambda_6 = 1,$$
$$\lambda_i \geqslant 0, \ i = 1, \ldots, 6,$$

where $\lambda_i, i = 1, \ldots, 6$ are the dual variables. The above problem has five solutions $\lambda = (\lambda_1, \ldots, \lambda_6)$, namely,

$$(0, 0, 1, 1, 0, 0), (1, 1, 1, 0, 0, 0), (1, 0, 0, 0, 0, 1), \ (0, 1, 0, 0, 1, 0), (0, 0, 0, \frac{1}{2}, \frac{1}{2}, \frac{1}{2}).$$

These solutions are the minimal balanced collection of weights. The inequality $\sum_{K \in C} \lambda_K v(K) \leq v(M)$ in Theorem 2.3 can be written as $v(M) \geqslant \max(30\lambda_4 + 20\lambda_5 + 20\lambda_6)$, with the maximum to be found over the set of the minimal balanced collection of weights. Substituting for the values of λ_4, λ_5, and λ_6 from the various solutions, we get the inequality $v(M) \geqslant \max\{30, 0, 20, 20, 35\} = 35$, which is satisfied because $v(M) = 60$. Therefore, the core is nonempty (which we already knew).

2.3.2   The Shapley Value

Shapley (1953) proposed a solution concept for cooperative games that assigns to each player a value that corresponds to a weighted sum of her contribution to all the coalitions she can join. Whereas the core focuses on the stability of the outcome and may be empty, the Shapley value always exists and selects a unique and fair imputation. Denote by $\phi(v) = (\phi_1(v), \ldots, \phi_m(v))$ the Shapley value, where $\phi_i(v)$ is Player i's component.

To define his value, Shapley states three axioms:

Symmetry: If $v(S \cup \{i\}) = v(S \cup \{j\})$ for every subset $S \subset M$ that contains neither i nor j, then $\phi_i(v) = \phi_j(v)$. This property is interpreted as fair or equal treatment for identical players.

Efficiency: The sum of the value components is equal to the grand coalition's gain, i.e., $\sum_{i=1}^{m} \phi_i(v) = v(M)$.

Linearity: If two cooperative games defined over the same set of players and described by CFs $v(\cdot)$ and $w(\cdot)$ are combined, then the allocated gains should correspond to the gains obtained from $v(\cdot)$ and the gains obtained from $w(\cdot)$, that is, $\phi_i(v + w) = \phi_i(v) + \phi_i(w)$ for all $i \in M$.

Theorem 2.4 *The unique m-vector satisfying the symmetry, efficiency, and linearity axioms is given by*

$$\phi_i(v) = \sum_{K \ni i} \frac{(m - k)!\,(k - 1)!}{m!} \left(v(K) - v(K \setminus \{i\}) \right), \ \forall i \in M. \tag{2.10}$$

The difference $v(K) - v(K \setminus \{i\})$ measures the marginal contribution of Player i to coalition K. Owen (2013) (p. 271) gives an intuitive interpretation of the term $\frac{(m-k)!(k-1)!}{m!}$. Suppose that $M = \{1, \ldots, m\}$ players enter a room in some order. Whenever a player enters the room and players $K \setminus \{i\}$ are already there, she is paid her marginal contribution $MC_i(K) = v(K) - v(K \setminus \{i\})$. Suppose all $m!$ orders are equally likely. Then, there are $(k - 1)!$ different orders in which the players in $K \setminus \{i\}$ can precede i, and $(m - k)!$ orders in which they can follow i. Therefore, we have a total of $(k - 1)!(m - k)!$ orders for that case of $m!$ total orders. Ex ante, the payoff of a player i is

$$\phi_i(v) = \sum_{K \ni i} \frac{(m - k)!\,(k - 1)!}{m!} MC_i(K), \ \forall i \in M,$$

that is, her Shapley value.

Example 2.7 In Example 2.2, about three neighboring towns, we had

$$v(\{1\}) = 0, \quad v(\{2\}) = 0, \quad v(\{3\}) = 0,$$
$$v(\{1, 2\}) = 30, \quad v(\{1, 3\}) = 20, \quad v(\{2, 3\}) = 20,$$
$$v(\{1, 2, 3\}) = 60.$$

Applying the formula in (2.10), we get

$$\phi(v) = \left(\frac{130}{6}, \frac{130}{6}, \frac{100}{6} \right).$$

As they are identical in all respects, it comes as no surprise that Players 1 and 2 obtain the same payoff.

Recalling that the core in this example is given by

$$C = \{(x_1, x_2, x_3) \mid 0 \leq x_1 \leq 40, \ 0 \leq x_2 \leq 40, \ 0 \leq x_3 \leq 30, \ x_1 + x_2 + x_3 = 60\},$$

it is clear that the Shapley value belongs to the core.

How general is the result obtained above, namely, that the Shapley value is in the core? As the Shapley value always exists and the core may be empty, we infer that this result cannot always hold true. Convex games, however, have the following nice property.

Theorem 2.5 (Shapley 1971) *If the game is convex, then the core is nonempty and the Shapley value is the core's center of gravity.*

Note that the convexity property is a sufficient, not necessary, condition.

2.4   The Nash Bargaining Solution

Following an axiomatic approach, Nash (1953) proposed a solution to two-player bargaining problems. Denote by u_i the utility (payoff) of Player i. To start with, two items are needed to define the Nash bargaining solution (NBS): (i) the set of feasible solutions (S), and (ii) the status quo (or disagreement) point (u^0, v^0), that is, the players' payoffs in case negotiation/cooperation fails. We suppose that S is a convex and compact set containing (u^0, v^0).

A bargaining procedure is a mapping

$$\phi : (S, u^0, v^0) \rightarrow (u^*, v^*),$$

where (u^*, v^*) is the pair of players' outcomes. To determine these outcomes, Nash introduced the following six axioms:

A1. Feasibility: $(u^*, v^*) \in S$.
A2. Individual rationality: $u^* \geq u^0$, $v^* \geq v^0$.
A3. Pareto optimality: If $(u, v) \in S$ with $u \geq u^*$ and $v \geq v^*$, then $(u, v) = (u^*, v^*)$.
A4. Invariance with respect to the linear transformation of utilities: Let T be obtained from S by the transformation

$$u' = \alpha_1 u + \beta_1 \quad (\alpha_1 > 0), \quad v' = \alpha_2 v + \beta_2 \quad (\alpha_2 > 0).$$

If $\phi(S, u^0, u^0) = (u^*, v^*)$, we require that

$$\phi(T, \alpha_1 u^0 + \beta_1, \alpha_2 v^0 + \beta_2) = (\alpha_1 u^* + \beta_1, \alpha_2 v^* + \beta_2).$$

A5. Symmetry: If

$$(i) : (u, v) \in S \Rightarrow (v, u) \in S,$$
$$(ii) : u^0 = v^0,$$
$$(iii) : \phi(S, u^0, v^0) = (u^*, v^*),$$

then $u^* = v^*$.

A6. Independence of irrelevant alternatives: If $(u^*, v^*) \in T \subset S$ and $(u^*, v^*) = \phi(S, u^0, v^0)$, then $(u^*, v^*) = \phi(T, u^0, v^0)$.

Theorem 2.6 (Nash 1953) *Suppose there exist points $(u, v) \in S$ with $u > u^0$ and $v > v^0$ and that the maximum of*

$$g(u, v) = (u - u^0)(v - v^0)$$

over this set is attained at (u^*, v^*). Then, the point (u^*, v^*) is uniquely determined, and $\phi(S, u^0, v^0) = (u^*, v^*)$ is the unique function that satisfies axioms A1–A6.

As the objective function $g(u, v)$ is continuous, and S is compact and convex, there exists an optimal solution to the optimization problem stated in the above theorem. Further, the objective function is strictly quasi-concave, and therefore, the solution is unique.

Axioms A1–A5 are intuitive. Indeed, A1 and A2 state that it does not help to consider a solution that is not feasible or leaves a player worse off than her status quo utility. (Individual rationality is a cornerstone of game theory.) A3 excludes an interior solution, that is, a solution dominated by another feasible one lying on the Pareto surface; otherwise, some gains would be wasted. One interpretation of A4 is that it does not matter whether bargaining is done in dollars or euros. Symmetry is synonymous with fairness or the egalitarian principle, which is an oft-sought property in negotiations/arbitration. If we let $c = \max(u + v)$, then the Nash bargaining outcome (u^*, v^*) is given by

$$u^* = u^0 + \frac{1}{2}(c - u^0 - v^0),$$
$$v^* = v^0 + \frac{1}{2}(c - u^0 - v^0),$$

that is, each player gets her status quo gain plus half of the surplus generated by the bargaining procedure. The solution is fair because

$$u^* - u^0 = v^* - v^0,$$

meaning that both players equally improve their gains with respect to the disagreement point.

Axiom A6 did not get the same positive reception as the five others. It states that a preferred solution in a larger set will still be a preferred solution in a smaller subset. To illustrate why this axiom is controversial, consider the following example

(p. 471, Ordeshook 1986). Two players must share a dollar, and if they fail to reach an agreement, then each person gets nothing. The arbitrated solution to this game is $(0.50, 0.50)$. Now, suppose that Player 2 cannot (for some reason) receive more than 0.50. Axiom A6 requires that the solution remain $(0.50, 0.50)$. One can argue that this solution is not fair, as Player 1's opportunities for gain appear to exceed Player 2's opportunities. Kalai and Smorodinsky (1975) proposed an alternative solution to the NBS, where a monotonicity axiom is included instead of axiom A6.

In the classical framework of the NBS, the status quo is considered as a given point. However, it could also correspond to the result of the implementation of specific (threat) actions. When the disagreement point is variable, then one can construct a game in which each player chooses a threat and receives a payoff according to the outcome of bargaining.

Example 2.8 Consider a supply chain formed of a supplier, Player s, who sells her product through a retailer, Player r. The manufacturer decides her margin m_s and rate of advertising a_s. The retailer controls her margin m_r. The retail price is the sum of margins, i.e.,

$$p = m_s + m_r.$$

The consumer demand q depends on the retail price p and the supplier's advertising, and is given by

$$q = a_s(\alpha - \beta p),$$

where α and β are strictly positive parameters. The manufacturer's advertising cost is convex increasing and given by $\frac{1}{2}wa_s^2$, where $w > 0$. The players maximize their profits given by

$$\pi_s(m_r, m_s, a_s) = m_s a_s (\alpha - \beta p) - \frac{1}{2}wa_s^2,$$

$$\pi_r(m_r, m_s, a_s) = m_r a_s (\alpha - \beta p).$$

Collectively, the supply chain is better off when the players coordinate their strategies and maximize their joint payoff. That is,

$$\max \pi = \pi_s + \pi_r = (m_s + m_r) a_s (\alpha - \beta p) - \frac{1}{2}wa_s^2.$$

Assuming an interior solution, the maximization yields a total payoff equal to $\frac{\alpha^4}{32w\beta^2}$. The sharing of this payoff depends on the status quo. One option is to let the disagreement point be a Nash equilibrium outcome. The rationale is straightforward: if the players do not coordinate their strategies, then they will play a noncooperative game, and the Nash equilibrium is then considered the security outcome. The equilibrium conditions for a Nash equilibrium are

$$\frac{\partial \pi_s}{\partial a_s} = \frac{\partial \pi_s}{\partial m_s} = \frac{\partial \pi_r}{\partial m_r} = 0.$$

Straightforward computations give the unique equilibrium outcomes

$$\left(\pi_s^0, \pi_r^0\right) = \left(\frac{\alpha^4}{162w\beta^2}, \frac{\alpha^4}{81w\beta^2}\right).$$

To get the Nash bargaining solution, we solve the following optimization problem:

$$g\left(\pi_m, \pi_r\right) = \left(\pi_m - \frac{\alpha^4}{162w\beta^2}\right)\left(\pi_r - \frac{\alpha^4}{81w\beta^2}\right),$$

$$\text{subject to} : \pi_m + \pi_r = \frac{\alpha^4}{32w\beta^2},$$

and obtain $\left(\pi_m^*, \pi_r^*\right) = \left(\frac{65\alpha^4}{5184w\beta^2}, \frac{97\alpha^4}{5184w\beta^2}\right)$. Figure 2.2 gives a visual illustration of this example.

In Example 2.8, we chose the Nash equilibrium outcomes as the status quo point, but it could have been something else. In the supply chain literature (see the surveys in Jørgensen and Zaccour 2014; De Giovanni and Zaccour 2019), the supplier is often assigned the role of leader in a game à la Stackelberg (von Stackelberg 1934). The reason is that the upstream firm must first announce its wholesale price before the downstream firm can choose its retail price. Therefore, one could argue that the relevant noncooperative solution is a Stackelberg equilibrium and, consequently, let the status quo point be the resulting outcomes.

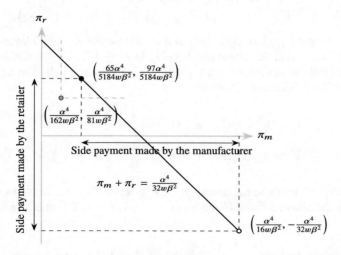

Fig. 2.2 Illustration of the Nash bargaining solution. The gray circle indicates the disagreement point, the black circle indicates the Nash bargaining solution, and the white circle indicates the joint optimization payoff of the players

The following example illustrates how the status quo can be chosen endogenously, that is, supported by the choice of a particular strategy by each player.

Example 2.9 Consider the following matrix game

$$
\begin{array}{cc}
 & \text{Player 2} \\
 & b_1 \qquad b_2 \\
\end{array}
$$

$$
\text{Player 1}\quad
\begin{array}{c}
a_1 \\
a_2 \\
\end{array}
\begin{array}{cc}
-1/2, -1/4 & 3/4, 1/4 \\
1/4, 3/4 & 0, 0 \\
\end{array}
$$

where the entries $\left(u_{ij}, v_{ij}\right), i, j = 1, 2$ are the payoffs of Player 1 and Player 2, respectively. The maximum payoff that the players can collectively achieve is one. Note that this game has two Nash equilibria in pure strategies, i.e., (a_2, b_1) and (a_1, b_2), and one in mixed strategies, i.e., $(r, q) = \left(\frac{3}{5}, \frac{1}{2}\right)$, where r is the probability of Player 1 choosing a_1, and q the probability of Player 2 choosing b_1. The corresponding payoff pair is $\left(\frac{5}{40}, \frac{6}{40}\right)$.

What would the NBS outcomes be? The answer depends on the choice of the status quo. If the players choose as their status quo (or threat) strategies

- (a_1, b_1), then $\left(u^0, v^0\right) = (u_{11}, v_{11}) = (-1/2, -1/4)$ and $(u^*, v^*) = \left(\frac{3}{8}, \frac{5}{8}\right)$;
- (a_1, b_2), then $\left(u^0, v^0\right) = (u_{12}, v_{12}) = (3/4, 1/4)$ and $(u^*, v^*) = \left(\frac{3}{4}, \frac{1}{4}\right)$;
- (a_2, b_1), then $\left(u^0, v^0\right) = (u_{21}, v_{21}) = (1/4, 3/4)$ and $(u^*, v^*) = \left(\frac{1}{4}, \frac{3}{4}\right)$;
- (a_2, b_2), then $\left(u^0, v^0\right) = (u_{22}, v_{22}) = (0, 0)$ and $(u^*, v^*) = \left(\frac{1}{2}, \frac{1}{2}\right)$;
- (r, q), then $\left(u^0, v^0\right) = \left(u_{rq}, v_{rq}\right) = \left(\frac{5}{40}, \frac{6}{40}\right)$ and $(u^*, v^*) = \left(\frac{39}{80}, \frac{41}{80}\right)$.

Clearly, Player 1 prefers $(3/4, 1/4)$ as a status quo point, while Player 2 prefers $(1/4, 3/4)$. But in fact, the status quo is going to be $(-1/2, -1/4)$, that is, none of the Nash equilibrium outcomes nor the symmetric outcome $(0, 0)$. The argument is as follows: The NBS outcomes are given by

$$
u^* = u^0 + \frac{1}{2}\left(1 - u^0 - v^0\right) = \frac{1}{2} + \frac{1}{2}\left(u^0 - v^0\right),
$$

$$
v^* = v^0 + \frac{1}{2}\left(1 - u^0 - v^0\right) = \frac{1}{2} - \frac{1}{2}\left(u^0 - v^0\right).
$$

Clearly, Player 1 wants to maximize $\left(u^0 - v^0\right)$, while Player 2 wants to minimize $\left(u^0 - v^0\right)$. Considering the differences $\left(u_{ij} - v_{ij}\right), i = 1, 2, r$ and $j = 1, 2, q$ we then have

$$
\text{Player 1}: \max\left\{-\frac{1}{4}, \frac{1}{2}, -\frac{1}{2}, 0, -\frac{1}{40}\right\} = 1/2 \rightarrow \text{choose } a_1,
$$

$$
\text{Player 2}: \min\left\{-\frac{1}{4}, \frac{1}{2}, -\frac{1}{2}, 0, -\frac{1}{40}\right\} = -1/2 \rightarrow \text{choose } b_1.
$$

The conclusion is that the best threats for players 1 and 2 are a_1 and b_1, respectively. The above reasoning can easily be generalized to the case where Player 1 has $K > 2$ strategies, Player 2 has $L > 2$ strategies, and the payoff to share is a constant c. Indeed, if the status quo strategy pair is (a_{kl}, b_{kl}) with corresponding payoffs (u_{kl}, v_{kl}), then the players get the following outcomes in the NBS:

$$u^* = \frac{c}{2} + \frac{1}{2}\left(u_{kl} - v_{kl}\right),$$

$$v^* = \frac{c}{2} - \frac{1}{2}\left(u_{kl} - v_{kl}\right).$$

To choose their threat strategies, Player 1 maximizes $(u_{kl} - v_{kl})$ and Player 2 minimizes $(u_{kl} - v_{kl})$, for $k = 1, \ldots, K$ and $l = 1, \ldots, L$.

2.4.1 Price of Anarchy

By virtue of joint optimization, cooperation leads to a collective outcome that is at least the sum of the individual payoffs under a noncooperative mode of play. In any cooperative game, the dividend of cooperation (DoC) is given by

$$DoC = v(M) - \sum_{i \in M} v(\{i\}).$$

Similarly, in the Nash bargaining solution we have $DoC = (u^* + v^*) - \left(u^0 + v^0\right)$. Recall that if the DoC is equal to zero, then the cooperative game is inessential.

If the individual noncooperative outcomes correspond to the equilibrium payoffs and if the equilibrium is not unique, then measuring the benefits of cooperation becomes trickier. In algorithmic game theory, the authors typically compute the so-called price of anarchy (PoA) to assess the cost of lack of a coordination between the players. This concept was proposed first by Koutsoupias and Papadimitriou (1999) to measure a network's efficiency, i.e., how the system payoff decreases when the players behave selfishly, with the benchmark being the "centralized" or cooperative payoff.

Consider a normal-form game $\Gamma = (M, \{S_i\}_{i \in M}, \{u_i\}_{i \in M})$, with M being the set of players, S_i the strategy set of Player i, and $u_i : S \to \mathbb{R}$ her utility function, where $S = S_1 \times \cdots \times S_m$ is the set of strategy profiles. Let the measure of efficiency of any strategy profile $s \in S$ be the sum of players' payoffs with this strategy profile, i.e., $\sum_{i \in M} u_i(s)$. Let S_{ne} be the set of Nash equilibria in game Γ.

The PoA in game Γ is the ratio between the optimal cooperative strategy profile and the "worst equilibrium", that is,

$$PoA = \frac{\max_{s \in S} \sum_{i \in M} u_i(s)}{\min_{s \in S_{ne}} \sum_{i \in M} u_i(s)}. \qquad (2.11)$$

If in game Γ, players have costs instead of payoffs, then the price of anarchy is given by

$$PoA = \frac{\max_{s \in S_{ne}} \sum_{i \in M} u_i(s)}{\min_{s \in S} \sum_{i \in M} u_i(s)}. \qquad (2.12)$$

There exists a related measure called the price of stability (PoS), which is the ratio between the payoffs in the "best equilibrium" and the cooperative solution. In the case of payoffs, we have

$$PoS = \frac{\max_{s \in S} \sum_{i \in M} u_i(s)}{\max_{s \in S_{ne}} \sum_{i \in M} u_i(s)}, \qquad (2.13)$$

and in the case of costs, we have

$$PoS = \frac{\min_{s \in S_{ne}} \sum_{i \in M} u_i(s)}{\min_{s \in S} \sum_{i \in M} u_i(s)}. \qquad (2.14)$$

One can easily notice that $1 \leq PoS \leq PoA$.

We illustrate the PoA and PoS in two types of games, namely, the prisoner's dilemma and congestion games.

Example 2.10 Consider the prisoner's dilemma game introduced in Chap. 1, Example 1.1.

Prisoner 2

Confess Deny

Confess (8, 8)   (0, 15)

Prisoner 1

Deny   (15, 0)   (1, 1)

The unique Nash equilibrium is (Confess, Confess) with a total cost of 16. The cooperative solution is (Deny, Deny) with a total cost of 2. Then, $PoA = 16/2 = 8$. Because the equilibrium is unique, the price of anarchy coincides with the price of stability, $PoS = 8$.

Example 2.11 Consider a two-player congestion game. Figure 2.3 represents the directed graph, where any player has two strategies: strategy A, which goes from O to F through A, and strategy B, which goes from O to F through B. If a player chooses strategy A, her cost will be $1 + x^2$, where x is the number of players using the route $O \rightarrow A \rightarrow F$. If a player chooses strategy B, her costs will be $2 + x^2$. The following matrix defines the players' costs:

Fig. 2.3 The directed graph
for a congestion game with
partially quadratic costs

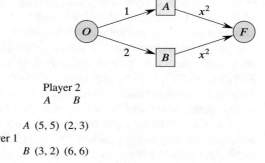

Player 2
     A B

                      A $(5, 5)$ $(2, 3)$
         Player 1
                      B $(3, 2)$ $(6, 6)$

There exist two Nash equilibria in pure strategies (A, B) and (B, A) and one equilibrium in mixed strategies $\left(\left(\frac{2}{3}, \frac{1}{3}\right); \left(\frac{2}{3}, \frac{1}{3}\right)\right)$ with players' costs $(2, 3)$, $(3, 2)$, and $(4, 4)$, respectively. There are two cooperative solutions corresponding to the strategy profiles, namely, (A, B) and (B, A), which are also Nash equilibria in pure strategies. Here, $PoA = 8/5$, while $PoS = 1$, because the "best" Nash equilibrium coincides with the cooperative solution.

Example 2.12 We extend the previous example to three players, represented by the following cost matrices, where Player 1 chooses a row, Player 2 a column, and Player 3 a matrix:

	Player 3 : X		Player 3 : Y	
	A	B	A	B
A	$(9, 9, 9)$	$(5, 3, 5)$	$(5, 5, 3)$	$(2, 6, 6)$
B	$(3, 5, 5)$	$(6, 6, 2)$	$(6, 2, 6)$	$(10, 10, 10)$

There are three Nash equilibria in pure strategies (A, B, X), (B, A, X), (A, A, Y) and one equilibrium in mixed strategies $\left(\left(\frac{4}{7}, \frac{3}{7}\right), \left(\frac{4}{7}, \frac{3}{7}\right), \left(\frac{4}{7}, \frac{3}{7}\right)\right)$ with the players' costs given by $(5, 3, 5)$, $(3, 5, 5)$, $(5, 5, 3)$, and $(5.7551, 5.7551, 5.7551)$, respectively. The strategy profiles (A, B, X), (B, A, X), and (A, A, Y) yield a collective (cooperative) cost of 13. The PoA is $(5.7551 \cdot 3)/(5 + 3 + 5) \approx 1.358$, while the PoS is equal to one.

2.5   Additional Readings

A broad range of textbooks and lecture notes covering cooperative games is available. Owen (2013) is a classic one (the first edition dates back to 1968). One of the early books dedicated to cooperative games is Driessen (1988), where classical solution concepts are described in details and illustrated with examples. Moulin

(2014) discusses cooperative games concepts in the context of microeconomics, while Curiel (2013) gives an extensive account of applications of cooperative games to combinatorial optimization problems, which are extensively studied in operations research in one-decision-maker setups. Peters (2008) covers a large number of topics in game theory, with seven chapters dedicated to cooperative games. Peleg and Sudhölter (2003), Chakravarty et al. (2015) deal exclusively and extensively with cooperative games. Readers interested in the computational issues arising in cooperative games would find interesting material in Chakravarty et al. (2015). The basic models of cooperative game theory, including the core, the Shapley value, stable sets, and the bargaining problem and its solutions, are described in Osborne and Rubinstein (1994). Myerson (2013) describes different aspects of bargaining in two-person games, cooperative games with transferable and nontransferable utilities, and models of cooperation under uncertainty. Bilbao (2000) covers cooperative game models defined on combinatorial structures. Readers can study properties, solutions, and applications of partition function games in Kóczy (2018), extending the ideas of cooperative game theory when the value of cooperation depends on outsiders' actions.

2.6 Exercises

Exercise 2.1 Show that every essential cooperative game is strategically equivalent to exactly one game in 0–1 normalized form.

Exercise 2.2 One hundred individuals each have a right shoe and 101 individuals each have a left shoe. Denote by K a coalition and by $v(K)$ the characteristic function of the cooperative game with 201 players. $v(K)$ is equal to one dollar times the number of complete pairs (i.e., with both one right shoe and one left shoe) that coalition K can realize. Determine the core of this game.

Exercise 2.3 Consider a three-player cooperative game where the utility function of Player i is given by

$$u_i = \alpha_i s_i - \frac{1}{2}\beta_i s_i^2 - \frac{1}{2}\sigma \left(\sum_{i=1}^{3} s_i \right), \quad i \in M = \{1, 2, 3\},$$

where $s_i \geq 0$ is the strategy of Player i, and α_i, β_i, and σ are positive parameters.
   Show that $v^\gamma(K) = v^\delta(K)$ for all $K \subset M$.

Exercise 2.4 (treasure hunt game) A group of m people have found a treasure, and each pair among them can carry no more than one piece. The cooperative game (M, v) is given by the set of players $M = \{1, \ldots, m\}$, $m > 1$, and the characteristic function, $v(K) = \frac{k}{2}$ if k is even, and $v(K) = \frac{k-1}{2}$ if k is odd.

1. Show that, for $m = 2$, the core coincides with the imputation set.

2. Show that, for $m \geqslant 4$ and even, the core is the unique point $(\frac{1}{2}, \ldots, \frac{1}{2})$.
3. Show that when m is odd, the core is the empty set.

Exercise 2.5 This exercise is based on Chander and Tulkens (1997). Consider a model with multilateral externalities and two kinds of commodities: a standard private good, whose quantities are denoted by $x_1, x_2, \ldots, x_m$, and an environmental good (pollution emissions), whose quantities are denoted by $e_1, e_2, \ldots, e_m$. Suppose that the private good can be produced by the agents according to the following production technology and the transfer function, respectively, $x_i = f_i(e_i)$ and $e = \sum_i e_i$, where e_i and x_i denote agent i's emissions and output, respectively. Here, $f_i(e_i)$ is assumed to be increasing, differentiable and concave in e_i. Each agent's preferences are represented by a quasi-linear utility function $u_i(x_i, e) = x_i - d_i(e) = f_i(e_i) - d_i(e)$, where $d_i(e)$ is agent i's disutility, which is assumed to be a positive, increasing, differentiable, and convex function of the level of the externality $e = \sum_{i \in M} e_i$.

1. Compute the values of $v^\gamma(K)$ for all $K \subset M$.
2. Show that the γ core is nonempty.
3. Compute the values of $v^\delta(K)$ for all $K \subset M$.
4. Show that the δ core is nonempty.
5. Show that δ core $\subseteq \gamma$-core.

Exercise 2.6 In Example 2.8, the assumption is that the status quo is given as the Nash equilibrium outcomes.

1. Suppose that the status quo corresponds to the Stackelberg equilibrium outcomes, with the supplier as leader and the retailer as follower. Determine the NBS outcome.
2. Suppose that the status quo corresponds to the Stackelberg equilibrium outcomes, with the retailer as leader and the supplier as follower. Determine the NBS outcome.
3. Compare and comment on the results.

Exercise 2.7 A two-player congestion game is depicted in Fig. 2.4. The cost of a route, i.e., from A to F and from B to F, is equal to the number of players choosing this route.

Fig. 2.4 The directed graph for a congestion game with partially linear costs

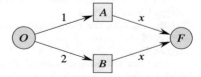

The following matrix gives the players' costs:

$$
\begin{array}{cc}
 & \text{Player 2} \\
 & A \qquad B
\end{array}
$$

$$
\text{Player 1} \quad
\begin{array}{c}
A \\
B
\end{array}
\begin{array}{cc}
(3, 3) & (2, 3) \\
(3, 2) & (4, 4)
\end{array}
$$

Compute the prices of anarchy and stability.

Exercise 2.8 We extend the previous exercise to three players represented by the following cost matrices, where Player 1 chooses a row, Player 2 a column, and Player 3 a matrix:

$$
\begin{array}{cc}
\text{Player 3 : } X & \text{Player 3 : } Y \\
A \qquad B & A \qquad B
\end{array}
$$

$$
\begin{array}{cc}
A \ (4, 4, 4) \ (3, 3, 3) & A \ (3, 3, 3) \ (2, 4, 4) \\
\\
B \ (3, 3, 3) \ (4, 4, 2) & B \ (4, 2, 4) \ (5, 5, 5)
\end{array}
$$

Compute the prices of anarchy and stability.

References

Aumann, R. J. and Maschler, M. (1965). The bargaining set for cooperative games. In Dresher, M., Shapley, L. S., and Tucker, A. W., editors, *Advances in Game Theory. (AM-52), Volume 52*, pages 443–476. Princeton University Press.

Aumann, R. J. and Peleg, B. (1960). Von Neumann-Morgenstern solutions to cooperative games without side payments. *Bulletin of the American Mathematical Society*, 66(3):173–179.

Bilbao, J. (2000). *Cooperative Games on Combinatorial Structures*. Theory and Decision Library C. Springer US.

Bondareva, O. N. (1963). Some applications of linear programming methods to the theory of cooperative games. *Problemy Kibernetiki*, 10:119–139.

Chakravarty, S., Mitra, M., and Sarkar, P. (2015). *A Course on Cooperative Game Theory*. Cambridge University Press.

Chander, P. (2007). The gamma-core and coalition formation. *International Journal of Game Theory*, 35(4):539–556.

Chander, P. and Tulkens, H. (1997). The core of an economy with multilateral environmental externalities. *International Journal of Game Theory*, 26(3):379–401.

Curiel, I. (2013). *Cooperative Game Theory and Applications: Cooperative Games Arising from Combinatorial Optimization Problems*. Theory and Decision Library C. Springer US.

Davis, M. and Maschler, M. (1965). The kernel of a cooperative game. *Naval Research Logistics Quarterly*, 12(3):223–259.

De Giovanni, P. and Zaccour, G. (2019). A selective survey of game-theoretic models of closed-loop supply chains. *4OR*, 17(1):1–44.

Driessen, T. (1988). *Cooperative Games, Solutions and Applications*. Theory and decision library: Series C, Game theory, mathematical programming, and operations research. Springer.

Germain, M., Toint, P., Tulkens, H., and de Zeeuw, A. (2003). Transfers to sustain dynamic core-theoretic cooperation in international stock pollutant control. *Journal of Economic Dynamics and Control*, 28(1):79–99.

Gillies, D. B. (1953). *Some Theorems on N-person Games*. Princeton University Press.

Gromova, E. V. and Petrosyan, L. (2017). On an approach to constructing a characteristic function in cooperative differential games. *Automation and Remote Control*, 78(9):1680–1692.

Jørgensen, S. and Zaccour, G. (2014). A survey of game-theoretic models of cooperative advertising. *European Journal of Operational Research*, 237(1):1–14.

Kalai, E. and Smorodinsky, M. (1975). Other solutions to Nash's bargaining problem. *Econometrica*, 43(3):513–518.

Kóczy, L. (2018). *Partition Function Form Games: Coalitional Games with Externalities*. Theory and Decision Library C. Springer International Publishing.

Koutsoupias, E. and Papadimitriou, C. (1999). Worst-case equilibria. In Meinel, C. and Tison, S., editors, *STACS 99*, pages 404–413, Berlin, Heidelberg. Springer Berlin Heidelberg.

Moulin, H. (2014). *Cooperative Microeconomics: A Game-Theoretic Introduction*. Princeton Legacy Library. Princeton University Press.

Myerson, R. (2013). *Game Theory: Analysis of Conflict*. Harvard University Press.

Nash, J. (1953). Two-person cooperative games. *Econometrica*, 21(1):128–140.

Naumova, N. I. and Solovieva, N. (2017). *Bondareva-Shapley theorem (in Russian)*, pages 52–58. In: Malozemov, V.N. (ed) Selected lectures on extremum problems, Saint Petersburg: BBM Publishing house.

Ordeshook, P. C. (1986). *Game Theory and Political Theory: An Introduction*. Cambridge University Press.

Osborne, M. and Rubinstein, A. (1994). *A Course in Game Theory*. The MIT Press. MIT Press.

Owen, G. (2013). *Game Theory*. Emerald Group Publishing Limited.

Peleg, B. (1965). An inductive method for constructing mimmal balanced collections of finite sets. *Naval Research Logistics Quarterly*, 12(2):155–162.

Peleg, B. and Sudhölter, P. (2003). *Introduction to the Theory of Cooperative Games*. Theory and Decision Library. Kluwer Academic Publishers.

Peters, H. (2008). *Game Theory: A Multi-Leveled Approach*. Springer Berlin Heidelberg.

Petrosyan, L. and Zaccour, G. (2003). Time-consistent Shapley value allocation of pollution cost reduction. *Journal of Economic Dynamics and Control*, 27(3):381–398.

Reddy, P. V. and Zaccour, G. (2016). A friendly computable characteristic function. *Mathematical Social Sciences*, 82:18–25.

Schmeidler, D. (1969). The nucleolus of a characteristic function game. *SIAM Journal on Applied Mathematics*, 17(6):1163–1170.

Shapley, L. S. (1953). A value for n-person games. In Kuhn, H. W. and Tucker, A. W., editors, *Contributions to the Theory of Games (AM-28), Volume II*, pages 307–318. Princeton University Press.

Shapley, L. S. (1967). On balanced sets and cores. *Naval Research Logistics Quarterly*, 14(4):453–460.

Shapley, L. S. (1971). Cores of convex games. *International Journal of Game Theory*, 1(1):11–26.

Thrall, R. M. and Lucas, W. F. (1963). N-person games in partition function form. *Naval Research Logistics Quarterly*, 10(1):281–298.

von Neumann, J. and Morgenstern, O. (1944). *Theory of Games and Economic Behavior*. Princeton University Press.

von Stackelberg, H. (1934). *Marktform und Gleichgewicht*. Springer.

Zaccour, G. (2003). Computation of characteristic function values for linear-state differential games. *Journal of Optimization Theory and Applications*, 117(1):183–194.

Part II
Deterministic Dynamic Games

Chapter 3
Multistage Games

In this chapter, we provide a brief introduction to the theory of multistage games. We recall some properties of dynamical systems described in state-space formalism, and introduce the idea of information structure, which defines what the players know when they make their decisions. Next, we characterize information, contingent equilibrium solutions of the game through either a coupled maximum principle or dynamic programming. Further, we present a class of multistage games described by linear dynamics and quadratic objectives, which are attractive from a computational viewpoint. This chapter is primarily based on Chap. 6 of Haurie et al. (2012).

3.1 State-Space Description of Multistage Games

Games that are not played once but evolve over multiple rounds or stages are referred to as multistage games or dynamic games. We recall from Chap. 1 that extensive-form games model strategic situations in which the players act sequentially. The tree structure of such games captures how the information unfolds during the game as a result of the players' actions. However, the tree representation ceases to be practical when the game is played over a large number of stages (think of a chess game) and when the players have large action sets. This shortcoming can be mitigated if the information carried by the tree structure is conveyed by one variable, referred to as a state variable, which evolves over time. This approach provides a way to model games played over multiple rounds using dynamical systems, which are defined and studied in a state space. A state space contains state variables that provide an exhaustive summary of how players' actions (also called inputs) have impacted the system in the past. By knowing the current state and the future time profiles of players' actions, we can predict the future behavior of a dynamic system.

Denote by $\mathbb{T} = \{0, 1, 2, \ldots, T\}$ the set of decision instants or stages of the game, and by $x_t \in X \subset \mathbb{R}^q$ the state vector at time $t \in \mathbb{T}$, and by $u_t^i \in U^i \subseteq \mathbb{R}^{\ell_i}$ the control vector of Player i. Let

$$\mathbf{x}_t := \{x_0, x_1, \ldots, x_t\}, \tag{3.1}$$

$$u_t := \{u_t^1, u_t^2, \ldots, u_t^m\}, \tag{3.2}$$

$$\mathbf{u}^i := \{u_0^i, u_1^i, \ldots, u_{T-1}^i\}, \tag{3.3}$$

$$\mathbf{u} := \{u_0, u_1, \ldots, u_{T-1}\}. \tag{3.4}$$

To define a multistage game we need to specify the following elements:

1. A set of players $M := \{1, 2, \ldots, m\}$;
2. A state transition equation governing the dynamic interaction environment of the players

$$x_{t+1} = f(t, x_t, u_t), \tag{3.5}$$

where $f : \mathbb{T} \times \mathbb{R}^q \times \mathbb{R}^{\ell_1} \times \cdots \times \mathbb{R}^{\ell_m} \to \mathbb{R}^q$ denotes the state transition function from stage t to $t + 1$. The initial state $x_0 \in \mathbb{R}^q$ is assumed to be known;

3. The history of the game at a stage t is defined by

$$\mathcal{H}_t := \{t, \{x_0, x_1, \ldots, x_t\}, \{u_0, u_1, \ldots, u_{t-1}\}\}, \tag{3.6}$$

which describes the sequence of control and state variable values until stage t;

4. An information structure, which represents the information that the players use when determining their controls at time t; usually, the information structure is a subset of the history $\mathcal{H}_t$;
5. Any other relevant restrictions on the system's variables;
6. A payoff function (also called utility function or performance criterion) for each player.

The payoff function depends on the planning horizon T of the game. When T is finite, the performance criterion of Player $i \in M$ is given by

$$J_i(x_0, \mathbf{u}) = \sum_{t=0}^{T-1} \phi_i(t, x_t, u_t) + \Phi_i(T, x_T), \tag{3.7}$$

where $\phi_i : \mathbb{T} \times \mathbb{R}^q \times \mathbb{R}^{\ell_1} \times \cdots \times \mathbb{R}^{\ell_m} \to \mathbb{R}$ is the instantaneous payoff function of Player i and $\Phi_i : \mathbb{T} \times \mathbb{R}^q \to \mathbb{R}$ is the terminal reward function of Player i, also called the salvage value.

When the planning horizon is infinite, the sum of instantaneous rewards may tend to infinity, which renders the comparison of different paths meaningless. Several approaches have been proposed to deal with this situation, but here we only list the two most often used while assuming that the system is stationary, i.e., the reward and state-transition functions do not depend explicitly on time t.

Discounted reward: The performance criterion of Player i is given by the discounted sum of instantaneous rewards

$$J_i(x_0, \mathbf{u}) := \sum_{t=0}^{\infty} \rho_i^t \, \phi_i(x_t, u_t), \tag{3.8}$$

where $\rho_i \in (0, 1)$ is the discount factor for Player i. If the transition reward is a uniformly bounded function, then the infinite discounted sum converges to a finite value, and the payoffs obtained in this manner can be compared. We notice that this criterion gives a diminishing weight to future rewards, which means that the performance is mostly influenced by what happens in the early periods, and hence, it discriminates against the future payoff streams.

Average reward: Another criterion that emphasizes the rewards obtained in the distant future is the limit of the average rewards, also known as the Cesaro limit, i.e.,

$$J_i(x_0, \mathbf{u}) := \liminf_{T \to \infty} \frac{1}{T} \sum_{t=0}^{T} \phi_i(x_t, u_t). \tag{3.9}$$

Notice that, as T tends to infinity, all rewards gained over a finite number of periods will tend to have a negligible influence on the criterion; only what happens in the very distant future, actually at infinity, matters. In essence, this criterion is not affected by what happens in the early periods.

When the sum of infinite streams of rewards does not converge, there exist other methods for comparing the player's performance criterion. We do not discuss these methods in detail here but refer readers to Haurie et al. (2012) and the bibliography provided there. A commonly used method for comparing agent payoffs is provided by the overtaking optimality criterion. When using this concept, we cannot say that a player maximizes this criterion; however, we can use it to determine whether one stream of rewards is better than another.

3.2 Information Structures

The most complete information that could be available to a player at any stage is the game history $\mathcal{H}_t$. However, as the total amount of information available at stage t tends to increase with t, using the entire history for decision-making may be impractical. In most cases of interest, not all information accumulated up to stage t turns out to be relevant for making a decision at that point. Consequently, the players could rely on only a subset of $\mathcal{H}_t$, of fixed and finite dimension, to choose their controls at every stage t. We consider the following information structures:

Open-loop information structure: At each stage t, the players use only the knowledge of the initial state x_0 and the stage t to determine their controls. A strategy for Player $i \in M$ is a mapping $\sigma^i : \mathbb{T} \setminus \{T\} \times \mathbb{R}^q \to U^i$, where the control variable of Player i at stage t is given by

$$u_t^i = \sigma^i(t, x_0).   \qquad (3.10)$$

The players may use this information structure if they do not have access to more information about the state, other than the initial condition x_0.

Feedback information structure: At each stage t, the players have access to the information about the current state x_t of the system and do not retain information from the previous stages. A strategy for Player i is a mapping $\sigma^i : \mathbb{T} \setminus \{T\} \times \mathbb{R}^q \to U^i$ from the current state x_t and the stage t to Player i's control set, that is,

$$u_t^i = \sigma^i(t, x_t).   \qquad (3.11)$$

Due to the no-memory nature of the information structure, these strategies are also referred to as Markov strategies.

Closed-loop information structure: At each stage t, the players have access to the history of the game $\mathcal{H}_t$. In a deterministic context, knowledge of the state x_t for $t > 0$ implicitly contains all the information about the players' controls, and therefore, the history can be written as

$$\mathcal{H}_t^x := (x_0, x_1, \ldots, x_t).   \qquad (3.12)$$

The strategy of Player i at stage t is a mapping $\sigma_t^i : \mathcal{H}_t^x \to U^i$, where the control variable of Player i at stage t is given by

$$u_t^i = \sigma_t^i(\mathcal{H}_t^x).   \qquad (3.13)$$

We saw in Chap. 1 that the solution of the game varies with the information available to the players. Similarly, in multistage games, the outcome of the game depends on the information structure.

Remark 3.1   1. The open-loop structure is often considered unrealistic in the context of dynamic games, mainly because it does not lead to subgame perfect equilibrium solutions; see also Chap. 1 in the context of extensive-form games.
2. In a deterministic setting, the game history (3.6) is equivalent to (3.12). However, if the players' strategies include threats based on the game history, then the outcomes under closed-loop and feedback information structures could be very different. This is because, in the feedback information structure, the players forget opponents' past actions and may not be able to implement adequate punishments. Further, if the dynamical system or the objective functionals contain any stochastic elements, then the game history (3.6) is not equivalent to (3.12).

3.3 Strategies and Equilibrium Solutions

In this section, we define equilibrium solution in a multistage game under open-loop and feedback information structures and finite horizon setting.

3.3.1 Open-Loop Nash Equilibrium

As the initial state x_0 is a given parameter, an open-loop strategy of Player i is in fact a profile of control actions $\mathbf{u}^i$, that is, there is really no need to distinguish between a strategy and a control. We first define a normal-form representation of the game with open-loop strategies and then define the corresponding Nash equilibrium.

In a finite-horizon setting, the utility function (or performance criterion) of Player $i \in M$ and the state equations are given by

$$J_i(0, x_0, (\mathbf{u}^i, \mathbf{u}^{-i})) = \sum_{t=0}^{T-1} \phi_i(t, x_t, (u_t^i, u_t^{-i})) + \Phi_i(T, x_T), \qquad (3.14)$$

$$x_{t+1} = f(t, x_t, (u_t^i, u_t^{-i})), \qquad (3.15)$$

where $u_t^i \in U^i, u_t^{-i} \in \prod_{\substack{j \in M \\ j \neq i}} U^j$ and the initial state $x_0 \in \mathbb{R}^q$ are assumed to be given.

When the players use their open-loop strategies, the state trajectory, starting from the initial position $(0, x_0)$, is generated by the strategy profile $(\mathbf{u}^i, \mathbf{u}^{-i})$ and obtained by solving (3.15). So, the performance criterion (3.14) along with (3.15) define the normal form of the open-loop multistage game starting at the initial position $(0, x_0)$.

Definition 3.1 The strategy profile $\mathbf{u}^* := (\mathbf{u}^{i*}, \mathbf{u}^{-i*})$ is an open-loop Nash equilibrium at the initial position $(0, x_0)$ if the following condition is satisfied for every player $i \in M$:

$$J_i(0, x_0, (\mathbf{u}^{i*}, \mathbf{u}^{-i*})) \geq J_i(0, x_0, \mathbf{u}^i, \mathbf{u}^{-i*})), \ \forall \mathbf{u}^i \in \mathbf{U}^i. \qquad (3.16)$$

The open-loop Nash equilibrium strategies have the property that, along the equilibrium state trajectory, they continue to retain the equilibrium property, also referred to as weak time consistency; see also Başar and Olsder (1999). This property is established in the following theorem.

Theorem 3.1 *Let the strategy profile* $\mathbf{u}^* := (\mathbf{u}^{i*}, \mathbf{u}^{-i*})$ *be an open-loop Nash equilibrium at the initial position* $(0, x_0)$, *and let* $\mathbf{x}^* := (x_0^*, x_1^*, \ldots, x_T^*)$, $x_0^* = x_0$, *be the corresponding equilibrium state trajectory generated by these controls starting from* $(0, x_0)$. *Then, the restriction* $\mathbf{u}_{[\tau, T-1]}^*$ *of these control sequences to the stages* $\tau, \tau + 1, \ldots, T - 1$ *is an open-loop Nash equilibrium for the starting point* (τ, x_τ^*), *where $\tau = 0, 1, \ldots, T - 1$ and x_τ^* is an intermediate state along the equilibrium state trajectory.*

Proof Let us assume that, at an intermediate position (τ, x_τ^*) with $0 < \tau < T - 1$, the restriction $\mathbf{u}^*_{[\tau,T-1]}$ is not an open-loop Nash equilibrium. Then, there exist, a player $i \in M$ and controls $\mathbf{u}^i_{[\tau,T-1]} := \{u^i_\tau, \ldots, u^i_{T-1}\}$ such that

$$J_i\left(\tau, x_\tau^*, (\mathbf{u}^i_{[\tau,T-1]}, \mathbf{u}^{-i*}_{[\tau,T-1]})\right) > J_i\left(\tau, x_\tau^*, \mathbf{u}^*_{[\tau,T-1]}\right),$$

where $J_i\left(\tau, x_\tau^*, (\mathbf{u}^i_{[\tau,T-1]}, \mathbf{u}^{-i*}_{[\tau,T-1]})\right) = \sum_{t=\tau}^{T-1} \phi_i(x_t, (u^i_t, u^{-i*}_t)) + \Phi_i(T, x_T)$ and $\{x_\tau, \ldots, x_T\}$ is the state trajectory generated by the restricted controls $(\mathbf{u}^i_{[\tau,T-1]}, \mathbf{u}^{-i*}_{[\tau,T-1]})$. Now, concatenating the controls $\mathbf{u}^{i*}|_{[0,\tau-1]}$ to $\mathbf{u}^i|_{[\tau,T-1]}$ as $\mathbf{u}^{i\dagger} = (\mathbf{u}^{i*}|_{[0,\tau-1]}, \mathbf{u}^i|_{[\tau,T-1]})$, we obtain

$$J_i(0, x_0, (\mathbf{u}^{i\dagger}, \mathbf{u}^{-i*})) > J_i(0, x_0, (\mathbf{u}^{i*}, \mathbf{u}^{-i*})),$$

which contradicts the assumption that $\mathbf{u}^*$ is an open-loop Nash equilibrium at the starting point $(0, x_0)$.   □

Though an open-loop Nash equilibrium is weakly time consistent, it fails to satisfy the subgame perfectness property in the sense of Selten (1975); see also Sect. 1.3.5. This means that, if a player deviates from using the equilibrium control temporarily and next decides to play the equilibrium controls, then the remainder of the equilibrium control sequence will no longer retain its equilibrium property. In particular, a temporary deviation by a player results in the state variable evolving away from the equilibrium state trajectory, which implies that the previously defined open-loop equilibrium strategies lose their equilibrium property for the rest of the game. Subgame perfection requires that, starting from any state, the strategy generates controls that remain in equilibrium for the remainder of the game.

3.3.1.1 Coupled Maximum Principle for Multistage Games

From Definition 3.1, we see that, to determine Player i's open-loop Nash equilibrium strategy, we must solve a dynamic optimization problem given by (3.16). (Put differently, given $\mathbf{u}^{-i*}$, Player i's best-reply open-loop strategy is given by $\mathbf{u}^{i*}$.) Consequently, we can use discrete-time optimal-control techniques (see Sethi and Thompson 2000), based on Pontryagin's maximum principle, to characterize open-loop Nash equilibrium solutions. To this end, we define the Hamiltonian function associated with Player i's optimal control problem (3.16) as follows:

$$H_i(t, \lambda^i_{t+1}, x_t, (u^i_t, u^{-i*}_t)) = \phi_i(t, x_t, (u^i_t, u^{-i*}_t)) + (\lambda^i_{t+1})' f(t, x_t, (u^i_t, u^{-i*}_t)).$$
$$(3.17)$$

Assumption 3.1 Assume that $f(t, x, u)$ and $\phi_i(t, x, u)$ are continuously differentiable in state x and continuous in controls u for each $t = 0, 1, \ldots, T - 1$ and $\Phi_i(t, x)$ is continuously differentiable in x. Assume that, for each $i \in M$, U^i is compact and convex. Assume also that, for each t, x, the function $H_i(t, \lambda^i, x, (u^i, u^{-i}))$ is concave in u^i.

The necessary conditions for the existence of open-loop equilibrium strategies are provided in the next theorem.

Theorem 3.2 *Let Assumption 3.1 holds true. Let* $\mathbf{u}^*$ *be an open-loop Nash equilibrium strategy profile, generating the state trajectory* $\mathbf{x}^*$ *from the initial point* $(0, x_0)$ *for the game described by (3.14)–(3.15). Then, there exist real-valued functions* $\lambda^i : \mathbb{T} \to \mathbb{R}^q$ *such that, for any* $i \in M$, *the following conditions hold true for* $t \in \mathbb{T} \backslash \{T\}$:

$$u_t^{i*} = \arg\max_{u_t^i \in U^i} H_i(t, \lambda_{t+1}^i, x_t^*, (u_t^i, u_t^{-i*})), \tag{3.18}$$

$$\lambda_t^i = \frac{\partial}{\partial x_t} H_i(t, \lambda_{t+1}^i, x_t^*, (u_t^i, u_t^{-i*})), \tag{3.19}$$

$$\lambda_T^i = \frac{\partial}{\partial x_T} \Phi_i(T, x_T^*). \tag{3.20}$$

Proof The proof follows from the fact that each player's best reply to the other players' equilibrium controls is an optimal control problem. Then, under Assumption 3.1, the maximum principle in discrete time holds true; see Başar and Olsder (1999) or Fan and Wang (1964) for a complete proof. □

The terminal condition (3.20) is referred to as a transversality condition. The necessary conditions for the existence of an open-loop equilibrium in an infinite-horizon multistage game naturally carry over from the finite-horizon case except for the transversality condition (3.20). When $T \to \infty$, it is usually required that the costate variable λ_∞^i not grow unbounded; see Halkin (1974), Michel (1982) for more details.

3.3.2 Feedback-Nash Equilibrium

In this section, we characterize the Nash equilibrium strategies of the players under a feedback information structure. The approach is based on the dynamic programming method introduced by Bellman for control systems; see Bellman (2003). We consider the normal-form representation of the multistage game in feedback strategies at the starting point (τ, x_τ), with $\tau \in \{0, 1, \ldots, T - 1\}$ and $x_\tau \in \mathbb{R}^q$. The payoff function of Player $i \in M$ is given by

$$J_i(\tau, x_\tau, (\sigma^i, \sigma^{-i})) = \sum_{t=\tau}^{T-1} \phi_i(t, x_t, (\sigma^i(t, x_t), \sigma^{-i}(t, x_t))) + \Phi_i(x_T), \qquad (3.21)$$

and the state variables evolve according to

$$x_{t+1} = f(t, x_t, (\sigma^i(t, x_t), \sigma^{-i}(t, x_t))), \ t \in \{\tau, \tau+1, \ldots, T-1\}. \qquad (3.22)$$

Here, Player $i \in M$ uses feedback strategy $\sigma^i : \mathbb{T}\backslash\{T\} \times \mathbb{R}^q$ and, as a result, the state variable evolves according to (3.22), and the players receive the payoffs in (3.21). We denote the set of admissible feedback strategies by Σ^i for Player $i \in M$, and the set of joint admissible feedback strategies, $\Sigma = \Sigma^1 \times \cdots \times \Sigma^m$.

Definition 3.2 The strategy profile $\sigma^* := (\sigma^{i*}, \sigma^{-i*}) \in \Sigma$ is a feedback-Nash equilibrium at the position (τ, x_τ), with $\tau \in \{0, 1, \ldots, T-1\}$ and $x_\tau \in \mathbb{R}^q$, if the following condition is satisfied for every Player $i \in M$:

$$J_i(\tau, x_\tau, (\sigma^{i*}, \sigma^{-i*})) \geq J_i(\tau, x_\tau, (\sigma^i, \sigma^{-i*})), \ \forall \sigma^i \in \Sigma^i. \qquad (3.23)$$

We note from Definition 3.2 that the equilibrium property should hold for any admissible starting point, not just the initial point $(0, x_0)$, and as a result, the feedback-Nash equilibrium differs from the open-loop Nash equilibrium. In other words, the subgame perfectness property is embedded in the definition of the feedback-Nash equilibrium. In multistage games, this property is also sometimes referred to as strong time consistency; see Başar and Olsder (1999).

Given that the open-loop strategies are defined at the initial point $(0, x_0)$ of the multistage game, it is possible to transform a multistage game into a static game by eliminating the state variables. Consequently, the existence conditions for open-loop Nash equilibria are similar to those introduced in Sect. 1.2.3 for concave games. Unfortunately, these methods are not applicable to feedback strategies, and as a result, it is difficult to formulate existence conditions for a feedback-Nash equilibrium in multistage games. The standard approach for calculating a feedback-Nash equilibrium is to formulate Definition 3.2 as dynamic programming equations, and then use a verification theorem to confirm that the solutions of these equations are equilibria. A verification theorem shows that, if we can obtain a solution to the dynamic programming equations, then this solution constitutes a feedback-Nash equilibrium. This implies that the existence of a feedback-Nash equilibrium can be established in specific cases where an explicit solution of the dynamic programming equations can be obtained.

3.3.2.1 Dynamic Programming for Multistage Games

We use the backward induction or dynamic programming algorithm that is used in Sect. 1.3.5 to establish subgame perfectness in extensive-form games and extend it to multistage games. Let $\sigma^* := (\sigma^{i^*}, \sigma^{-i^*}) \in \Sigma$ be a feedback-Nash equilibrium solution of the multistage game, at the starting point (τ, x_τ), described by (3.21)–(3.22), and let $(x_\tau^*, x_{\tau+1}^*, \ldots, x_T^*)$ be the associated equilibrium state trajectory. The value function of Player $i \in M$ is given by

$$V_i(\tau, x_\tau) = \sum_{t=\tau}^{T-1} \phi_i(t, x_t^*, (\sigma^{i^*}(t, x_t^*), \sigma^{-i^*}(t, x_t^*))) + \Phi_i(x_T^*), \qquad (3.24)$$

$$V_i(T, x_T^*) = \Phi_i(x_T^*), \qquad (3.25)$$

$$x_{t+1}^* = f(t, x_t^*, (\sigma^{i^*}(t, x_t^*), \sigma^{-i^*}(t, x_t^*))), \qquad (3.26)$$

$$x_\tau^* = x_\tau. \qquad (3.27)$$

The value function is the payoff that Player i will receive if the feedback equilibrium strategy is played from initial point (τ, x_τ). The next lemma establishes a decomposition of the equilibrium conditions over time.

Lemma 3.1 *For every Player $i \in M$, the value function $V_i(t, x_t)$ satisfies the following recurrent equation, backward in time, also known as the Bellman equation:*

$$V_i(t, x_t^*) = \max_{u_t^i \in U^i} \left\{ \phi_i(t, x_t^*, (u_t^i, \sigma^{-i^*}(t, x_t^*))) + V_i(t+1, f(t, x_t^*, (u_t^i, \sigma^{-i^*}(t, x_t^*)))) \right\}, \quad (3.28)$$

$$t = T - 1, \ldots, 0,$$

$$V_i(T, x_T^*) = \Phi_i(x_T^*). \qquad (3.29)$$

Proof We claim that the feedback-Nash equilibrium control of Player i at (t, x_t^*) satisfies

$$\sigma^{i^*}(t, x_t^*) = \arg\max_{u_t^i \in U^i} \left\{ \phi_i(t, x_t^*, (u_t^i, \sigma^{-i^*}(t, x_t^*))) + V_i(t+1, x_{t+1}^*) \right\}.$$

This means that the feedback equilibrium control of Player i at point (t, x_t^*) must be the best reply to the feedback control of the players from $M \setminus \{i\}$ at point (t, x_t^*), combined with the use of a feedback equilibrium pair at all future stages $\{t+1, \ldots, T-1\}$. To prove this, we assume that the statement of the claim is not true. Then, there exists a control $\hat{u}_t^i \in U^i$ such that

$$V_i(t, x_t^*) = \phi_i(t, x_t^*, (u_t^i, \sigma^{-i^*}(t, x_t^*))) + V_i(t+1, f(t, x_t^*, (\sigma^{i^*}(t, x_t^*), \sigma^{-i^*}(t, x_t^*))))$$

$$< \phi_i(t, x_t^*, (\hat{u}_t^i, \sigma^{-i^*}(t, x_t^*))) + V_i(t+1, f(t, x_t^*, (\hat{u}_t^i, \sigma^{-i^*}(t, x_t^*)))). \qquad (3.30)$$

Next, we define a new strategy $\hat{\sigma}^i$ for Player i as $\hat{\sigma}^i(\tau, x_\tau) = \sigma^{i^*}(\tau, x_\tau)$ for $\tau \neq t$ and $\hat{\sigma}_i(t, x_t) = \hat{u}_t^i$. Then, it is easy to verify that the payoff generated from the initial point (t, x_t^*) by the strategy profile $(\hat{\sigma}^i, \sigma^{-i^*})$ is given by the expression on the right-hand side of (3.30). This contradicts the equilibrium property at starting point (t, x_t^*) for the strategy $(\sigma^{i*}, \sigma^{-i*})$. Therefore, the claim is true and, as a result, the recurrence relation (3.28) holds true.   □

The above lemma provides necessary and sufficient conditions for a feedback-Nash equilibrium. Further, it reveals the structure of this equilibrium, that is, that the equilibrium condition must hold in a set of local (static) games, defined at each possible initial point (τ, x_τ).

For any starting point (τ, x_τ), we introduce the local (static) game where the players' actions are $(u_\tau^i, u_\tau^{-i}) \in U^i \times U^{-i}$ and their payoffs are given by

$$h_i(\tau, x_\tau, (u_\tau^i, u_\tau^{-i})) = \phi_i(\tau, x_\tau, (u_\tau^i, u_\tau^{-i})) + V_i(\tau + 1, f(\tau, x_\tau, (u_\tau^i, u_\tau^{-i}))). \tag{3.31}$$

Then, from Lemma 3.1, it is clear that the feedback-Nash equilibrium control $(\sigma^{i*}(\tau, x_\tau), \sigma^{-i*}(\tau, x_\tau))$ evaluated at (τ, x_τ) is a Nash equilibrium for the local game described by the payoffs in (3.31).

The lemma suggests the following recursive procedure for obtaining the feedback-Nash equilibrium.

1. At stage $T - 1$, for any starting point $(T - 1, x_{T-1})$, solve the local (static) game where the payoff of Player $i \in M$ is given by

$$h_i(T - 1, x_{T-1}, (u_{T-1}^i, u_{T-1}^{-i})) = \phi_i(T - 1, x_{T-1}, (u_{T-1}^i, u_{T-1}^{-i}))$$
$$+ \Phi_i(T, f(T - 1, x_{T-1}, (u_{T-1}^i, u_{T-1}^{-i}))). \tag{3.32}$$

Assume that a Nash equilibrium exists for this local game. Then, the equilibrium is a function of the initial data $(T - 1, x_{T-1})$, and we define it as $(\sigma^{i*}(T - 1, x_{T-1}), \sigma^{-i*}(T - 1, x_{T-1}))$. This implies that, for each player $i \in M$, we have

$$\sigma^{i*}(T - 1, x_{T-1}) = \arg\max_{u_{T-1}^i \in U^i} h_i(T - 1, x_{T-1}, (u_{T-1}^i, \sigma^{-i*}(T - 1, x_{T-1}))).$$

The equilibrium payoff of Player i is defined as

$$V_i(T - 1, x_{T-1}) = h_i(T - 1, x_{T-1}, (\sigma^{i*}(T - 1, x_{T-1}), \sigma^{-i*}(T - 1, x_{T-1}))).$$

2. At stage $T - 2$, for any starting point $(T - 2, x_{T-2})$, solve the local (static) game where the payoff of Player $i \in M$ is given by

$$h_i(T - 2, x_{T-2}, (u_{T-2}^i, u_{T-2}^{-i})) = \phi_i(T - 2, x_{T-2}, (u_{T-2}^i, u_{T-2}^{-i}))$$
$$+ V_i(T - 1, f(T - 2, x_{T-2}, (u_{T-2}^i, u_{T-2}^{-i}))).$$

Assume that a Nash equilibrium exists for this local game. Then, the equilibrium is a function of the initial data $(T - 2, x_{T-2})$, and we define it as $(\sigma^{i*}(T - 2, x_{T-2}), \sigma^{-i*}(T - 2, x_{T-2}))$. This implies that, for each player $i \in M$, we have

$$\sigma^{i^*}(T-2, x_{T-2}) = \arg\max_{u^i_{T-2} \in U^i} \ h_i(T-2, x_{T-2}, (u^i_{T-2}, \sigma^{-i^*}(T-2, x_{T-2}))).$$

The equilibrium payoff of Player i is defined as

$$V_i(T-2, x_{T-2}) = h_i(T-2, x_{T-2}, (\sigma^{i^*}(T-2, x_{T-2}), \sigma^{-i^*}(T-2, x_{T-2}))).$$

3. At time $T-3$, etc., proceed recursively, defining local games and their solutions for all preceding periods, until period 0 is reached.

In the next theorem, we show that this procedure defines a feedback-Nash equilibrium strategy vector.

Theorem 3.3 *Suppose there exist value functions $V_i(t, x_t)$ and feedback strategies $(\sigma^{i^*}, \sigma^{-i^*})$ that satisfy the local game equilibrium conditions defined in equations (3.24)–(3.29) for $t = \{0, 1, \ldots, T-1\}$ and $x_t \in \mathbb{R}^q$. Then, the strategy profile $(\sigma^{i^*}, \sigma^{-i^*})$ constitutes a (subgame) perfect equilibrium of the dynamic game with a feedback information structure. Moreover, the value function $V_i(t, x_t)$ represents the equilibrium payoff of Player i for the game starting at point (t, x_t).*

Proof To prove the theorem, we proceed by backward induction. First, we start with the initial point $(T-1, x_{T-1})$ with $x_{T-1} \in \mathbb{R}^q$. From the first step of the recursive procedure, at each of these points, the feedback strategy profile $(\sigma^{i^*}, \sigma^{-i^*})$ cannot be improved upon by a unilateral change by Player i. This is a consequence of the equilibrium condition for the local game (3.32). Therefore, the feedback strategy $(\sigma^{i^*}, \sigma^{-i^*})$ is an equilibrium solution at any $(T-1, x_{T-1})$.

At stage $T-2$, consider a starting position $(T-2, x_{T-2})$ with $x_{T-2} \in \mathbb{R}^q$. If Player i uses a control according to an arbitrary strategy $\sigma^i \in \Sigma^i$ at stage $T-2$, while other players from $M \setminus \{i\}$ use their equilibrium strategies σ^{-i^*}, then Player i receives the payoff

$$\begin{aligned} h_i(T-2, x_{T-2}, &(\sigma^i(T-2, x_{T-2}), \sigma^{-i^*}(T-2, x_{T-2}))) \\ &= \phi_i(T-2, x_{T-2}, (\sigma^i(T-2, x_{T-2}), \sigma^{-i^*}(T-2, x_{T-2}))) \\ &\quad + h_i(T-1, \hat{x}_{T-1}, (\sigma^i(T-1, \hat{x}_{T-1}), \sigma^{-i^*}(T-1, \hat{x}_{T-1}))), \end{aligned}$$
$$(3.33)$$

where $\hat{x}_{T-1} = f(T-2, x_{T-2}, (\sigma^i(T-2, x_{T-2}), \sigma^{-i^*}(T-2, x_{T-2})))$.

Since the strategy $(\sigma^{i^*}, \sigma^{-i^*})$ provides an equilibrium control at $(T-1, \hat{x}_{T-1})$, we have

$$V_i(T-1, \hat{x}_{T-1}) \geq h_i(T-1, \hat{x}_{T-1}, (\sigma^i(T-1, \hat{x}_{T-1}), \sigma^{-i^*}(T-1, \hat{x}_{T-1}))).$$
$$(3.34)$$

Using (3.33) in (3.34), we get

$$
\begin{aligned}
h_i(T - 2, & x_{T-2}, (\sigma^i(T - 2, x_{T-2}), \sigma^{-i^*}(T - 2, x_{T-2}))) \\
& \leq \phi_i(T - 2, x_{T-2}, (\sigma^i(T - 2, x_{T-2}), \sigma^{-i^*}(T - 2, x_{T-2}))) \\
& \quad\quad\quad\quad\quad\quad\quad\quad\quad\quad\quad\quad + V_i(T - 1, \hat{x}_{T-1}).
\end{aligned}
\tag{3.35}
$$

From (3.28), the right-hand side of the above equation is, by construction, less than or equal to $V_i(T - 2, x_{T-2})$. This implies that there is no other feedback strategy σ^i that can do better than the strategy σ^{i^*} in the last two stages. The proof follows by repeating this verification process in the remaining stages $T - 3, T - 4, \ldots, 0$.

The subgame perfectness of the feedback-Nash equilibrium follows from the fact that, at every stage, the equilibrium control, which is obtained as a solution of (3.28), depends only on the starting point (t, x_t).   □

Remark 3.2 The above result is a verification theorem. Jointly with Lemma 3.1, it shows that the dynamic programming and the Bellman equations are both necessary and sufficient conditions for a feedback-Nash equilibrium to exist. This means that we need to solve the system of dynamic programming equations (3.24)–(3.27) for all the players, which requires that the value functions $V_i(t, x_t)$ be determined for every $x_t \in X$, for each $i \in M$. In practice, this can be achieved if X is a finite set or the system of dynamic programming equations (3.24)–(3.27) has an analytical solution that could be obtained through the method of undetermined coefficients.

Remark 3.3 We can extend the method for computing the feedback-Nash equilibrium to infinite horizon stationary discounted games using the approach given in Theorem 3.3; see Haurie et al. (2012), Sect. 6.6.

3.4   Linear-Quadratic Multistage Games

We consider multistage games with linear state-transition equation and stage-additive quadratic payoffs, which are also known as linear-quadratic dynamic games (LQDGs). There are two major reasons for the popularity of this class of games. First, LQDGs yield closed-form expressions for equilibrium strategies, and importantly, theorems characterizing the existence and uniqueness of equilibria are available. Second, while being tractable and involving simple functional forms, they still capture interactions between the control and state variables of the different players and nonconstant returns to scale, two features that occur quite naturally in many applications. The linear dynamics, which are a priori restrictive, could be, at least in some instances, considered an acceptable approximation of a possibly nonlinear specification. In brief, the conceptual and methodological grounds of this class of games are well established and provide an attractive ready-to-use framework for many applications. For complete coverage of LQDGs, see, e.g., the books of Başar and Olsder (1999), Engwerda (2005).

In an LQGD, the payoff of Player $i \in M$ is given by

$$J_i(0, x_0, (\mathbf{u}^i, \mathbf{u}^{-i})) = \frac{1}{2} x_T' Q_T^i x_T + p_T^i{}' x_T$$

$$+ \sum_{t=0}^{T-1} \left(\frac{1}{2} x_t' Q_t^i x_t + p_t^i{}' x_t + \frac{1}{2} u_t^i{}' R_t^{ii} u_t + \sum_{j \neq i} \frac{1}{2} u_t^j{}' R_t^{ij} u_t^j \right),$$

(3.36)

and the state transition equation is given by

$$x_{t+1} = A_t x_t + \sum_{i \in M} B_t^i u_t^i, \quad t = 0, 1, \ldots, T - 1,$$

(3.37)

where the initial condition x_0 is given and where A_t, B_t^i, Q_t^i, R_t^{ij}, p_t^i are matrices and vectors of appropriate dimensions. We assume that the matrices $\{R_t^{ii}, \ t \in \mathbb{T} \setminus \{T\}\}$ are negative definite, and $\{Q_t^i, \ t \in \mathbb{T}\}$ are symmetric and negative semidefinite. Note that there are no restrictions on the number of players, or state and control variables.

3.4.1 Open-Loop Nash Equilibrium

We provide the open-loop Nash equilibrium solution for LQDGs described by (3.36)–(3.37), using the coupled maximum principle; see Sect. 3.3.1.1. The Hamiltonian function associated with Player i's optimal control problem (3.16) is given by

$$H_i(t, \lambda_{t+1}^i, x_t, (u_t^i, u_t^{-i*})) = \left(\frac{1}{2} x_t' Q_t^i x_t + p_t^i{}' x_t + \frac{1}{2} u_t^i{}' R_t^{ii} u_t + \sum_{j \neq i} \frac{1}{2} u_t^j{}' R_t^{ij} u_t^j \right)$$

$$+ (\lambda_{t+1}^i)' \left(A_t x_t + B_t^i u_t^i + \sum_{j \neq i} B_t^j u_t^{j*} \right).$$

(3.38)

The necessary conditions for a strategy profile $\mathbf{u}^* := (\mathbf{u}^{i*}, \mathbf{u}^{-i*})$ to be an open-loop Nash equilibrium are then given by (3.18)–(3.20). By the negative definiteness of R_t^{ii}, we have that the Hamiltonian (3.38) is strictly concave with respect to u_t^i. Assuming interior solutions, the necessary condition (3.18) results in

$$R_t^{ii} u_t^{i*} + B_t^i{}' \lambda_{t+1}^i = 0 \quad \Rightarrow \quad u_t^{i*} = - \left(R_t^{ii} \right)^{-1} B_t^i{}' \lambda_{t+1}^i.$$

(3.39)

Using (3.39) in (3.37), (3.19), and (3.20), we obtain the following two-point boundary value (TPBV) equations:

$$x_{t+1}^* = A_t x_t^* - \sum_{j=1}^m B_t^j (R_t^{jj})^{-1} B_t^{j'} \lambda_{t+1}^j, \tag{3.40}$$

$$\lambda_t^i = A_t' \lambda_{t+1}^i + Q_t^i x_t^* + p_t^i, \tag{3.41}$$

with the boundary conditions x_0 and $\lambda_T^i = Q_T^i x_T^* + p_T^i$. Due to the linear dynamics and quadratic objectives, the co-state variables at every stage $t \in \mathbb{T}$ can be taken as affine functions of the state variable, that is,

$$\lambda_t^i = K_t^i x_t^* + k_t^i, \tag{3.42}$$

where parameters $K_t^i \in \mathbb{R}^{q \times q}$, $i \in M$, $t \in \mathbb{T}$ are to be determined. Using (3.42) in (3.40), the state equation is given by

$$\left(I + \sum_{j=1}^m B_t^j (R_t^{jj})^{-1} B_t^{j'} K_{t+1}^j \right) x_{t+1}^* = A_t x_t^* - \sum_{j=1}^m B_t^j (R_t^{jj})^{-1} B_t^{j'} k_{t+1}^j.$$

Denoting $\Lambda_t = I + \sum_{j=1}^m B_t^j (R_t^{jj})^{-1} B_t^{j'} K_{t+1}^j$ for all t, and assuming that these matrices are invertible for all t, we get

$$x_{t+1}^* = \Lambda_t^{-1} A_t x_t^* - \Lambda_t^{-1} \sum_{j=1}^m B_t^j (R_t^{jj})^{-1} B_t^{j'} k_{t+1}^j. \tag{3.43}$$

Using (3.42) and (3.43) in (3.41), we obtain

$$K_t^i x_t^* + k_t^i = A_t' \left(K_{t+1}^i x_{t+1}^* + k_{t+1}^i \right) + Q_t^i x_t^* + p_t^i$$

$$= \left(A_t' K_{t+1}^i \Lambda_t^{-1} A_t + Q_t^i \right) x_t^* + p_t^i + A_t' k_{t+1}^i - A_t' K_{t+1}^i \Lambda_t^{-1} \sum_{j=1}^m B_t^j (R_t^{jj})^{-1} B_t^{j'} k_{t+1}^j.$$

The above equation has to hold for an arbitrary x_t^*. Then, identifying the linear terms in x_t^* and constant terms on both sides of the above equation, we obtain the following set of backward recursive equations for each $i \in M$ and $t = 0, 1, \ldots, T - 1$:

$$\Lambda_t = I + \sum_{j=1}^m B_t^j (R_t^{jj})^{-1} B_t^{j'} K_{t+1}^j, \tag{3.44}$$

$$K_t^i = A_t' K_{t+1}^i \Lambda_t^{-1} A_t + Q_t^i, \tag{3.45}$$

$$k_t^i = p_t^i + A_t' k_{t+1}^i - A_t' K_{t+1}^i \Lambda_t^{-1} \sum_{j=1}^m B_t^j (R_t^{jj})^{-1} B_t^{j'} k_{t+1}^j, \tag{3.46}$$

where $K_T^i = Q_T^i$ and $k_T^i = p_T^i$.

Remark 3.4 It can be shown that the invertibility of the matrices $\{\Lambda_t, \ t \in \mathbb{T}\backslash\{T\}\}$ implies the unique solvability of the TPBV equations described by (3.40)–(3.41); see Jank and Abou-Kandil (2003).

Using the above calculations we have the following theorem.

Theorem 3.4 *For an m-person multistage linear quadratic game described by (3.36)–(3.37), let the matrices $\{Q_t^i, \ i \in M, \ t \in \mathbb{T}\}$ be negative semidefinite and $\{R_t^{ii}, \ i \in M, \ t \in \mathbb{T}\backslash\{T\}\}$ be negative definite. Let the set of matrices $\{\Lambda_t, \ t \in \mathbb{T}\backslash\{T\}\}$ recursively defined by (3.44) be invertible. Then, there exists a unique open-loop Nash equilibrium solution defined by*

$$
u_t^{i*} = -(R_t^{ii})^{-1} B_t^{i'} K_{t+1}^i \Lambda_t^{-1} \left(A_t x_t^* - \sum_{j=1}^m B_t^j (R_t^{jj})^{-1} B_t^{j'} k_{t+1}^j \right),
$$

for $i \in M$ and $t \in \mathbb{T}\backslash\{T\}$, where $\{K_t^i, k_t^i, \ i \in M, \ t \in \mathbb{T}\}$ are the solutions of the backward recursive equations (3.45)–(3.46).

Proof The proof involves first transforming the dynamic game into a static game by eliminating the state variables. If Q_t^i is negative semidefinite, and R_t^{ii} is negative definite, then the objective function of Player i is strictly concave in $\mathbf{u}^i$ for all $\mathbf{u}^{-i}$, and the existence of an open-loop Nash equilibrium follows from Rosen (1965). Further, the strict concavity of the Hamiltonian (3.38) with respect to u_t^i implies that the equilibrium is unique. For a more detailed proof, see Jank and Abou-Kandil (2003). □

3.4.1.1 An Example

Consider the following two-player, multistage dynamic oligopoly game. The players are producers of a homogeneous commodity sold in a competitive market. Denote by q_t^i the output of Player i at time t ($t = 0, 1, \ldots, T$). The price of the product is given by the following inverse-demand law:

$$
P_t = \alpha_t - \beta_t Q_t,
$$

where $Q_t = q_t^1 + q_t^2$ is the total quantity of the product available on the market. Player i ($i = 1, 2$) is described by the following data:

- A production capacity X_t^i, $t = 0, 1, 2, \ldots, T$, which accumulates over time according to the following difference equation:

$$
X_{t+1}^i = X_t^i + I_t^i,
$$

where I_t^i is the physical investment in the production capacity at time t ($t = 0, 1, \ldots, T - 1$) and X_0^i is the initial production capacity of Player i;

- Production- and investment-cost functions, given by $C_i(q_t^i) = \frac{\gamma^i}{2}(q_t^i)^2$ and $F_i(I_t^i) = \frac{\delta^i}{2}(I_t^i)^2$, respectively;
- A payoff function for Player i,

$$J_i(\cdot) = \rho^T \left(P_T \, q_T^i - C_i(q_T^i) \right) + \sum_{t=0}^{T-1} \rho^t \left(P_t \, q_t^i - C_i(q_t^i) - F_i(I_t^i) \right),$$

where $\rho \in (0, 1)$ is the common discount factor;
- A capacity constraint $q_t^i \leq X_t^i$, $i = 1, 2$, $t = 0, 1, \ldots, T$.

We analyze the dynamic duopoly game with the assumption that the players produce at their full capacity, that is, $q_t^i = X_t^i$, $i = 1, 2$, $t = 0, 1, \ldots, T$. This assumption enables us to model this game as a linear-quadratic dynamic game. The state and control variables are given by

$$x_t = \begin{bmatrix} X_t^1 \\ X_t^2 \end{bmatrix}, \quad u_t^1 = I_t^1, \quad u_t^2 = I_t^2.$$

The payoff function for Player 1 is written in these variables as

$$J_1(\cdot) = \rho^T \left((\alpha_T - \beta_T Q_T) \, q_T^1 - \frac{\gamma^1}{2}(q_T^1)^2 \right)$$

$$+ \sum_{t=0}^{T-1} \rho^t \left((\alpha_t - \beta_t Q_t) \, q_t^1 - \frac{\gamma^1}{2}(q_t^1)^2 - \frac{\delta^1}{2}(I_t^1)^2 \right).$$

Using the production at the capacity assumption $q_t^i = X_t^i$, we get

$$J_1(\cdot) = \frac{1}{2} \begin{bmatrix} X_T^1 \\ X_T^2 \end{bmatrix}' \left(-\rho^2 \begin{bmatrix} 2\beta_T + \gamma^1 & \beta_T \\ \beta_T & 0 \end{bmatrix} \right) \begin{bmatrix} X_T^1 \\ X_T^2 \end{bmatrix} + \begin{bmatrix} \rho^2 \alpha_T \\ 0 \end{bmatrix}' \begin{bmatrix} X_T^1 \\ X_T^2 \end{bmatrix}$$

$$+ \sum_{t=0}^{T-1} \frac{1}{2} \begin{bmatrix} X_t^1 \\ X_t^2 \end{bmatrix}' \left(-\rho^t \begin{bmatrix} 2\beta_t + \gamma^1 & \beta_t \\ \beta_t & 0 \end{bmatrix} \right) \begin{bmatrix} X_t^1 \\ X_t^2 \end{bmatrix} + \begin{bmatrix} \rho^t \alpha_t \\ 0 \end{bmatrix}' \begin{bmatrix} X_t^1 \\ X_t^2 \end{bmatrix} + \frac{1}{2} I_t^1 \left(-\rho^t \delta^1 \right) I_t^1.$$

Similarly, writing the payoff function for Player 2 as above, the problem parameters associated with the LQDG are as follows:

$$A_t = \begin{bmatrix} 1 & 0 \\ 0 & 1 \end{bmatrix}, \quad B_t^1 = \begin{bmatrix} 1 \\ 0 \end{bmatrix}, \quad B_t^2 = \begin{bmatrix} 0 \\ 1 \end{bmatrix}, \quad t = 0, 1, \ldots, T-1,$$

$$Q_t^1 = -\rho^t \begin{bmatrix} 2\beta_t + \gamma^1 & \beta_t \\ \beta_t & 0 \end{bmatrix}, \quad Q_t^2 = -\rho^t \begin{bmatrix} 0 & \beta_t \\ \beta_t & 2\beta_t + \gamma^2 \end{bmatrix}, \quad t = 0, 1, \ldots, T,$$

$$p_t^1 = \begin{bmatrix} \rho^t \alpha_t \\ 0 \end{bmatrix}, \quad p_t^2 = \begin{bmatrix} 0 \\ \rho^t \alpha_t \end{bmatrix}, \quad t = 0, 1, \ldots, T,$$

$$R_t^{11} = -\rho^t \delta^1, \quad R_t^{12} = 0, \quad R_t^{21} = 0, \quad R_t^{22} = -\rho^t \delta^2, \quad t = 0, 1, \ldots, T-1.$$

Table 3.1 Open-loop Nash equilibrium investment strategies and the corresponding capacity trajectories

t	$\rho = 0.9$				$\rho = 0.5$			
	$q_t^1 = X_t^1$	$q_t^2 = X_t^2$	I_t^1	I_t^2	$q_t^1 = X_t^1$	$q_t^2 = X_t^2$	I_t^1	I_t^2
0	10	6	0.535	0.436	10	6	0.293	0.320
1	10.535	6.436	0.374	0.221	10.293	6.320	0.258	0.222
2	10.908	6.657	0.288	0.140	10.550	6.543	0.235	0.167
3	11.196	6.797	0.241	0.112	10.785	6.709	0.219	0.136
4	11.437	6.909	0.214	0.103	11.004	6.846	0.209	0.119
5	11.652	7.011	0.198	0.101	11.213	6.965	0.201	0.111
6	11.849	7.112	0.184	0.102	11.414	7.076	0.195	0.106
7	12.034	7.214	0.169	0.101	11.609	7.182	0.186	0.104
8	12.203	7.315	0.143	0.096	11.794	7.285	0.168	0.098
9	12.345	7.411	0.095	0.073	11.962	7.384	0.122	0.078
10	12.440	7.485			12.084	7.462		

We assume that the parameters in the inverse-demand law vary over time as follows:

$$\alpha_t = \alpha_{t-1}(1 + \epsilon), \ \beta_t = \frac{1}{1 + \epsilon} \beta_{t-1}, \ 0 < \epsilon < 1, \ t = 0, 1, \ldots, T.$$

For a numerical illustration, the parameters are set as

$$\alpha_0 = 100, \ \beta_0 = 2, \ \epsilon = 0.01, \ X_0^1 = 10, \ X_0^2 = 6,$$
$$\gamma^1 = 4, \ \gamma^2 = 8, \ \delta^1 = 20, \ \delta^2 = 12, T = 10.$$

Notice that the parameters satisfy the conditions stated in the Theorem 3.4 and Remark 3.4. So, the unique open-loop Nash equilibrium can be obtained by solving the Riccati equations (3.44)–(3.46). Table 3.1 illustrates the open-loop investment strategies and the resulting capacity trajectories of the players when the discount factor is set to $\rho = 0.9$ and $\rho = 0.5$.

3.4.2 Feedback-Nash Equilibrium

We provide the feedback-Nash equilibrium solution for the LQDG obtained by solving the dynamic programming equations; see Sect. 3.3.2.1. Due to the linear-quadratic structure of the game, we assume that the value functions are quadratic and given by

$$V_i(t, x_t) = \frac{1}{2} x_t' S_t^i x_t + r_t^{i'} x_t + w_t^i, \tag{3.47}$$

where $S_t^i \in \mathbb{R}^{q \times q}$, $r_t^i \in \mathbb{R}^q$, and $w_t^i \in \mathbb{R}$ are the coefficients to be determined. Again, due to the linear dynamics and quadratic payoffs, the equilibrium strategies are supposed to be affine functions of the state variable, that is, for every $i \in M$,

$$\sigma^{i*}(t, x_t) := Z_t^i x_t + z_t^i, \tag{3.48}$$

where $Z_t^i \in \mathbb{R}^{\ell_i \times q}$ and $z_t^i \in \mathbb{R}^{\ell_i}$ are parameters to be determined. The local game associated with the dynamic programming equation (3.31) is then given by

$$\begin{aligned} h_i(t, x_t, (u_t^i, u_t^{-i})) = {} & \frac{1}{2} x_t' Q_t^i x_t + p_t^{i'} x_t + \frac{1}{2} u_t^{i'} R_t^{ii} u_t^i + \sum_{j \neq i} \frac{1}{2} u_t^{j'} R_t^{ij} u_t^j \\ & + \frac{1}{2} \Big(A_t x_t + \sum_{i \in M} B_t^i u_t^i \Big)' S_{t+1}^i \Big(A_t x_t + \sum_{i \in M} B_t^i u_t^i \Big) \\ & + r_{t+1}^{i'} \Big(A_t x_t + \sum_{i \in M} B_t^i u_t^i \Big) + w_{t+1}^i. \end{aligned}$$

The local-game payoff of Player i must be maximized with respect to u_t^i. This implies that the gradient in u_t^i must be equal to zero, that is,

$$\Big(R_t^{ii} + B_t^{i'} S_{t+1}^i B_t^i \Big) u_t^i + B_t^{i'} r_{t+1}^i + \sum_{j \neq i} B_t^{i'} S_{t+1}^i B_t^j u_t^j + B_t^{i'} S_{t+1}^i A_t x_t = 0.$$

Using the structure of the feedback control (3.48) in the above equation, we get

$$\Big(R_t^{ii} + B_t^{i'} S_{t+1}^i B_t^i \Big) \Big(Z_t^i x_t + z_t^i \Big) + B_t^{i'} r_{t+1}^i + \sum_{j \neq i} B_t^{i'} S_{t+1}^i B_t^j \Big(Z_t^j x_t + z_t^j \Big) + B_t^{i'} S_{t+1}^i A_t x_t = 0.$$

As the starting point (t, x_t) is arbitrary, we obtain the following two equations for each $i \in M$:

$$\Big(R_t^{ii} + B_t^{i'} S_{t+1}^i B_t^i \Big) Z_t^i + \sum_{j \neq i} B_t^{i'} S_{t+1}^i B_t^j Z_t^j = -B_t^{i'} S_{t+1}^i A_t, \tag{3.49}$$

$$\Big(R_t^{ii} + B_t^{i'} S_{t+1}^i B_t^i \Big) z_t^i + \sum_{j \neq i} B_t^{i'} S_{t+1}^i B_t^j z_t^j = -B_t^{i'} r_{t+1}^i. \tag{3.50}$$

Clearly, the coefficients associated with the linear-feedback equilibrium controls (3.48) are obtained by solving (3.49)–(3.50) for all $i \in M$.

Next, from Lemma 3.1, using the form of the value function (3.47) in the recurrence relation (3.28), and using the equilibrium controls (3.48), we get

$$\frac{1}{2}x_t'S_t^i x_t + r_t^{i'} x_t + w_t^i = \frac{1}{2}x_t'Q_t^i x_t + p_t^{i'} x_t + \frac{1}{2}\left(Z_t^i x_t + z_t^i\right)' R_t^{ii}\left(Z_t^i x_t + z_t^i\right)$$

$$+ \sum_{j\neq i}\frac{1}{2}\left(Z_t^j x_t + z_t^j\right)' R_t^{ij}\left(Z_t^j x_t + z_t^j\right)$$

$$+ \frac{1}{2}\left(A_t x_t + \sum_{i\in M} B_t^i\left(M_t^i x_t + z_t^i\right)\right)' S_{t+1}^i\left(A_t x_t + \sum_{i\in M} B_t^i\left(Z_t^i x_t + z_t^i\right)\right)$$

$$+ r_{t+1}^{i'}\left(A_t x_t + \sum_{i\in M} B_t^i\left(Z_t^i x_t + z_t^i\right)\right) + w_{t+1}^i.$$

Identifying the quadratic and linear terms in x_t and the constant terms on both sides of the above equation, we obtain the following set of backward recursive equations for each $i \in M$ and $t \in \mathbb{T}\backslash\{T\}$:

$$S_t^i = Q_t^i + \bar{A}_t' S_{t+1}^i \bar{A}_t + \sum_{j\in M} Z_t^{j'} R_t^{ij} Z_t^j, \tag{3.51}$$

$$r_t^i = p_t^i + \bar{A}_t' r_{t+1}^i + \bar{A}_t' S_{t+1}^i \bar{z}_t + \sum_{j\in M} Z_t^{j'} R_t^{ij} z_t^j, \tag{3.52}$$

$$w_t^i = w_{t+1}^i + \sum_{j\in M} z_t^{j'} R_t^{ij} z_t^j + r_{t+1}^{i'} \bar{z}_t + \frac{1}{2}\bar{z}_t' S_{t+1}^i \bar{z}_t, \tag{3.53}$$

where $\bar{A}_t = A_t + \sum_{i\in M} B_t^i Z_t^i$, $\bar{z}_t = \sum_{i\in M} B_t^i z_t^i$, and the boundary conditions are given by $S_T^i = Q_T^i, r_T^i = p_T^i, w_T^i = 0$ for $i \in M$.

Theorem 3.5 *Let the set of matrices* $\{S_t^i, r_t^i, w_t^i, \ i \in M, \ t \in \mathbb{T}\}$ *solve the backward recursive coupled Riccati equations* (3.51)–(3.53). *Let the set of matrices* $\{R_t^{ii} + B_t^{i'} S_{t+1}^i B_t^i, \ i \in M, \ t \in \mathbb{T}\backslash\{T\}\}$ *be negative definite. Then, the feedback-Nash equilibrium controls of Player* $i \in M$ *for* $t \in \mathbb{T}\backslash\{T\}$ *are given by*

$$\sigma^{i*}(t, x_t) := Z_t^i x_t + z_t^i,$$

where Z_t^i *and* z_t^i *are obtained by solving* (3.49)–(3.50). *Further, the equilibrium payoff of Player* i *is given by* $V_i(0, x_0)$.

Proof The proof follows from the verification of Theorem 3.3. The negative definiteness of the matrix $R_t^{ii} + B_t^{i'} S_{t+1}^i B_t^i$ guarantees that the right-hand side of the dynamic programming equation (3.28) is maximized in the player's own strategy. Hence, the equilibrium is obtained as a solution of the dynamic programming equation. $\square$

Remark 3.5 The game solution defined by equations (3.49)–(3.50) and (3.51)–(3.53) is valid for any dimensionality of the problem, i.e., these formulas are the same for any number of players, and any dimension of the state and control variables.

Remark 3.6 We can extend the method for computing feedback-Nash equilibria to infinite-horizon stationary discounted games. In this setting, it is assumed that

Table 3.2 Feedback-Nash equilibrium investment strategies and the corresponding capacity trajectories

t	$\rho = 0.9$				$\rho = 0.5$			
	$q_t^1 = X_t^1$	$q_t^2 = X_t^2$	I_t^1	I_t^2	$q_t^1 = X_t^1$	$q_t^2 = X_t^2$	I_t^1	I_t^2
0	10	6	0.608	0.460	10	6	0.313	0.329
1	10.608	6.460	0.417	0.228	10.313	6.329	0.273	0.226
2	11.025	6.688	0.313	0.141	10.587	6.555	0.247	0.169
3	11.338	6.829	0.256	0.111	10.833	6.724	0.228	0.137
4	11.594	6.940	0.223	0.102	11.061	6.860	0.216	0.119
5	11.817	7.042	0.202	0.100	11.277	6.980	0.206	0.110
6	12.019	7.142	0.184	0.100	11.483	7.090	0.198	0.105
7	12.202	7.242	0.159	0.097	11.681	7.195	0.185	0.102
8	12.361	7.339	0.116	0.085	11.867	7.297	0.161	0.094
9	12.477	7.424	0.061	0.060	12.028	7.391	0.111	0.073
10	12.538	7.483			12.139	7.464		

the matrices entering the state dynamics (3.5) and payoff functions (3.36) are stage invariant, that is, $A(t) = A$, $B_t^i = B^i$, $Q_t^i = Q^i$, $R_t^{ij} = R^{ij}$, $i, j \in M$; see Haurie et al. (2012), Sect. 6.7.2, for further details.

3.4.2.1 An Example

We revisit the dynamic duopoly game studied in Sect. 3.4.1.1 under feedback information structure. For a numerical illustration, the parameters are set exactly in the same way as the open-loop case. For these parameter values, we observe that the matrices $\{R_t^{ii} + B_t^{i'} S_{t+1}^i B_t^i, \ i \in M, \ t \in \mathbb{T}\backslash\{T\}\}$ are negative definite. The feedback-Nash equilibrium strategies are obtained by solving the coupled Riccati difference equations (3.51)–(3.53). Table 3.2 illustrates the feedback investment strategies and the resulting capacity trajectories of the players when the discount factor is set to $\rho = 0.9$ and $\rho = 0.5$.

From Tables 3.1 and 3.2, we observe that players consistently produce more with feedback-Nash equilibrium strategies, compared to the open-loop Nash equilibrium strategies. This behavior is somewhat expected, as feedback intensifies competition with respect to precommitment (open loop); see also Reynolds (1987).

3.5 Additional Readings

Dynamic games, in both discrete and continuous time, can be seen as offsprings of game theory and dynamic optimization. The necessary game-theoretic concepts were covered in Chap. 1. There are many books covering optimal control theory and dynamic programming, e.g., Kamien and Schwartz (2012), Sethi and Thompson (2000), Seierstad and Sydsæter (1987), Bertsekas (2000). For an introduction (and some advanced material) on multistage games, see, e.g., Başar and Olsder (1999), Haurie et al. (2012), Krawczyk and Petkov (2018). For comprehensive coverage of linear-quadratic differential games see Engwerda (2005).

3.6 Exercises

Exercise 3.1 Consider an industry made up of $m > 2$ firms competing in a two-period Cournot game. Each firm is endowed with an initial production capacity X_t^i, $i = 1, \ldots, m$. At period $t + 1$, the capacity is given by

$$X_{t+1}^i = X_t^i + I_t^i,$$

where $I_t^i \geq 0$ is the investment made by firm i at period t. This means that it takes one period before an investment becomes productive. The investment cost of firm $i = 1, \ldots, m$ is given by the convex function

$$c_i(I_t^i) = \frac{1}{2} \alpha_i (I_t^i)^2,$$

where α_i is a positive parameter. Denote by q_t^i the quantity produced by firm i, and Q_t the total quantity produced by m firms at period t. The production costs of firm $i = 1, \ldots, m$ are assumed to be quadratic and given by

$$g_i(q_t^i) = \frac{1}{2} \beta_i (q_t^i)^2,$$

where β_i is a positive parameter. The inverse demand law is given by

$$p_t = a_t - b_t Q_t,$$

where a_t and b_t are positive demand parameters that satisfy

$$a_{t+1} = (1 + \epsilon)a_t, \ b_{t+1} = \frac{1}{1 + \epsilon} b_t, \ 0 < \epsilon < 1, \ t \in \mathbb{T} \setminus \{T\}.$$

1. Assuming that the firms produce at full capacity, model the competition as a linear-quadratic difference game.
2. Compute open-loop and feedback-Nash equilibrium investments and quantities for each firm when $T = 5$, $m = 3$, $\alpha_1 = \alpha_2 = 1.5$, $\alpha_3 = 0.8$, $\beta_1 = 0.5$, $\beta_2 = \beta_3 = 1$, $a_0 = 4.5$, $b_0 = 0.4$, $\epsilon = 0.01$, $K_0^1 = 2$, $K_0^2 = 1.5$. Verify if the parameter values satisfy the conditions stated in Theorems 3.4 and 3.5.

3. Suppose that the firms form a cartel and optimize their joint profit. Determine the optimal quantities, investments, and joint profit. Compare this solution to the open-loop and feedback-Nash equilibria and discuss the results (in particular, the impact of the cartel's formation on consumers).

Exercise 3.2 Suppose that the m firms in the above example compete à la Cournot in two different markets.

1. Extend the model to account for the additional market.
2. Write down the necessary equations for computing open-loop and feedback-Nash equilibrium investment strategies.

Exercise 3.3 Consider a three-player dynamic game of pollution control studied in Germain et al. (2003). Let $M = \{1, 2, 3\}$ be the set of players representing countries, and $\mathbb{T} = \{0, 1, 2, 3\}$ be the set of time periods. Countries produce emissions of some pollutant at any time period except the terminal one, that is,

$$x_{t+1} = (1 - \delta)x_t + \sum_{i=1}^{3} u_t^i, \ x_0 \text{ given,}$$

where $\delta \in (0, 1)$ is the rate of pollution absorption by nature. The damage cost is an increasing function in the pollution stock and has a quadratic form $D_i(x_t) = \alpha_i x_t^2$, $i \in M$, where α_i is a strictly positive parameter. The cost of emissions is also given by a quadratic function $C_i(u_t^i) = \frac{1}{2}\gamma_i \left(u_t^i - e\right)^2$, where e and γ_i are strictly positive constants.

1. The instantaneous and terminal costs of Player $i \in M$ are given by $\rho^t(C_i(u_t^i) + D_i(x_t))$ and $\rho^T D_i(x_T)$, where $\rho \in (0, 1)$ denotes the discount factor. Model this multistage pollution game as a linear-quadratic difference game by setting up the equations for the computation of the open-loop and feedback-Nash equilibria.
2. Find the open-loop and feedback-Nash equilibria using the following parameters:

$$\alpha_1 = 0.1, \ \alpha_2 = 0.2, \ \alpha_3 = 0.3,$$
$$\gamma_1 = 0.1, \ \gamma_2 = 0.2, \ \gamma_3 = 0.3,$$
$$\delta = 0.6, \ \rho = 0.9, \ e = 30, \ x_0 = 5.$$

Verify if the parameter values satisfy the conditions stated in Theorems 3.4 and 3.5.
3. Assume that, in an effort to minimize their costs, the countries cooperate and jointly minimize the total stock of pollutants. Formulate the associated optimal control problem and calculate the optimal emissions using the parameter values in (b). Compare the optimal emissions and pollution levels with those obtained at equilibrium.

References

Başar, T. and Olsder, G. (1999). *Dynamic Noncooperative Game Theory: Second Edition*. Classics in Applied Mathematics. Society for Industrial and Applied Mathematics.

Bellman, R. (2003). *Dynamic Programming*. Dover Books on Computer Science Series. Dover Publications.

Bertsekas, D. (2000). *Dynamic Programming and Optimal Control*. Number v. 2 in Athena Scientific optimization and computation series. Athena Scientific.

Engwerda, J. (2005). *LQ dynamic optimization and differential games*. John Wiley & Sons.

Fan, L. and Wang, C. (1964). *The Discrete Maximum Principle: A Study of Multistage Systems Optimization*. Wiley.

Germain, M., Toint, P., Tulkens, H., and de Zeeuw, A. (2003). Transfers to sustain dynamic core-theoretic cooperation in international stock pollutant control. *Journal of Economic Dynamics and Control*, 28(1):79–99.

Halkin, H. (1974). Necessary conditions for optimal control problems with infinite horizons. *Econometrica*, 42(2):267–272.

Haurie, A., Krawczyk, J. B., and Zaccour, G. (2012). *Games and Dynamic Games*. World Scientific, Singapore.

Jank, G. and Abou-Kandil, H. (2003). Necessary conditions for optimal control problems with infinite horizons. *IEEE Transactions on Automatic Control*, 8(2):267–271.

Kamien, M. and Schwartz, N. (2012). *Dynamic Optimization: The Calculus of Variations and Optimal Control in Economics and Management*. Dover books on mathematics. Dover Publications.

Krawczyk, J. B. and Petkov, V. (2018). *Multistage Games*, pages 157–213. In: Başar T., Zaccour G. (eds) Handbook of Dynamic Game Theory. Springer International Publishing, Cham.

Michel, P. (1982). On the transversality condition in infinite horizon optimal problems. *Econometrica*, 50(4):975–985.

Reynolds, S. S. (1987). Capacity investment, preemption and commitment in an infinite horizon model. *International Economic Review*, 28(1):69–88.

Rosen, J. B. (1965). Existence and uniqueness of equilibrium points for concave n-person games. *Econometrica*, 33(3):520–534.

Seierstad, A. and Sydsæter, K. (1987). *Optimal Control Theory with Economic Applications*. Advanced Textbooks in Economics. Elsevier Science.

Selten, R. (1975). Reexamination of the perfectness concept for equilibrium points in extensive games. *International Journal of Game Theory*, 4(1):25–55.

Sethi, S. and Thompson, G. (2000). *Optimal Control Theory: Applications to Management Science and Economics*. Springer Nature Book Archives Millennium. Springer.

Chapter 4
Sustainability of Cooperation in Dynamic Games

This chapter introduces the concepts of time consistency and cooperative equilibrium in cooperative dynamic games. To simplify the exposition, we retain a deterministic setup. In the next chapters, we extend the results to dynamic games played over event trees (DGPETs).

4.1 Dynamic Cooperative Games

In Chap. 2, we saw that solving a cooperative static game with a transferable utility is, schematically, a two-step process:

Step 1: Determine the best outcome that the grand coalition can achieve by optimizing the (possibly weighted) sum of the players' payoffs.

Step 2: Allocate the optimal outcome to the players, using one of the many solutions available for cooperative game, e.g., Shapley value and core.

The strategy profile corresponding to the collectively optimal solution determined in the first step can be interpreted as the *operational clause* of a cooperative agreement (or contract), while the sharing defined in the second step represents the *financial clause*. In a one-shot game, both clauses are fulfilled simultaneously.

An additional concern comes up in a dynamic setting, namely, the durability (or sustainability) of an agreement over time. It is an empirical fact that some long-term contracts fail to remain in place till their maturity. Examples of breakdowns include divorce, the United States leaving the Paris Agreement, and Canada abandoning the

Kyoto Protocol. Haurie (1976) gives two reasons why a given agreement that is individually rational at an initial instant of time may fail to remain in place till its maturity date T[1]:

1. If the players accept a renegotiation of the original agreement at an intermediary date $\tau, 0 < \tau < T$, it is not sure that they will wish to continue with that agreement. In fact, they may not go on with the original agreement if it is not a solution of the cooperative game that starts out at time τ.
2. If a player obtains a higher payoff by leaving the agreement at time τ than by continuing to implement her cooperative strategy, then she will indeed deviate from cooperation. Such individual deviation may occur because the cooperative solution is not an equilibrium.

If a long-term agreement can break down, then the following questions are quite natural:

1. Why do rational players engage in intertemporal collaboration instead of cooperating one period at a time?
2. Can cooperation be sustained?

The answer to the first question is quite simple. If a problem is inherently dynamic, meaning that today's decisions have an impact on tomorrow's gain, then solving a T-period dynamic optimization problem always yields at least the same outcome that would be obtained by solving T static problems. If the difference between the two results is negligible, then we would conclude that the intertemporal feature is not important.

To answer the second, of how to achieve sustainable cooperation, the literature on dynamic games has followed two approaches. The first aims at ensuring that the agreed-upon solution of the cooperative game, e.g., the Shapley value, the Nash bargaining solution, is *time consistent*. This means that, at each instant of time, the players prefer to stick to the agreement rather than switch to a noncooperative mode of play. In a nutshell, the property of time consistency can be implemented by defining appropriate payments over time, say, $\beta_{i0}, \ldots, \beta_{iT}$. The second approach seeks, through the implementation of trigger or incentive strategies, to make the cooperative solution an equilibrium of an associated noncooperative game.

4.1.1 Elements of the Game

To simplify the exposition of the main ideas related to sustainability, we consider a deterministic environment. Let the dynamic game played on $\mathbb{T} = \{0, 1, \ldots, T\}$ be defined by the following:

---

[1] An agreement is initially individually rational if the total gain under cooperation exceeds the gain that a player can secure by acting alone. By definition, an imputation satisfies this requirement.

1. A set of players $M = \{1, 2, \ldots, m\}$.
2. For each player $i \in M$, a vector of control (or decision) variables $u_t^i \in U^i \subseteq \mathbb{R}^{\ell_i}$ at $t = 0, \ldots, T - 1$, where U^i is the set of admissible control values for Player i. Let $\mathbf{U} = \prod_{i \in M} U^i$.
3. A vector of state variables $x_t \in X \subset \mathbb{R}^q$ at time $t \in \mathbb{T}$, where X is the set of admissible states and where the evolution over time of the state is given by

$$x_{t+1} = f(x_t, u_t), \quad x_0 \text{ given}, \tag{4.1}$$

   where $u_t \in \mathbf{U}$, $t = 1, \ldots, T$ and x_0 is the initial state at $t = 0$.
4. A payoff functional for Player $i \in M$,

$$J_i(\mathbf{u}; x_0) = \sum_{t=0}^{T-1} \rho^t \phi_i(x_t, u_t) + \rho^T \Phi_i(x_T), \tag{4.2}$$

   where $\rho \in (0, 1)$ is the discount factor; $u_t = (u_t^1, \ldots, u_t^m)$, and $\mathbf{u}$ is given by

$$\mathbf{u} = \{u_t^i \in U^i : t \in \mathbb{T}\backslash\{T\}, \ i \in M\}; \tag{4.3}$$

   $\phi_i(x_t, u_t)$ is the reward to Player i at $t = 0, \ldots, T - 1$, and $\Phi_i(x_T)$ is the reward to Player i at terminal time T.
5. An information structure that defines the information that is available to Player $i \in M$ when she selects her control vector u_t^i at time $t \in \mathbb{T}\backslash\{T\}$.
6. A strategy σ^i for Player $i \in M$, which is an m_i-dimensional vector-decision rule that defines the control $u_t^i \in U^i$ as a function of the information available at time $t = 0, \ldots, T - 1$.

Remark 4.1 As in the games played over event trees, we retain a finite horizon, but all concepts and results in this chapter can be easily extended to an infinite horizon.

Remark 4.2 To keep it as simple as possible, we assumed that the control set is time invariant. The other options are to have a control set that varies with time, i.e., U_t^i, or to depend on the state and time, i.e., $U_t^i(x_t)$. Adopting either of these options complicates solving some of the problems involved, but does not affect the conceptual frameworks that are introduced.

Suppose that the players agree to maximize their joint payoff, given by

$$J(\mathbf{u}; x_0) = \sum_{i \in M} J_i(\mathbf{u}; x_0). \tag{4.4}$$

The result is the grand coalition payoff to be shared between the players.

Remark 4.3 A more general formulation is to suppose that the players optimize a weighted sum of their objectives, i.e., $J = \sum_{i \in M} \alpha_i J_i(\mathbf{u}; x_0)$, where the weights reflect bargaining power and satisfy $\alpha_i \geq 0$ and $\sum_{i \in M} \alpha_i = 1$. For parsimony, we assume that all players have the same weight.

Denote by $u_t^* = \left(u_t^{1*}, \ldots, u_t^{m*}\right), t = 0, \ldots, T - 1$, the control paths that solve the optimal control problem in (4.4), subject to the state equations in (4.1). Denote by $J^*(x_0)$ the outcome of the joint optimization problem. If this solution is implemented throughout the game, then Player i's outcome, before any side payment, is given by

$$J_i^*(x_0) = \sum_{t=0}^{T-1} \rho^t \phi_i \left(x_t^*, u_t^*\right) + \rho^T \Phi_i(x_T^*), \tag{4.5}$$

where x_t^* is the solution to the state equation

$$x_{t+1} = f\left(x_t, u_t^*\right), \quad x_0 \text{ given.} \tag{4.6}$$

In the terminology used in the introduction, the control paths $u_t^* = \left(u_t^{1*}, \ldots, u_t^{m*}\right)$, $t = 0, \ldots, T - 1$ represent the operational clause of the agreement. By now, each player knows what action she must implement at each period t. If all players abide by the agreement, then they realize the joint payoff

$$J^*(x_0) = \sum_{i \in M} J_i^*(x_0).$$

Let $v(K; x_0)$ be the characteristic function (CF) value of coalition K in the entire cooperative game starting at time 0 in state x_0. We add x_0 as an argument of $v(K)$, because, later on, we will also need to consider CF values of games starting at other initial state values. The CF value of the grand coalition is

$$v(M; x_0) = J^*(x_0).$$

Remark 4.4 To be more rigorous, we should write $v(K; x_0, 0)$ instead of $v(K; x_0)$ to specify the starting time for the CF value of coalition K. Indeed, the value $x_0 \in X$ can materialize at any period, not only at the initial time. To keep the notation simple, we omit the time argument.

To compute the values $v(K; x_t)$ for all $K \subset M$ and $t \in \mathbb{T}$, the players must select a CF concept, e.g., $\alpha, \beta, \gamma,$ or δ CF. The conceptual developments to follow apply to any choice of CF, but for clarity, we suppose that the players adopt the γ CF. Recall that, in this case, $v(K; x_t)$ is defined as the partial equilibrium outcome of the noncooperative game between coalition K and left-out players acting individually.

In particular, $v(\{i\}; x_t)$ is the Nash equilibrium outcome of Player i in the m-player noncooperative game. In case of multiple equilibria, we suppose that the players select one via some mechanism.[2]

Denote by x_t^{nc} and $u_t^{nc}, t = 0, \ldots, T-1$ the state and control trajectories in the noncooperative game, respectively. If the game is played noncooperatively throughout the entire horizon, then Player i gets the following outcome:

$$J_i^{nc}(x_0) = \sum_{t=0}^{T-1} \rho^t \phi_i\left(x_t^{nc}, u_t^{nc}\right) + \rho^T \Phi_i(x_T^{nc}). \tag{4.7}$$

Observe that $J_i^{nc}(x_0) = v(\{i\}; x_0)$.

The next step is to select an imputation, or a subset of imputations, from the set

$$Y(x_0) = \left\{ (y_1(x_0), \ldots, y_m(x_0)) \; \middle| \; y_i(x_0) \geq v(\{i\}; x_0), \forall i \in M \right.$$

$$\left. \text{and } \sum_{i=1}^{m} y_i(x_0) = v(M; x_0) \right\}.$$

Suppose that the players agree on imputation $y^*(x_0) = \left(y_1^*(x_0), \ldots, y_m^*(x_0)\right)$ to share $J^*(x_0)$. The discussion to follow is valid for any choice of $y(x_0) \in Y(x_0)$.

4.2 Time-Consistent Solution

The imputation $y^*(x_0)$ defines what the players would get if they cooperated throughout the whole game. It does not tell us how the total individual payoffs are distributed over time, nor what would happen if the agreement broke down. To define time consistency and other concepts, we need to determine the following outcomes:

1. $J_i^*\left(x_\tau^*\right)$: Cooperative payoff-to-go, *before any side payment,* of Player i, at time $\tau \in \mathbb{T}$ and state value x_τ^*.
2. $J_i^c\left(x_\tau^*\right)$: Cooperative payoff-to-go, *after any side payment,* of Player i, at time $\tau \in \mathbb{T}$ and state value x_τ^*.
3. $J_i^{nc}\left(x_\tau^*\right)$: Noncooperative (Nash equilibrium) payoff-to-go of Player i, at time $\tau \in \mathbb{T}$ and state value x_τ^*.

The value of $J_i^*(x_\tau^*)$ is obtained from (4.5) by a restriction of the time interval to $[\tau, T]$. The amount $J_i^c\left(x_\tau^*\right)$ corresponds to what Player i will actually get in the game on $[\tau, T]$, with $J_i^c(x_0) = y_i^*(x_0)$. The difference between $J_i^c\left(x_\tau^*\right)$ and $J_i^*(x_\tau^*)$ can assume any sign, depending on whether the player is receiving or paying a certain amount.

---

[2] Selection of an equilibrium is beyond the scope of this book.

Remark 4.5 As the objective is to avoid a breakdown of the agreement before its maturity, say at time $\tau > 0$, the payoffs-to-go $J_i^* \left(x_\tau^* \right)$, $J_i^c \left(x_\tau^* \right)$, and $J_i^{nc} \left(x_\tau^* \right)$ are quite naturally computed along the optimal collective trajectory x_τ^*, that is, assuming that cooperation has prevailed from the initial date till period τ.

Remark 4.6 The noncooperative payoff $J_i^{nc} \left(x_\tau^* \right)$ is not obtained by restricting the equilibrium payoff computed on $\mathbb{T}$ to $[\tau, T]$. Indeed, the restriction of (4.7) to the time interval $[\tau, T]$ would give Player i the following outcome:

$$J_i \left(x_\tau^{nc} \right) = \sum_{t=\tau}^{T-1} \rho^{t-\tau} \phi_i \left(x_t^{nc}, u_t^{nc} \right) + \rho^{T-\tau} \Phi_i (x_T^{nc}).$$

Unless the Nash equilibrium is efficient (Pareto optimal), there is no reason to believe that x_τ^{nc} and x_τ^* are the same. Hence, determining $J_i \left(x_\tau^{nc} \right)$ requires solving a noncooperative game on $[\tau, T]$ with an initial state value given by x_τ^*.

Definition 4.1 A cooperative solution is time consistent at initial state x_0 if, at any x_τ^* for all $\tau > 0$, it holds that

$$J_i^c(x_\tau^*) \geq J_i^{nc}(x_\tau^*), \quad i \in M, \tag{4.8}$$

where $\{x_\tau^*, \tau \in \mathbb{T}\}$ denotes the cooperative state trajectory.

The condition in (4.8) states that the cooperative agreement designed at the start of the game is time consistent if each player's cooperative payoff-to-go dominates her noncooperative payoff-to-go at any intermediate date τ. Two comments are in order.

1. In (4.8), $J_i^{nc}(x_\tau^*)$ is the fallback (or benchmark) outcome that Player i can secure if the agreement breaks down at τ. Two implicit assumptions are made. First, when making the comparison, each player is assuming that either *all* the players cooperate or there is no agreement. That is, it suffices for one player to leave the agreement for it to become void. The option of a subset of players continuing to play cooperatively from τ onward is not on the menu. Second, if the agreement breaks down at τ, then noncooperation prevails till T, with no possibility of resuming cooperation at a future date t, with $\tau < t < T$.
2. Whereas $J_i^{nc}(x_\tau^*)$ is endogenously obtained by solving for a Nash equilibrium, the after-side-payment cooperative outcomes $J_i^c(x_\tau^*)$, $i \in M$ are exogenously determined, taking constraints into account. Then, the question is how to construct the vector $J^c(x_\tau^*) = \left(J_1^c(x_\tau^*), \ldots, J_m^c(x_\tau^*) \right)$ to satisfy the condition in (4.8).

A stronger condition for dynamic individual rationality is that the cooperative payoff-to-go dominates (at least weakly) the noncooperative payoff-to-go *along any state trajectory*. This amounts to relaxing the assumption that the players have been following the cooperative state trajectory until the comparison point, as is the case in time consistency. This is the *agreeability* concept, introduced in Kaitala and Pohjola (1990).

Definition 4.2 A cooperative solution is agreeable at initial state x_0 if, at any x_τ for all $\tau > 0$, the following inequality holds:

$$J_i^c(x_\tau) \geq J_i^{nc}(x_\tau), \quad i \in M.$$

Clearly, agreeability implies time consistency. Jørgensen et al. (2003, 2005) analyzed the relationships between the two concepts in the case of linear-state and linear-quadratic differential games, respectively.

4.2.1 Imputation Distribution Procedure

Denote by β_{it} the payment that Player i receives at time t, and let $\beta_i = (\beta_{i0}, \ldots, \beta_{iT})$ be the payment schedule to Player i and $\beta = (\beta_1, \ldots, \beta_m)$.

Definition 4.3 The payment schedule $\beta_i = (\beta_{i0}, \ldots, \beta_{iT})$ is a payoff distribution procedure if it satisfies the following condition:

$$\sum_{t=0}^{T} \rho^t \beta_{it} = J_i^c(x_0), \quad \text{for all } i \in M. \tag{4.9}$$

The condition in (4.9) means that the sum of discounted payments to each player must be equal to what she is entitled to receive in the entire cooperative game. The idea of a *payoff distribution procedure* (PDP) was introduced in Petrosyan (1997).

Definition 4.4 The PDP $\beta = (\beta_1, \ldots, \beta_m)$ is time consistent at initial state x_0 if, at any x_τ^* for all $\tau > 0$, the following condition holds:

$$\sum_{t=\tau}^{T} \rho^{t-\tau} \beta_{it} \geq J_i^{nc}\left(x_\tau^*\right), \quad \text{for all } i \in M. \tag{4.10}$$

The condition in (4.10) states that, if the players reconsider the agreement at any intermediate date τ, the total discounted payment to each player if they continue to cooperate is at least equal to her noncooperative payoff-to-go.

Two comments are in order regarding the definition of a PDP. First, the values of $\beta_{it}, t \in \mathbb{T}$ can assume any sign and are not constrained to be nonnegative, i.e., some players may have to pay, instead of receiving money, at some periods of time. If this is undesirable in a practical problem, then we can implement a regularization procedure to alleviate the problem (see Petrosyan and Danilov 1979). Second, there is an infinite number of time functions that qualify as a PDP. In particular, the payment β_{it} may not be directly related to the revenues and costs of Player i at time t. Put differently, a PDP may have neither a particular relationship with the data of the problem beyond the fact that it ensures the agreement's sustainability, nor any particular economic interpretation. A natural question is, then, could the payments be related to a cooperative game solution or a bargaining outcome? To answer this question, we introduce the concept of an *imputation distribution procedure* (IDP).

Definition 4.5 Let $y^* (x_0) = \left(y_1^* (x_0), \ldots, y_m^* (x_0) \right)$ be an imputation agreed upon by the m players. An imputation distribution procedure (IDP) is a vector $\alpha = (\alpha_1, \ldots, \alpha_m)$, with $\alpha_i = (\alpha_{i0}, \ldots, \alpha_{iT})$ satisfying

$$\sum_{t=0}^{T} \rho^t \alpha_{it} = y_i^* (x_0), \text{ for all } i \in M. \tag{4.11}$$

The above definition states that the discounted stream of payments to Player i must be equal to her imputation for the whole cooperative game. The next definition gives the condition for an IDP to be time consistent.

Definition 4.6 An IDP $\alpha = (\alpha_1, \ldots, \alpha_m)$ is time consistent at the initial state x_0 if, at any x_τ^*, the following conditions hold:

$$\sum_{t=0}^{\tau-1} \rho^t \alpha_{it} + \rho^\tau y_i \left(x_\tau^* \right) = y_i^* (x_0), \text{ for } \tau = 1, \ldots, T - 1, \forall i \in M, \tag{4.12}$$

$$\text{and } \alpha_{iT} = y_i \left(x_T^* \right), \forall i \in M, \tag{4.13}$$

where $y_i^* (x_0)$ and $y_i \left(x_\tau^* \right)$ belong to the same solution of a cooperative game and subgame starting at time τ and state x_τ^*, respectively.

We make the following comments:

1. The imputations $y_i^* (x_0)$ and $y_i \left(x_\tau^* \right)$ correspond to the cooperative payoff-to-go from periods 0 and τ, respectively. The two numbers are computed using the same solution concept, e.g., the Shapley value or the core.
2. For a PDP to be time consistent, we needed the condition $\sum_{t=\tau}^{T} \rho^{t-\tau} \beta_{it} \geq J_i^{nc} \left(x_\tau^* \right)$ (see (4.10)). Because an imputation is by definition individually rational, there is no need to explicitly add this condition for an IDP.
3. To interpret the condition in (4.12), suppose that the players have played the game cooperatively from the initial date to period τ. At τ, the state of the system is x_τ^*. If the players continue to cooperate from τ till the end of the game, then they will choose one imputation using the same solution concept as initially, say $y \left(x_\tau^* \right) = \left(y_1 \left(x_\tau^* \right), \ldots, y_m \left(x_\tau^* \right) \right)$. For the IDP to be time consistent, Player i's cooperative payoff-to-go at τ plus what she had received till τ must be equal to her imputation or cooperative payoff-to-go at 0.

Proposition 4.1 *The time-consistent IDP $\alpha_i = (\alpha_{i0}, \ldots, \alpha_{iT})$, $i \in M$ is given by*

$$\alpha_{it} = y_i \left(x_t^* \right) - \rho y_i \left(x_{t+1}^* \right), \text{ for } t = 0, \ldots, T - 1, \text{ and} \tag{4.14}$$
$$\alpha_{iT} = y_i \left(x_T^* \right).$$

Proof Substitute for $\alpha_{it} = y_i\left(x_t^*\right) - \rho y_i\left(x_{t+1}^*\right)$ in $\sum_{t=0}^{\tau-1} \rho^t \alpha_{it} + \rho^\tau y_i\left(x_\tau^*\right)$ to get

$$
\sum_{t=0}^{\tau-1} \rho^t \left(y_i\left(x_t^*\right) - \rho y_i\left(x_{t+1}^*\right)\right) + \rho^\tau y_i\left(x_\tau^*\right)
$$
$$
= \rho^0 \left(y_i\left(x_0^*\right) - \rho y_i\left(x_1^*\right)\right) + \rho^\tau y_i\left(x_\tau^*\right)
$$
$$
+ \rho \left(y_i\left(x_1^*\right) - \rho y_i\left(x_2^*\right)\right)
$$
$$
+ \rho^2 \left(y_i\left(x_2^*\right) - \rho y_i\left(x_3^*\right)\right) + \dots
$$
$$
+ \rho^{\tau-1} \left(y_i\left(x_{\tau-1}^*\right) - \rho y_i\left(x_\tau^*\right)\right)
$$
$$
= y_i\left(x_0^*\right).
$$

Further, the equality in (4.13) is clearly satisfied.   □

The condition in (4.11) is an individual budget-balance equation. By construction, an IDP also satisfies the collective budget balance

$$
\sum_{i=1}^{m} \sum_{t=0}^{T} \rho^t \beta_{it} = \sum_{i=1}^{m} \sum_{t=0}^{T} \rho^t \alpha_{it} = J^*\left(x_0\right). \tag{4.15}
$$

The following lemma shows that a time-consistent IDP is balanced at each period of time, that is, at any t, the sum of allocations to the players is equal to the total cooperative payoff realized at that period.

Lemma 4.1 *A time-consistent IDP satisfies the following property:*

$$
Period\ budget\ balance : \sum_{i=1}^{m} \alpha_{i\tau} = \sum_{i=1}^{m} \phi_i\left(x_\tau^*, u_\tau^*\right)
$$

for any $\tau \in \mathbb{T}$.

Proof Write (4.12) for $\tau = 1$:

$$
\alpha_{i0} + \rho y_i\left(x_1^*\right) = y_i^*\left(x_0\right)
$$

and sum it over $i \in M$ to get

$$
\sum_{i \in M} \alpha_{i0} = \sum_{i \in M} y_i^*\left(x_0\right) - \rho \sum_{i \in M} y_i\left(x_1^*\right)
$$
$$
= J^*(x_0) - \rho J^*(x_1^*) = \sum_{i \in M} \phi_i(x_0, u_0^*) + \rho J^*(x_1^*) - \rho J^*(x_1^*)
$$
$$
= \sum_{i \in M} \phi_i(x_0, u_0^*).
$$

Let the IDP period budget-balance condition be true for any $t = 0, \ldots, \tau - 1$. We prove it for $t = \tau$. For $t = \tau + 1$, we have this for a time-consistent IDP:

$$\sum_{i \in M} \sum_{t=0}^{\tau} \rho^t \alpha_{it} = \sum_{i \in M} y_i^* (x_0) - \rho^{\tau+1} \sum_{i \in M} y_i \left(x_{\tau+1}^* \right)$$

$$= \sum_{t=0}^{\tau} \rho^t \sum_{i \in M} \phi_i (x_t^*, u_t^*) + \rho^{\tau+1} J^* (x_{\tau+1}^*) - \rho^{\tau+1} \sum_{i \in M} y_i \left(x_{\tau+1}^* \right)$$

$$= \sum_{t=0}^{\tau-1} \rho^t \sum_{i \in M} \alpha_{it} + \rho^\tau \sum_{i \in M} \phi_i (x_\tau^*, u_\tau^*) + \rho^{\tau+1} \left(J^* (x_{\tau+1}^*) - v(M; x_{\tau+1}^*) \right),$$

and we obtain that

$$\sum_{i \in M} \alpha_{i\tau} = \sum_{i \in M} \phi_i (x_\tau^*, u_\tau^*).$$

$\square$

To wrap up, designing a time-consistent IDP requires the implementation of the following algorithm, where k is the number of players in coalition K :

Step 1:   Compute the characteristic function values $v(K; x_t^*)$ of the cooperative game for all $K \subseteq M$ and each $t \in \mathbb{T} = \{0, 1, \ldots, T\}$.

> **Step 1a**:   Solve the joint-optimization problem in (4.4) to obtain the cooperative state trajectory $x^* = \left(x_0^*, \ldots, x_T^* \right)$.
>
> **Step 1b**:   For any t, with the initial state value x_t^*, and for any coalition K, determine the Nash equilibrium outcomes $J_i^{nc} \left(x_t^*; K \right), i \in M$ of the $(m - k + 1)$-player noncooperative game, with K acting as one player (i.e., optimizing the sum of the members' payoffs), and the $m - k$ left-out players acting individually.
> Set $v(K; x_t^*) = \sum_{i \in K} J_i^{nc} \left(x_t^*; K \right)$. If the equilibrium is not unique, then select one through some mechanism.

Step 2:   Choose a solution concept, i.e., the Shapley value, the core, etc. For each cooperative subgame starting at time $t = 0, \ldots, T$ in state x_t^*, select an imputation $y \left(x_t^* \right) = \left(y_1 \left(x_t^* \right), \ldots, y_m \left(x_t^* \right) \right)$ following the solution concept. (If the solution concept is set-valued, then the players must agree on one.)
Set $\sum_{t=0}^{T} \rho^t \alpha_{it} = y_i^* (x_0)$, for all $i \in M$.

Step 3:   Use (4.12) to obtain the IDP $\alpha_i = (\alpha_{i0}, \ldots, \alpha_{iT}), i \in M$ as follows:

$$\alpha_{it} = y_i \left(x_t^* \right) - \rho y_i \left(x_{t+1}^* \right), \text{ for } t = 0, \ldots, T - 1, \text{ and}$$
$$\alpha_{iT} = y_i \left(x_T^* \right).$$

4.2.2 An Example

To illustrate the construction of an IDP, we consider the model of pollution control in Germain et al. (2003). Denote by $M = \{1, 2, 3\}$ the set of players (countries), and by $\mathbb{T} = \{0, 1, \ldots, T\}$ the set of periods, with $T = 9$. Let $u_t = (u_t^1, u_t^2, u_t^3)$ be the vector of countries' emissions of some pollutant at time t, and denote by x_t the stock of pollution at period t. The evolution of this stock is governed by the following difference equation:

$$x_t = (1 - \delta)x_{t-1} + \sum_{i \in M} u_{t-1}^i, \quad t = 1, \ldots, T, \quad x_0 \text{ given}, \qquad (4.16)$$

where $\delta \in (0, 1)$ is the rate of pollution absorption by Mother Nature.

The damage cost of country i is an increasing convex function in the pollution stock having the form

$$D_i(x_t) = \alpha_i x_t^2, \quad i \in M,$$

where α_i is a strictly positive parameter. The cost of emissions is also given by a quadratic function,

$$C_i(u_t^i) = \frac{\gamma_i}{2} \left(u_t^i - e\right)^2, \quad i \in M,$$

where e and γ_i are strictly positive constants. The total discounted cost $J_i(\mathbf{x}, \mathbf{u})$, where $\mathbf{x} = (x_0, x_1, \ldots, x_T)$ and $\mathbf{u} = (u_0, u_1, \ldots, u_{T-1})$, and $u_t = (u_t^i : i \in M)$ to be minimized by Player $i \in M$, is given by

$$J_i(\mathbf{x}, \mathbf{u}) = \sum_{t=0}^{T-1} \rho^t \left(C_i(u_t^i) + D_i(x_t)\right) + \rho^T D_i(x_T),$$

where $\rho \in (0, 1)$ is the common discount rate, subject to (4.16), given initial stock x_0 before the game starts, and the constraint $u_t^i \in [0, e]$ for all $i \in M$, and time $t \in \mathbb{T}$.

We adopt the following parameter values:

$$\alpha_1 = 0.06, \quad \alpha_2 = 0.07, \quad \alpha_3 = 0.08,$$
$$\gamma_1 = 5.35, \quad \gamma_2 = 5.85, \quad \gamma_3 = 6.00,$$
$$\delta = 0.15, \quad e = 10, \quad \rho = 0.96.$$

Now, we apply the algorithm introduced above.

Step 1a: Compute the cooperative state trajectory.

Solving the grand coalition (or joint) optimization problem

$$\min_{u^1, u^2, u^3} \sum_{i \in M} J_i(\mathbf{x}, \mathbf{u}) = \sum_{i \in M} \sum_{t=0}^{T-1} \rho^t \left(C_i(u_t^i) + D_i(x_t)\right) + \sum_{i \in M} \rho^T D_i(x_T),$$

we obtain the following cooperative state trajectory:

x_0^*	x_1^*	x_2^*	x_3^*	x_4^*	x_5^*	x_6^*	x_7^*	x_8^*	x_9^*
0	9.457	15.314	19.138	21.964	24.579	27.739	32.394	39.969	52.806

Step 1b: Compute the characteristic function values.

Grand coalition. The CF value for the grand coalition for the entire duration of the game is given by

$$v(M; x_0) = \sum_{i \in M} J_i^*(x_0) = 4662.7.$$

This is the value to share between the players.

To determine the other values, we use the γ-CF, i.e., the players in coalition $K \subset M$ minimize the sum of their costs, while each left-out player minimizes her own cost.

One-player coalitions. The value $v(\{i\}, x_0), i \in M$ corresponds to Player i's cost in the Nash equilibrium of the three-player noncooperative game. Solving for this equilibrium, we get the following state trajectory: and payoffs $J_i^{nc}(x_0)$.

x_0^{nc}	x_1^{nc}	x_2^{nc}	x_3^{nc}	x_4^{nc}	x_5^{nc}	x_6^{nc}	x_7^{nc}	x_8^{nc}	x_9^{nc}
0	16.501	28.901	38.513	46.372	53.347	60.236	67.859	77.161	89.324

Set $v(\{i\}, x_0) = J_i^{nc}(x_0)$. To compute $v(\{i\}; x_t^*), i \in M$, we solve for the Nash equilibrium in the subgame starting at $t = 1, \ldots T$, with the initial state value being x_t^* (that is, the cooperative state). The results are reported in Table 4.1 (columns $v(1), v(2), v(3)$).

Two-player coalitions. Consider coalition $K = \{1, 2\}$. The value $v(\{1, 2\}; x_0)$ is obtained by solving a two-player noncooperative game between Players 1 and 2 acting as one player, i.e., minimizing the sum of their costs, and Player 3 minimizing her own cost. Denote by $J_i^{nc}(x_t^*; \{1, 2\})$ the equilibrium payoff of Player i when $K = \{1, 2\}$. Set $v(\{1, 2\}; x_t^*) = \sum_{i \in K} J_i^{nc}(x_t^*)$.

The same type of computations are done for coalitions $\{1, 3\}$ and $\{2, 3\}$. The results are given in Table 4.1.

Remark 4.7 All computed Nash equilibria are unique.

Remark 4.8 It is easy to verify that the γ characteristic function is subadditive at each node, i.e., $v(S \cup T) \leq v(S) + v(T)$ for any disjoint coalitions $S, T \subset M$.

Step 2:   Choose an imputation. Suppose that the players agree on using the Shapley value to share the total cost $v(M; x_0)$. Using the characteristic function values from Table 4.1, we compute the Shapley value in all subgames (see Table 4.2).

The total cost of the grand coalition, which is equal to 4662.67, is allocated as follows:

$$Sh_0(x_0) = (Sh_{10}(x_0), Sh_{20}(x_0), Sh_{30}(x_0)) = (1234.45, 1550.53, 1877.70).$$

Table 4.1 Characteristic functions for the game and all subgames

t	$v(1)$	$v(2)$	$v(3)$	$v(12)$	$v(13)$	$v(23)$	$v(123)$
0	1743.390	2064.170	2421.920	3756.550	4053.130	4364.520	4662.670
1	1657.290	1959.690	2294.130	3542.420	3820.950	4113.230	4437.910
2	1521.090	1795.970	2097.020	3225.260	3477.670	3742.280	4090.500
3	1344.150	1584.480	1844.770	2828.960	3049.220	3279.760	3643.440
4	1138.280	1339.430	1554.580	2381.560	2565.990	2758.530	3128.630
5	915.817	1075.600	1244.120	1910.490	2057.660	2210.680	2577.170
6	691.033	809.970	933.516	1444.510	1555.320	1669.760	2020.110
7	481.676	563.462	647.093	1015.840	1093.620	1173.150	1490.340
8	306.140	357.510	409.304	655.266	705.523	756.259	1016.510
9	167.306	195.191	223.075	362.497	390.382	418.266	585.572

Table 4.2 The Shapley value of the game and all subgames

t	Sh_1	Sh_2	Sh_3
0	1234.450	1550.530	1877.700
1	1178.910	1476.250	1782.740
2	1091.430	1361.170	1637.900
3	977.431	1212.870	1453.140
4	845.052	1041.900	1241.680
5	702.178	858.573	1016.420
6	556.521	673.208	790.385
7	416.107	496.765	577.472
8	287.794	338.847	389.872
9	167.306	195.191	223.075

Note that the players' Shapley values follow the same order as the damage cost parameter values.

Step 3: Determine the IDP $\alpha_i = (\alpha_{i0}, \ldots, \alpha_{iT})$, $i \in M$. We specialize (4.14) to the case of the Shapley value, that is,

$$\alpha_{it} = Sh_{it}\left(x_t^*\right) - \rho Sh_{i,t+1}\left(x_{t+1}^*\right), \text{ for } t = 0, \ldots, T-1, \text{ and}$$
$$\alpha_{iT} = Sh_{iT}\left(x_T^*\right),$$

and obtain the results in Table 4.3.

Remark 4.9 We can easily verify that the Shapley value and its IDP are time consistent, i.e., condition (4.12) is satisfied for any time τ and $i \in M$. For example, for Player 1 and $\tau = 3$, we have

Table 4.3 IDP for the Shapley value

t	α_{1t}	α_{2t}	α_{3t}
0	102.690	133.326	166.265
1	131.144	169.526	210.353
2	153.095	196.822	242.884
3	166.181	212.645	261.129
4	170.961	217.667	265.919
5	167.918	212.293	257.650
6	157.058	196.315	236.012
7	139.825	171.472	203.195
8	127.180	151.463	175.720
9	167.306	195.191	223.075

$$\alpha_{10} + \rho\alpha_{11} + \rho^2\alpha_{12} + \rho^3 Sh_{13} = Sh_{10},$$
$$102.690 + 0.96 \cdot 131.144 + (0.96)^2 \cdot 153.095 + (0.96)^3 \cdot 977.431 = 1234.45.$$

4.3   Cooperative Equilibria

If the cooperative solution happens to be an equilibrium, then the durability of the agreement is no longer an issue, as it will be in the best interest of each player not to deviate (unilaterally) from the agreement. To endow the cooperative solution with an equilibrium property, one approach is to use trigger strategies that credibly and effectively punish any player deviating from the agreement.

Suppose that the players agree, before the game starts, to implement the cooperative controls u^* that maximize their joint payoff. Also, suppose that at any period τ, each player knows the *history* of the realized controls from the initial date to the previous period $\tau - 1$.

Definition 4.7 The collection of control vectors realized at time periods 0, 1, ..., $\tau - 1$ is called the history at time τ and is denoted by $\mathcal{H}_\tau^u = (u_0, u_1, \ldots, u_{\tau-1})$, where u_t is a control vector chosen at time t.

Notice that we do not include the sequence of observed states in the definition of history because, when the control profiles are known, the state trajectory can be obtained as the unique solution of state equation (4.6).

Now, we introduce history-dependent strategies that tell a player what to do at each moment of time, depending on what has happened before.

Definition 4.8 A strategy σ^i of Player $i \in M$ is a mapping that associates to each history $\mathcal{H}_\tau^u$, a control $u_\tau^i \in U^i$, that is,

$$\sigma_\tau^i : \mathcal{H}_\tau^u \to U^i. \tag{4.17}$$

Denote by Σ^i the set of strategies of Player i defined by (4.17), by $\sigma = (\sigma^1, \ldots, \sigma^m)$ a strategy profile, and by $\Sigma = \Sigma^1 \times \cdots \times \Sigma^m$ the set of possible strategy profiles.

For a given strategy profile and initial state, we can compute the payoff of Player i in all subgames as a function of the strategy profile, including in the whole game. To avoid adding new notations, the payoff of Player i in the subgame starting at time τ with history $\mathcal{H}_\tau^u$ is denoted by

$$J_i(\sigma | \mathcal{H}_\tau^u) = J_i(\mathbf{u}[\tau, x_\tau]; x_\tau),$$

where x_τ is a unique state defined by history $\mathcal{H}_\tau^u$, and $\mathbf{u}[\tau, x_\tau]$ is a trajectory of controls in the subgame starting at time τ and determined by profile σ. As the history of time 0 is $\mathcal{H}_0^u = \varnothing$, we have $J_i(\sigma | \mathcal{H}_0^u) = J_i(\sigma; x_0)$.

It is well known from repeated games (see Mailath and Samuelson 2006) that a Pareto-optimal (or cooperative) solution can be realized as an equilibrium if the game is repeated infinitely many times and the players are patient, that is, the discount rate is high enough. This result does not carry over to finite-horizon games. The reason is that, at the last period, it is in the best interest of the players to deviate to noncooperative strategies. Then, by a backward induction argument, one can show that the players end up deviating in all stages.[3] A less ambitious outcome is to have the cooperative solution as an ε-equilibrium, which we now define.

Definition 4.9 A strategy profile $\hat{\sigma}$ is an ε-equilibrium if, for each player $i \in M$ and each strategy $\sigma^i \in \Sigma^i$, the following inequality holds:

$$J_i(\hat{\sigma}; x_0) \geq J_i((\hat{\sigma}^{-i}, \sigma^i); x_0) - \varepsilon. \tag{4.18}$$

If the players adopt an ε-equilibrium, a player cannot gain more than ε through an individual deviation. Definition 4.9 does not guarantee that a player cannot gain more than ε if she deviates at any intermediate time. The next definition takes into account possible individual deviations in any time period on a realized trajectory.

Definition 4.10 A strategy profile $\hat{\sigma}$ is a subgame perfect ε-equilibrium if, for each player $i \in M$, any time τ, each strategy $\sigma^i \in \Sigma^i$, and each history $\mathcal{H}_\tau^u$, the following inequality holds:

$$J_i(\hat{\sigma} | \mathcal{H}_\tau^u) \geqslant J_i((\hat{\sigma}^{-i}, \sigma^i) | \mathcal{H}_\tau^u) - \varepsilon,$$

where $J_i(\hat{\sigma} | \mathcal{H}_\tau^u)$ is Player i's payoff in the subgame starting at time τ with history $\mathcal{H}_\tau^u$ when the players use strategy profile $\hat{\sigma}$.

---

[3] If you are not familiar with this result, a good exercise is to derive it using the prisoner's dilemma example in Chap. 1. Repeat the game twice and show by backward induction that the players deviate from the Pareto solution in both stages. Generalize the result to an arbitrary finite number of stages with a common discount factor for both players.

To strategically support cooperation in the finite-horizon dynamic game, we must construct an approximated equilibrium.[4] Before that, however, we need to define the trigger strategy that will be used in this construction. Let us suppose that (i) the players want to realize the cooperative controls $\mathbf{u}^*$ maximizing (4.4) and (ii) if Player i deviates from cooperation at time τ by implementing the control $u_\tau^i \neq u_\tau^{i*}$, then cooperation breaks down and all players switch to Nash equilibrium strategies in the subgame starting at subsequent time $\tau + 1$ in state $x_{\tau+1} = f\left(x_\tau^*, (u_\tau^{-i*}, u_\tau^i)\right)$.

If cooperation breaks down, then the payoff to any player $p \in M$ in the subgame starting at time $\tau + 1$ in state $x_{\tau+1}$ is $J_p^{nc}(x_{\tau+1})$, that is, the profit that Player p achieves when all players implement Nash equilibrium strategies. Denote by $\hat{\sigma} = (\hat{\sigma}^p : p \in M)$ a strategy profile that calls for Player p to implement the cooperative control u_t^{p*} at time t if in the history of this time period no deviations from the cooperative trajectory have been observed, and otherwise, to play the noncooperative control $u_t^{p,nc}$.

Below we will use the following notations:

$\hat{x}_{\tau+1}^i = f\left(x_\tau^*, (u_\tau^{-i*}, u_\tau^i)\right)$: The state at time $\tau + 1$, given by function f if at time τ Player i individually deviates from cooperation and uses the control u_τ^i, while all other players stick to the cooperative controls u_τ^{-i*}.

$\hat{\mathbf{u}}^{nc}[\tau + 1, \hat{x}_{\tau+1}^i] = (\hat{u}_t^{p,nc} : p \in M, t \geq \tau + 1)$: The collection of controls forming a Nash equilibrium in the subgame starting at time $\tau + 1$ with state $\hat{x}_{\tau+1}^i$, after Player i's deviation at time τ.

The trigger strategy of a player consists of two behavior types or modes:

Nominal mode. If the history of time τ coincides with the cooperative one:

$$\mathcal{H}_\tau^{u*} = (u_0^*, u_1^*, \ldots, u_{\tau-1}^*), \tag{4.19}$$

i.e., all players used their cooperative controls from time 0 until time $\tau - 1$, then Player p, $p \in M$, implements u_τ^{p*} at time τ.

Trigger mode. If the history of time τ is such that there exists a time period $t < \tau$ such that $u_t \neq u_t^*$, then Player p's strategy is the Nash equilibrium strategy calculated for the subgame starting from the next time period $t + 1$ and a corresponding state. Here, the history of time τ is such that there exists a time period $t < \tau$ and at least one deviating player $i \in M$, that is, the history $\mathcal{H}_t^u$ of time t is part of $\mathcal{H}_\tau^{u*}$, but the history $\mathcal{H}_{t+1}^u$ is different from $\mathcal{H}_{t+1}^{u*}$.

Formally, the trigger strategy of Player $p \in M$ is defined as follows:

$$\hat{\sigma}_p(\mathcal{H}_\tau^u) = \begin{cases} u_\tau^{p*}, & \text{if } \mathcal{H}_\tau^u = \mathcal{H}_\tau^{u*}, \\ \hat{u}_\tau^{p,nc}, & \text{if } \mathcal{H}_\tau^u \neq \mathcal{H}_\tau^{u*} \text{ and } \exists t < \tau \text{ such that} \\ & \mathcal{H}_t^u = \mathcal{H}_t^{u*}, \text{ and } \mathcal{H}_{t+1}^u \neq \mathcal{H}_{t+1}^{u*}, \end{cases} \tag{4.20}$$

---

[4] The development is in line with what has been done in repeated games, except that here we additionally have a vector of state variables evolving over time.

where $\hat{u}_\tau^{p,nc}$ is Player p's control at time τ. The control $\hat{u}_\tau^{p,nc}$ implements the punishing strategy in the subgame starting at time τ from a corresponding state. The control $\hat{u}_\tau^{p,nc}$ is calculated as part of the Nash equilibrium for the subgame starting at time τ.

To construct the trigger strategies, we need to find m punishing strategy profiles for each subgame on the cooperative state trajectory.

Theorem 4.1 *For any $\varepsilon \geq \tilde{\varepsilon}$, there exists a subgame perfect ε-equilibrium in trigger strategies with players' payoffs $J_1(\mathbf{u}^*; x_0), \ldots, J_m(\mathbf{u}^*; x_0)$, and*

$$\tilde{\varepsilon} = \max_{i \in M} \max_{\tau = 0, \ldots, T-1} \varepsilon_\tau^i, \tag{4.21}$$

where

$$\varepsilon_\tau^i = \max_{u_\tau^i \in U^i} \left\{ \phi_i(x_\tau^*, (u_\tau^{-i*}, u_\tau^i)) - \phi_i(x_\tau^*, u_\tau^*) \right. \tag{4.22}$$

$$+ \sum_{t=\tau+1}^{T-1} \rho^{t-\tau}\left(\phi_i(\hat{x}_t^i, \hat{u}_t^{nc}) - \phi_i(x_t^*, u_t^*)\right) + \rho^{T-\tau}\left(\Phi_i(\hat{x}_T^i) - \Phi_i(x_T^*)\right) \left. \right\},$$

where $\hat{u}_t^{nc}$ is a control profile at time t corresponding to a strategy profile $\hat{\sigma}$ determined by (4.20) and when the trigger mode of the strategy begins in the subgame starting at time $\tau + 1$ and in state $\hat{x}_{\tau+1}^i$ after the individual deviation of Player i is observed at time period τ. Therefore, the differences in the second line also depend on the control u_τ^i. The state $\hat{x}_t^i, t > \tau$ belongs to a state trajectory corresponding to $\hat{\mathbf{u}}^{nc}[\tau + 1, \hat{x}_{\tau+1}^i]$.

Proof Suppose that the players implemented their cooperative controls from the initial period till $\tau - 1$. The payoffs-to-go from period τ onward will correspond to one of the two following scenarios: (i) there is a deviation at τ, and then all players switch to the trigger mode of strategies $\hat{\sigma}$ from period $\tau + 1$; or (ii) there are no deviations in the history of time τ or at any subsequent period.

Consider the trigger strategy $\hat{\sigma}^i$ defined in (4.20) and the subgame starting from any time $\tau = 0, \ldots, T - 1$. Consider possible histories of this time period and compute the benefit of Player i first if, in the history, all players chose only cooperative controls. Her payoff in this subgame will be given by

$$J_i(\mathbf{u}^*[\tau, x_\tau^*]; x_\tau^*) = \phi_i(x_\tau^*, u_\tau^*) + \sum_{t=\tau+1}^{T-1} \rho^{t-\tau}\phi_i(x_t^*, u_t^*) + \rho^{T-\tau}\Phi_i(x_T^*), \tag{4.23}$$

where $\mathbf{u}^*[\tau, x_\tau^*] = (u_t^{p*} : p \in M, t \geq \tau)$ is a collection of cooperative controls in the subgame staring at time τ with state x_τ^*.

Then, consider an individual deviation of Player i. First, let the history of time τ be $\mathcal{H}_\tau^{u*}$. Suppose Player i deviates at time τ from the cooperative control profile. In this case, she may secure the following payoff in the subgame starting at time τ, given the information that the strategy profile $\hat{\sigma} = (\hat{\sigma}^p(\cdot) : p \in M)$ determined by

(4.20) will give the following maximal payoff to Player i:

$$\max_{u_\tau^i \in U^i} \left\{ \phi_i(x_\tau^*, (u_\tau^{-i*}, u_\tau^i)) + \sum_{t=\tau+1}^{T-1} \rho^{t-\tau} \phi_i(\hat{x}_t^i, \hat{u}_t^{nc}) + \rho^{T-\tau} \Phi_i(\hat{x}_T^i) \right\}, \qquad (4.24)$$

where a punishing Nash strategy starts to be implemented from time $\tau + 1$. Then, we may compute Player i's benefit from deviation at time τ as a difference between (4.24) and (4.23) to get ε_τ^i as follows:

$$\varepsilon_\tau^i = \max_{u_\tau^i \in U^i} \left\{ \phi_i(x_\tau^*, (u_\tau^{-i*}, u_\tau^i)) - \phi_i(x_\tau^*, u_\tau^*) \right. \qquad (4.25)$$

$$\left. + \sum_{t=\tau+1}^{T-1} \rho^{t-\tau}(\phi_i(\hat{x}_t^i, \hat{u}_t^{nc}) - \phi_i(x_t^*, u_t^*)) + \rho^{T-\tau}\left(\Phi_i(\hat{x}_T^i) - \Phi_i(x_T^*)\right) \right\},$$

which can be positive or zero. If ε_τ^i is equal to zero for any i, then no player can gain from individual deviation at time τ. If there exists at least one player j for whom ε_τ^i is positive, this means that the strategy profile is not stable against individual players' deviation at time τ.

Second, suppose that the history at time τ does not coincide with $\mathcal{H}_\tau^{u*}$. This means that all players have switched from nominal to trigger mode according to strategy (4.20). Player i will not benefit from the deviation at time τ because the players implement their Nash equilibrium strategies regardless of which player (or group of players) has deviated in the previous periods.

Calculating the maximum benefit from deviation for any subgame and any player given by (4.25), we obtain the value of $\tilde{\varepsilon}$ in the theorem statement, that is,

$$\tilde{\varepsilon} = \max_{i \in M} \max_{\tau=0,\dots,T-1} \varepsilon_\tau^i.$$

And for any $\varepsilon \geq \tilde{\varepsilon}$, the strategy profile determined by (4.20) is a subgame perfect ε-equilibrium by construction. $\qquad \Box$

Remark 4.10 The construction of a subgame perfect equilibrium requires the existence and uniqueness of the Nash equilibrium in any subgame, the conditions of which are discussed in Chap. 2. In case of multiple equilibria, one equilibrium may be selected for any subgame and used in the construction of strategy (4.20).

4.3.1 An Example

We illustrate, using the same example as in Sect. 4.2.2, the determination of a subgame perfect ε-equilibrium in the game.

First, we compute the players' costs in the entire game and in any subgame starting at time $\tau = 1, \ldots, 9$ when the cooperative strategy profile is implemented. The cooperative payoffs $J_i^c(x_\tau^*)$, where x_τ^*, $\tau = 1, \ldots, 9$, on a cooperative state trajectory, are given in Table 4.4. To implement a time-consistent solution as in, e.g., Sect. 4.2.2, the total cooperative payoff must be reallocated among the players following some procedure. Here, each player receives an amount given by her payoff function evaluated at the cooperative control and state values. In any subgame starting at τ, the initial state value is x_τ^*, that is, the value resulting from playing the game cooperatively from the initial date to τ. The cooperative state trajectory was calculated in Sect. 4.2.2 (see Step 1a).

Second, we find the subgame perfect ε-equilibrium in the game using Theorem 4.1. To do this, we must find the maximum benefit from individual deviation at any time period over the set of players, taking into account that the players adopt the trigger strategies defined by (4.20). The benefit of Player i for period τ is computed by formula (4.22). In the case of an individual deviation, notice that, to calculate ε_τ^i for any time period τ, we need to find, for any player, the payoff's maximizing strategy, given that, from the subsequent period, all the players will switch to their Nash equilibrium strategies. The Nash equilibrium strategies are calculated for the subgame starting from the corresponding state. The values of ε_τ^i are presented in Table 4.5. Therefore, by Theorem 4.1, we can state that for any $\varepsilon \geq 42.93$, there exists a subgame perfect ε-equilibrium in the game.

Table 4.5 shows that no player has a profitable deviation until $\tau = 5$. At time $\tau = 5$ and $\tau = 6$, only Player 1 has an incentive to deviate. Starting from time $\tau = 7$, any player benefits from individual deviation. We note that the value ε_τ^i is nonmonotonic in time for Player 1. The largest benefit for Player 1 is at time $\tau = 7$. The benefit to Player 2 at time 7 is larger than the one at time 8. The opposite is true for Player 3. The values of ε_τ^i are not calculated for $\tau = 9$, as the players do not choose controls at the terminal time.

Table 4.4 Cooperative payoffs $J_i^c(x_\tau^*)$, $i \in M$

τ	$J_1^c(x_\tau^*)$	$J_2^c(x_\tau^*)$	$J_3^c(x_\tau^*)$
0	1572.52	1528.92	1561.23
1	1488.71	1456.06	1493.13
2	1362.43	1343.11	1384.97
3	1202.77	1197.45	1243.22
4	1020.99	1029.50	1078.14
5	828.02	849.41	899.75
6	635.08	667.28	717.76
7	454.65	493.75	541.95
8	298.83	337.95	379.73
9	167.31	195.19	223.08

Table 4.5 The benefits from individual deviations ε_τ^i, $i \in M$

τ	ε_τ^1	ε_τ^2	ε_τ^3
0	0	0	0
1	0	0	0
2	0	0	0
3	0	0	0
4	0	0	0
5	24.05	0	0
6	42.62	0	0
7	42.93	20.56	4.15
8	21.16	16.83	14.12

4.4 Incentive Equilibria

In the previous section, we assumed that the players agree in a preplay arrangement to implement a punishing strategy if a deviation from the agreement is observed. Using trigger strategies, we constructed a cooperative equilibrium (in an ε sense). Although trigger strategies are effective and credible, they may embody large discontinuities, i.e., a slight deviation leads to a harsh retaliation, generating a very different path from the agreed-upon one. Recall that if a deviation is observed, then the players switch to a noncooperative mode of play for the remainder of the game, even if it is infinite.

In a two-player game,[5] one alternative approach to sustain cooperation is to use incentive strategies that are continuous in the information. In what follows, the information on which a player bases her decision is the other player's action.

As before, suppose that the players agree to maximize their joint payoff, given by

$$J(\mathbf{u}; x_0) = \sum_{i=1}^{2} \left(\sum_{t=0}^{T-1} \rho^t \phi_i(x_t, u_t) + \rho^T \Phi_i(x_T) \right), \qquad (4.26)$$

subject to the state equations in (4.1). Denote by $u_t^* = \left(u_t^{1*}, u_t^{2*} \right) \in U^1 \times U^2, t = 0, \ldots, T - 1$, the resulting control path and by $J^*(x_0)$ the joint optimal outcome.

An incentive strategy for Player i is defined by

$$\psi^i : U^j \to U^i, \ i, j = 1, 2; \ i \neq j,$$

that is, a mapping from Player j's control set into Player i's control set. Denote by Ψ_i the strategy set of Player i. A straightforward interpretation of an incentive strategy is that each player selects an action that depends on the other player's choice.

---

[5] The framework cannot be generalized to more than two players.

Suppose that the players agree before the game starts, as in a cooperative-equilibrium framework, on a control path to be implemented throughout the entire duration of the game. Although in general the players can agree on any path, it makes intuitively sense that they choose a joint maximizing payoff control path, which we denote by $u^* = (u^{1*}, u^{2*})$, where $u^{i*} = (u_0^{i*}, \ldots, u_{T-1}^{i*})$, $i = 1, 2$.

Definition 4.11 The pair of strategies (ψ_1, ψ_2) is an incentive equilibrium at u^*, if

$$J_1(u^{1*}, u^{2*}) \geq J_1(u^1, \psi^2(u^1)), \ \forall u^1 \in U^1,$$
$$J_2(u^{1*}, u^{2*}) \geq J_2(\psi^1(u^2), u^2), \ \forall u^2 \in U^2,$$
$$\psi^1(u^{2*}) = u^{1*} \text{ and } \psi^2(u^{1*}) = u^{2*}, \ t = 0, \ldots, T - 1.$$

The above definition states that if a player implements her part of the agreement, then the other player's best response is to do the same. In this sense, each player's incentive strategy represents a threat to implement a different control than the optimal one if the other player deviates from her optimal strategy. To determine these incentive strategies, we need to solve two optimal control problems, in each of which one player assumes that the other player is using her incentive strategy.

One important concern with incentive strategies is their credibility. These strategies are said to be credible if it is in each player's best interest to implement her incentive strategy if she detects a deviation by the other player from the agreed-upon solution. Otherwise, the threat is not believable, and a player can freely cheat on the agreement without facing any retaliation. A formal definition of credibility follows.

Definition 4.12 The incentive equilibrium strategy $(\psi^i \in \Psi_i, \ \forall i)$ is credible at $u^* \in U^1 \times U^2$ if the following inequalities are satisfied:

$$J_1(\psi^1(u^2), u^2) \geq J_1(u^{1*}, u^2), \ \forall u^2 \in U^2, \tag{4.27}$$
$$J_2(u^1, \psi^2(u^1)) \geq J_2(u^1, u^{2*}), \ \forall u^1 \in U^1. \tag{4.28}$$

To compute incentive strategies, we need to specify the sets of admissible ones, which is by no means trivial. Indeed, it is an instance of a more general problem: how players choose strategies in dynamic games. This involves choosing which observations the strategies should be based upon, and which particular functional forms should link the observations to the choice of actions. Here we follow the standard procedure by letting these choices be exogenous, that is, stated as rules of the game. For instance, the players agree on using linear strategies, that is,

$$\psi^i(u_t^{3-i}) = u_t^{i*} + p_t^i \left(u_t^{3-i} - u_t^{3-i*} \right), \ i = 1, 2.$$

The interpretation is as follows: Player i's action is equal to her control under cooperation, plus a term that depends on the other player's choice. If the other player deviates from the agreement by $u_t^{3-i} - u_t^{3-i*}$, then Player i deviates by $p_t^i \left(u^{3-i} - u^{3-i*} \right)$, where p_t^i is a penalty term to be determined endogenously. However, if Player $3 - i$

behaves gently and chooses the agreed-upon control, i.e., u_t^{3-i*}, then Player i would respond by implementing her part of the agreement.

Remark 4.11 Other choices of incentive strategies are possible. For example, one can have strategies that are nonlinear or discontinuous in the information (here, the other player's current decision). Moreover, increasing the information that is available to the players would allow for the construction of more elaborate strategies that depend on the current state as well as the current decisions, and even on the entire histories of state and decisions. The particular choice of incentive strategies will, among other things, be conditioned upon the informational assumptions that are appropriate in a given context.

The computation of an incentive equilibrium requires us to implement the following two-step algorithm:

Step 1:   Solve the joint-optimization problem to obtain $u_t^* = \left(u_t^{1*}, u_t^{2*}\right) \in U^1 \times U^2$ and $x_t^*, t = 0, \dots, T - 1$.

Step 2:   For Player $i = 1, 2$, solve the following dynamic optimization problem:

$$\max J_i\,(\mathbf{u}; x_0) = \sum_{t=0}^{T-1} \rho^t \phi_i\,(x_t, u_t) + \rho^T \Phi_i(x_T),$$

$$\text{subject to } :$$

$$x_{t+1} = f\,(x_t, u_t)\,, \quad x_0 \text{ given},$$

$$u_t^{3-i} = \psi^{3-i}(u_t^i).$$

In the above optimization problem, we have assumed that Player $3 - i$ will implement her incentive strategy. Consequently, in this second step, Player i computes her best reply to other player's incentive strategy. To solve this problem, we need to specify the functional form of ψ^i, $i = 1, 2$.

4.4.1 An Example

To illustrate how to determine the incentive strategies and (the) equilibrium, we use a two-player version of the example in Sect. 4.2.2, $M = \{1, 2\}$, $T = 10$.

In this example, the players minimize their payoff functions. To obtain the cooperative (desirable or agreed-upon) solution, we solve the following optimization problem:

$$\min J(\mathbf{x}, \mathbf{u}) = \sum_{i=1}^{2} \left\{ \sum_{t=0}^{T-1} \rho^t \left(\frac{\gamma_i}{2}(u_t^i - e)^2 + \alpha_i x_t^2 \right) + \rho^T \alpha_i x_T^2 \right\},$$

subject to the pollution dynamics

$$x_t = (1 - \delta)x_{t-1} + \sum_{i \in M} u^i_{t-1}, \ t = 1, \ldots, T, \ x_0 \text{ given.}$$

As before, we adopt the following parameter values:

$$\alpha_1 = 0.06, \ \alpha_2 = 0.07, \ \gamma_1 = 5.35, \ \gamma_2 = 5.85, \ \delta = 0.15, \ e = 10, \ \rho = 0.96, \ x_0 = 0.$$

The optimal control values and state trajectory are illustrated in Table 4.6. The next step is to determine the best-reply strategy of each player, assuming that the other indeed implements her incentive strategy. Given the linear-quadratic structure of the game, we assume that the players use linear incentive strategies, i.e.,

$$u^i_t = \psi^i(u^{3-i}_t) = u^{i*}_t + p^i_t \left(u^{3-i}_t - u^{3-i*}_t\right),$$
$$u^{3-i}_t = \psi^{3-i}(u^i_t) = u^{3-i*}_t + p^{3-i}_t \left(u^i_t - u^{i*}_t\right).$$

Then, Player $i = 1, 2$ solves the following optimization problem:

$$\min J_i(\mathbf{x}, \mathbf{u}) = \sum_{t=0}^{T-1} \rho^t \left(\frac{\gamma_i}{2}\left(u^i_t - e\right)^2 + \alpha_i x^2_t \right) + \rho^T \alpha_i x^2_T$$

subject to $x_t = (1 - \delta)x_{t-1} + u^i_{t-1} + u^{3-i*}_{t-1} + p^{3-i}_{t-1}\left(u^i_{t-1} - u^{i*}_{t-1}\right)$, $t = 1, \ldots, T$, and x_0 is given. We can rewrite the state dynamics as follows:

$$x_t = (1 - \delta)x_{t-1} + (1 + p^{3-i}_{t-1})u^i_{t-1} + u^{3-i*}_{t-1} - p^{3-i}_{t-1}u^{i*}_{t-1}.$$

Table 4.6 Cooperative control and state trajectories

t	u^{1*}_t	u^{2*}_t	x^*_t
0	4.6298	5.0888	0
1	3.9746	4.4896	9.7186
2	3.5721	4.1215	16.7249
3	3.3754	3.9416	21.9098
4	3.3647	3.9318	25.9403
5	3.5463	4.0979	29.3457
6	3.9543	4.4710	32.5881
7	4.6565	5.1132	36.1252
8	5.7657	6.1276	40.4760
9	7.4580	7.6752	46.2979
10			54.4865

Remark 4.12 In this problem, only Player i's control appears in the objective. Obviously, this need not be the case in general.

Player i's Hamiltonian is given by

$$H_{i,t} = \frac{\gamma_i}{2}\left(u_t^i - e\right)^2 + \alpha_i x_t^2 + \rho\lambda_{t+1}^i\left((1-\delta)x_t + (1+p_t^{3-i})u_t^i + u_t^{3-i*} - p_t^{3-i}u_t^{i*}\right),$$

where u_t^{i*}, $i = 1, 2$, is a cooperative control represented in the table above, and λ_t^i, $t = 0, \ldots, T$ is the costate variable appended by Player i to the state dynamics.
  The costate variables satisfy the system

$$\lambda_t^i = 2\alpha_i x_t + \rho(1-\delta)\lambda_{t+1}^i, \quad t = 0, \ldots, T-1,$$
$$\lambda_T^i = 2\alpha_i x_T.$$

Minimizing the Hamiltonian function, we obtain the control

$$u_t^i = e - \frac{\rho}{\gamma_i}\lambda_{t+1}^i(1 + p_t^{3-i}).$$

Equating the control u_t^i to the cooperative one u_t^{i*}, we find p_t^{3-i}:

$$p_t^{3-i} = \frac{\gamma_i(e - u_t^{i*})}{\rho\lambda_{t+1}^i} - 1.$$

Substituting the cooperative controls and solving the dynamic system with respect to λ_t^i, $i = 1, 2$, we obtain

$$p_t^{3-i} = \frac{\alpha_{3-i}}{\alpha_i} \text{ for all } t.$$

For our numerical example, we have

$$p_t^1 = p^1 = 0.857143, \quad p_t^2 = p^2 = 1.16667, \text{ for all } t,$$

and the following incentive strategies:

$$u_t^1 = u_t^{1*} + 0.857143\left(u_t^2 - u_t^{2*}\right),$$
$$u_t^2 = u_t^{2*} + 1.16667\left(u_t^1 - u_t^{1*}\right).$$

Credibilty

In this cost minimization problem, these incentive strategies are credible if the following two conditions hold:

$$J_1(\psi^1(u^2), u^2) \leq J_1(u^{1*}, u^2), \quad \forall u^2 \in U^2,$$
$$J_2(u^1, \psi^2(u^1)) \leq J_2(u^1, u^{2*}), \quad \forall u^1 \in U^1.$$

That is, if a player is cheated, she would be better off implementing her incentive strategy than sticking to the agreement. The above quantities are as follows:

$$J_1(\psi^1(u^2), u^2) = \sum_{t=0}^{T-1} \rho^t \left(\frac{\gamma_1}{2} \left(u_t^{1*} + p^1 \left(u_t^2 - u_t^{2*} \right) - e \right)^2 + \alpha_1 \hat{x}_t^2 \right) + \rho^T \alpha_1 \hat{x}_T^2,$$

$$\hat{x}_t = (1 - \delta)\hat{x}_{t-1} + u_{t-1}^{1*} + p^1 \left(u_{t-1}^2 - u_{t-1}^{2*} \right) + u_{t-1}^2,$$

$$J_1(u^{1*}, u^2) = \sum_{t=0}^{T-1} \rho^t \left(\frac{\gamma_1}{2} \left(u_t^{1*} - e \right)^2 + \alpha_1 \tilde{x}_t^2 \right) + \rho^T \alpha_1 \tilde{x}_T^2,$$

$$\tilde{x}_t = (1 - \delta)\tilde{x}_{t-1} + u_{t-1}^{1*} + u_{t-1}^2.$$

The first condition is satisfied if

$$J_1(\psi^1(u^2), u^2) - J_1(u^{1*}, u^2) = \sum_{t=0}^{T-1} \rho^t \left(\frac{\gamma_1}{2} p^1 \left(u_t^2 - u_t^{2*} \right) \left(p^1 \left(u_t^2 - u_t^{2*} \right) + 2 \left(u_t^{1*} - e \right) \right) \right.$$
$$\left. + \alpha_1 \left(\hat{x}_t^2 - \tilde{x}_t^2 \right) \right) + \rho^T \alpha_1 \left(\hat{x}_T^2 - \tilde{x}_T^2 \right)$$
$$\leq 0,$$

with the difference in state values given by

$$\hat{x}_t - \tilde{x}_t = (1 - \delta) \left(\hat{x}_{t-1} - \tilde{x}_{t-1} \right) + p^1 \left(u_{t-1}^2 - u_{t-1}^{2*} \right).$$

To examine if the difference $J_1(\psi^1(u^2), u^2) - J_1(u^{1*}, u^2)$ is nonpositive for any t and any $u_t^2, t = 0, \ldots, T - 1$ seems a challenging problem in a general form of cost functions, as the difference is a function of T variables $u_0^2, \ldots, u_{T-1}^2$. To develop an intuition about the credibility issue here, we examine this difference for $T = 1$ and $T = 2$.

First, we consider $T = 1$ and assume that $u_t^2 \in [0, e]$. For the numerical example, we have

$$J_1(\psi^1(u^2), u^2) - J_1(u^{1*}, u^2) = 198.422 - 40.9423u_0^2 + 2.10637(u_0^2)^2,$$

which is nonpositive when $u_0^2 \in [9.21664, 10]$ (see Fig. 4.1a). We can make the similar calculations for the second player and obtain

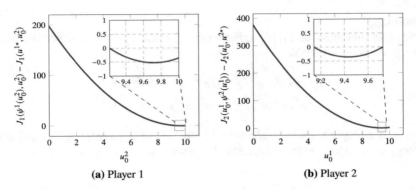

(a) Player 1   **(b)** Player 2

Fig. 4.1 Differences in payoffs with one variable

$$J_2(u^1, \psi^2(u^1)) - J_2(u^1, u^{2*}) = 376.159 - 79.8121u_0^1 + 4.22952(u_0^1)^2,$$

which is nonpositive when $u_0^1 \in [9.14343, 9.72685]$ (see Fig. 4.1b). The incentive
strategy is credible against the other player's deviations within this interval.

Second, we consider $T = 2$, and difference $J_1(\psi^1(u^2), u^2) - J_1(u^{1*}, u^2)$ is a
function of two variables, $u_0^2 \in [0, e]$, $u_1^2 \in [0, e]$. The graph of this function is given
in Fig. 4.2a. The difference $J_2(u^1, \psi^2(u^1)) - J_2(u^1, u^{2*})$ is a function of two variables, $u_0^1 \in [0, e]$, $u_1^1 \in [0, e]$. The graph of this function is given in Fig. 4.2b. In both
graphs, the red region is for negative values of the difference (the incentive strategy
is credible against the other player's deviations within this region), and the blue is
for positive values.

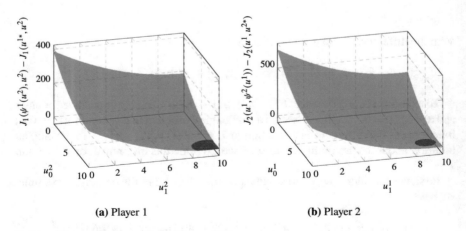

(a) Player 1   **(b)** Player 2

Fig. 4.2 Differences in payoffs with two variables

4.5 Additional Readings

In this section, we provide some references on time consistency and cooperative equilibria in deterministic dynamic games. The list is by no means exhaustive and aims only to offer a sampling of the work that has been done.

4.5.1 Time Consistency

Early contributions. Haurie (1976) showed that a bargaining solution agreed upon at the initial time may fail to remain in place as time goes by, that is, it is time inconsistent. The concept of time consistency and its implementation in cooperative differential games was proposed in Petrosyan (1977) and further discussed in Petrosyan and Danilov (1979), Petrosyan (1991, 1993b). In these publications in Russian, as well as in the subsequent books in English (Petrosyan 1993a; Petrosyan and Zenkevich (2016); Petrosyan (1997)), time consistency was termed dynamic stability. Petrosyan and Danilov (1979) introduced the idea of an imputation distribution procedure to make the cooperative solution time consistent in differential games. For an early discussion of IDP payments, see Petrosyan (1991). For a survey of time consistency in differential games with side payments, see Zaccour (2008) and Petrosyan and Zaccour (2018). Yeung and Petrosyan (2018) surveyed time consistency in differential games without transferable utility.

Applications. A series of contributions have dealt with the sustainability over time of cooperative agreements in environmental games. In a two-player downstream pollution game, Jørgensen and Zaccour (2001) defined a payment schedule from the polluted downstream country to the polluting upstream country, which leads to lower emissions and is Pareto optimal and time consistent. Petrosyan and Zaccour (2003) proposed a time-consistent Shapley value in an n-player global pollution-control problem. Cabo et al. (2006) designed a time-consistent agreement in an interregional differential game of pollution and trade. Smala Fanokoa et al. (2011) determined the conditions under which the cooperation of a country not vulnerable to pollution can be bought, and designed a time-consistent payment schedule to insure the sustainability of this cooperation. Wrzaczek et al. (2014) formulate a continuous-age-structure, overlapping-generations model of pollution control and define a time-consistent tax scheme to implement the optimal cooperative solution. Tur and Gromova (2020) construct a time-consistent IDP in a game of pollution control played by the three largest industrial enterprises of Bratsk, Russia.

Andrés-Domenech et al. (2015) proposed a time-consistent decomposition over time of the Nash bargaining solution to achieve a sustainable forest development in a two-player differential game. In Gromova et al. (2018), a nonrenewable resource extraction game model with random duration is proposed and an IDP for this game is constructed. Mazalov and Rettieva (2010) proposed an IDP to implement a time-consistent solution in a fishery divided between two regions, with a monopoly exploiting the resource in each region. Dahmouni et al. (2019) design a time-consistent sharing of joint-optimization outcomes to sustain cooperation between players in a fishery game where pollution externalities are present.

In marketing, Jørgensen and Gromova (2016) considered a three-player infinite-horizon differential game of advertising competition. If the players decide to jointly optimize their profits, then they can sustain their cooperation by implementing a time-consistent imputation from the core. In network applications, Zakharov and Shchegryaev (2015) study the problem of the time inconsistency of the Shapley value in transportation problems. Petrosyan and Sedakov (2016) examined the time consistency of the Shapley value in a class of dynamic network games with shocks. Petrosyan et al. (2021) construct time-consistent cooperative solutions for two-stage network games with pairwise interactions.

Games with a random terminal time: Time-consistent solutions in cooperative differential games of random duration are analyzed in Petrosyan and Shevkoplyas (2000). Marín-Solano and Shevkoplyas (2011) consider a deterministic differential game with a stochastic terminal time. In particular, they proposed an IDP when the players use nonconstant discounting. A class of cooperative multistage games with a random time horizon is considered in Gromova and Plekhanova (2019). They proposed a time-consistent Shapley value and imputation from the core.

Multicriteria dynamic games: Petrosyan and Tanaka (1996), Petrosyan et al. (1998) are among the first contributions to examine time consistency in cooperative mul-tistage games with vector payoffs. The problem of finding time-consistent Nash bargaining outcomes in dynamic multicriteria game is considered in Rettieva (2018) for a finite-horizon game, and in Rettieva (2020) for a random terminal horizon. In Kuzyutin and Nikitina (2017), Kuzyutin et al. (2018, 2019), a series of properties of time-consistent solutions of cooperative games is discussed, and an algorithm for finding such solutions is designed.

4.5.2 Cooperative Equilibria

Sustaining a Pareto outcome as an equilibrium has a long history in repeated games, and a well-known result in this area is the so-called folk theorem, which (informally) states that if the players are sufficiently patient, then any Pareto-optimal outcome can be achieved as a Nash equilibrium, see, e.g., Osborne and Rubinstein (1994). A similar theorem has been proved for stochastic games by Dutta (1995).

In multistage and differential games, memory strategies that are based on the past history of the control and the initial state have been used to render the Pareto-optimal solution an equilibrium. The first contributions appeared in the 1980s, e.g., Green and Porter (1984), Haurie and Tolwinski (1985), Tolwinski et al. (1986), Haurie and Pohjola (1987), Kaitala and Pohjola (1988). In Green and Porter (1984), the authors present a model of a noncooperatively supported cartel under demand uncertainty (the cartel corresponding to the industry-optimal outcome). In a two-player multistage game with an infinite number of stages, Haurie and Tolwinski (1985) show how a nominal cooperative control sequence can be implemented by using trigger strategies. Tolwinski et al. (1986) deal with a class of equilibria for differential games where a threat is included in a cooperative strategy. The threat is used as soon as a player deviates from the agreed-upon cooperative control. Haurie and Pohjola (1987) show how a Pareto-optimal solution in a differential game of capitalism à la Lancaster can be supported by strategic threats. The result obtained sensibly modifies the previous interpretation of capitalism's inherent inefficiency. In Kaitala and Pohjola (1988), the problem of managing a transboundary renewable resource is studied. It is again shown that the optimal outcome can be achieved by considering memory strategies and an appropriate threat.

In the above-cited references, memory strategies and threats are used to deter some players from cheating on the cooperative agreement. There are instances where there is no need to design such sophisticated strategies because the Pareto solution itself happens to be an equilibrium as in, e.g., Chiarella et al. (1984), Amir and Nannerup (2006), Seierstad (2014). Further, one can obtain conditions under which the Pareto solution is an equilibrium; see Dockner and Kaitala (1989) for an example of natural-resource extraction. Rincón-Zapatero et al. (2000) state some necessary and sufficient conditions for a feedback-Nash equilibrium to be efficient in a class of differential games. Martín-Herrán and Rincón-Zapatero (2005) develop a method for the characterization of Markov-perfect Nash equilibria to be Pareto efficient in nonlinear differential games.

Recently, Fonseca-Morales and Hernández-Lerma (2017) introduced several classes of deterministic and stochastic potential differential games in which open-loop Nash equilibria are also Pareto optimal. See Fonseca-Morales and Hernández-Lerma (2018) for a discussion of potential differential games.

4.5.3 Incentive Equilibria

The concept of incentive strategies has been around for a long time in dynamic games (and economics), but it has often been understood and used in a leader-follower (or principal-agent) sense. The idea is that the leader designs an incentive to induce the follower to act in a certain way, which is often meant to be (only) in the leader's best interest, but may also be in the best collective interest. In such a case, the incentive is one sided.

The incentive strategies presented in this chapter were introduced in Ehtamo and Hämäläinen (1986, 1989, 1993), with the main idea being to support the cooperative solution in two-player dynamic games. Martín-Herrán and Zaccour (2005, 2009) look at incentive strategies and equilibria in linear-state and linear-quadratic differential games, respectively. Jørgensen and Zaccour (2001, 2003) implement such strategies in marketing channels and (Buratto and Zaccour 2009) in licensing. Breton et al. (2008) determined an incentive equilibrium in an overlapping-generations environmental game. In De Frutos and Martín-Herrán (2015, 2020) nonlinear incentive functions are considered that depend on both players' controls and on the current state, with applications in environmental games.

4.6 Exercises

Exercise 4.1 Consider an emissions control game with two neighboring regions. Region 1 suffers from pollution damages due to industrial activities in both regions. Region 2 incurs no damages at all. Denote by y_t^i the rate of production in Region i, $i = 1, 2$, at time $t = 0, \ldots, T$ and let

$$U_i(y^i) = -\ln y^i$$

represent Region i's instantaneous utility of producing at the rate $y^i \geq 0$. Denote by e_t^i the rate of emissions defined by

$$e_t^i = \gamma_i e^{-\beta_i K_t^i} y_t^i,$$

where K_t^i is Region i's stock of abatement capital at time t. Capital K_t^i, $i = 1, 2$ evolves according to the following dynamics:

$$K_{t+1}^i = I_t^i + (1 - \delta) K_t^i, \; K_0^1 = K_0^1 > 0, \; K_0^2 = 0, \tag{4.29}$$

where $\delta \in (0, 1)$ is the constant depreciation rate. The stock of pollution, S_t, is governed by the following difference equation:

$$S_{t+1} = \sum_{i=1}^{2} e_t^i y_t^i + (1 - \nu) S_t, \; S_0 > 0 \text{ given}, \tag{4.30}$$

where $\nu \in (0, 1)$ is a constant natural decay rate of pollution.

Let I_t^i be the physical investment in abatement stock. The cost is given by $C_i(I^i)$, a strictly increasing convex function. Suppose, the investment rates are unconstrained. The objective of Region i is to maximize its social welfare function

$$W_i = \sum_{t=0}^{T} \left[U_i(y_t^i) - C_i(I_t^i) - D_i(S_t) \right], \tag{4.31}$$

where $D_i(S)$ is a function representing the damage incurred to Region i by the stock S. The damage function is increasing and strictly convex for (the vulnerable) Region 1 and is identically equal to zero for (the nonvulnerable) Region 2.

1. Adopt suitable functional forms for $D_i(S)$ and $C_i(I^i)$, and determine a Nash equilibrium in feedback strategies. Interpret the results.

2. Compute the jointly optimal solution. Interpret the results.
3. Compare the Nash equilibrium and the jointly optimal solution (the strategies and outcomes).
4. Use the Nash bargaining solution to share the total optimal payoff, assuming that the status quo is given by the Nash equilibrium.
5. Determine a time-consistent Nash bargaining solution.

Exercise 4.2 Consider a fully symmetric linear-state dynamic game played on $\mathbb{T} = \{0, 1, \ldots, T\}$ and defined by the following elements: (i) a set of players $M = \{1, 2, \ldots, m\}$; (ii) for each player $i \in M$, a control variable $u_t^i \in \mathbb{R}$ at $t = 0, \ldots, T - 1$; (iii) a state variable $x_t \in \mathbb{R}$ at time $t \in \mathbb{T}$. The evolution over time of the state is given by

$$x_{t+1} = \sum_{i \in M} \alpha u_t^i + (1 - \delta) x_t, \quad x_0 \text{ given},$$

where x_0 is the initial state at $t = 0$; (iv) a payoff functional for Player $i \in M$,

$$J_i (\mathbf{u}; x_0) = \sum_{t=0}^{T-1} \rho^t \left(a x_t + b \ln \left(1 + u_t^i \right) \right) + \rho^T s x_T,$$

where $\rho \in (0, 1)$ is the discount factor, a, b, s are positive constants; $u_t = \left(u_t^1, \ldots, u_t^m \right)$ and $\mathbf{u}$ is given by

$$\mathbf{u} = \{ u_t^i \in U^i \mid t \in \mathbb{T} \setminus \{T\}, \ i \in M \},$$

$\phi_i (x_t, u_t)$ is the reward to Player i at $t = 0, \ldots, T - 1$, and $\Phi_i(x_T)$ is the reward to Player i at terminal time T.

1. Determine a Nash equilibrium in feedback strategies.
2. Determine a Nash equilibrium in open-loop strategies.
3. Compare the strategies and outcomes of these two equilibria.
4. Compute the optimal control and state variables that maximize the total joint payoff.
5. Show that any time-consistent solution is also agreeable (see Definition 4.2).
6. What would change if the game was not symmetric?

Exercise 4.3 Consider an m-player dynamic game played on $\mathbb{T} = \{0, 1, \ldots, T\}$. Let $u_t^i \in \mathbb{R}$ be the control variable of Player i, and $x_t \in \mathbb{R}$ be the state variable, whose evolution is given by the following difference equation:

$$x_{t+1} = A x_t + \sum_{i \in M} B_i u_t^i, \quad x_0 \text{ given}, \tag{4.32}$$

where A and B are given parameters. Player $i \in M$ aims at maximizing the payoff functional

$$J_i (\mathbf{u}; x_0) = \sum_{t=0}^{T-1} \rho^t (C_i x_t + D_i u_t^i) + \rho^T F_i x_T, \qquad (4.33)$$

where $\rho \in (0, 1)$ is the discount factor, and C_i, D_i, and F_i are given parameters.

1. Suppose that the players adopt the γ characteristic function. Determine its values for all possible coalitions.
2. Suppose that the players adopt the δ characteristic function. Determine its values for all possible coalitions.
3. Compare the results obtained in 1 and 2.
4. Show that a time-consistent solution is also agreeable.

References

Amir, R. and Nannerup, N. (2006). Information structure and the tragedy of the commons in resource extraction. *Journal of Bioeconomics*, 8(2):147–165.

Andrés-Domenech, P., Martín-Herrán, G., and Zaccour, G. (2015). Cooperation for sustainable forest management: An empirical differential game approach. *Ecological Economics*, 117:118–128.

Breton, M., Sokri, A., and Zaccour, G. (2008). Incentive equilibrium in an overlapping-generations environmental game. *European Journal of Operational Research*, 185(2):687–699.

Buratto, A. and Zaccour, G. (2009). Coordination of advertising strategies in a fashion licensing contract. *Journal of Optimization Theory and Applications*, 142(1):31–53.

Cabo, F., Escudero, E., and Martín-Herrán, G. (2006). A time-consistent agreement in an interregional differential game on pollution and trade. *International Game Theory Review*, 08(03):369–393.

Chiarella, C., Kemp, M. C., Long, N. V., and Okuguchi, K. (1984). On the economics of international fisheries. *International Economic Review*, 25(1):85–92.

Dahmouni, I., Vardar, B., and Zaccour, G. (2019). A fair and time-consistent sharing of the joint exploitation payoff of a fishery. *Natural Resource Modeling*, 32(3):art. no. e12216.

De Frutos, J. and Martín-Herrán, G. (2015). Does flexibility facilitate sustainability of cooperation over time? A case study from environmental economics. *Journal of Optimization Theory and Applications*, 165(2):657–677.

De Frutos, J. and Martín-Herrán, G. (2020). Non-linear incentive equilibrium strategies for a transboundary pollution differential game. In Pineau, P.-O., Sigué, S., and Taboubi, S., editors, *Games in Management Science: Essays in Honor of Georges Zaccour*, pages 187–204, Cham. Springer International Publishing.

Dockner, E. J. and Kaitala, V. (1989). On efficient equilibrium solutions in dynamic games of resource management. *Resources and Energy*, 11(1):23–34.

Dutta, P. (1995). A folk theorem for stochastic games. *Journal of Economic Theory*, 66(1):1–32.

Ehtamo, H. and Hämäläinen, R. P. (1986). On affine incentives for dynamic decision problems. In Başar, T., editor, *Dynamic Games and Applications in Economics*, pages 47–63, Berlin, Heidelberg. Springer Berlin Heidelberg.

Ehtamo, H. and Hämäläinen, R. P. (1989). Incentive strategies and equilibria for dynamic games with delayed information. *Journal of Optimization Theory and Applications*, 63(3):355–369.

Ehtamo, H. and Hämäläinen, R. P. (1993). A cooperative incentive equilibrium for a resource management problem. *Journal of Economic Dynamics and Control*, 17:659–678.

Fonseca-Morales, A. and Hernández-Lerma, O. (2017). A note on differential games with Pareto-optimal Nash equilibria: Deterministic and stochastic models. *Journal of Dynamics & Games*, 4(3):195–203.

Fonseca-Morales, A. and Hernández-Lerma, O. (2018). Potential differential games. *Dynamic Games and Applications*, 8(2):254–279.

Germain, M., Toint, P., Tulkens, H., and de Zeeuw, A. (2003). Transfers to sustain dynamic core-theoretic cooperation in international stock pollutant control. *Journal of Economic Dynamics and Control*, 28(1):79–99.

Green, E. J. and Porter, R. H. (1984). Noncooperative collusion under imperfect price information. *Econometrica*, 52(1):87–100.

Gromova, E., Malakhova, A., and Palestini, A. (2018). Payoff distribution in a multi-company extraction game with uncertain duration. *Mathematics*, 6(9).

Gromova, E. V. and Plekhanova, T. M. (2019). On the regularization of a cooperative solution in a multistage game with random time horizon. *Discrete Applied Mathematics*, 255:40–55.

Haurie, A. (1976). A note on nonzero-sum differential games with bargaining solution. *Journal of Optimization Theory and Applications*, 18(1):31–39.

Haurie, A. and Pohjola, M. (1987). Efficient equilibria in a differential game of capitalism. *Journal of Economic Dynamics and Control*, 11(1):65–78.

Haurie, A. and Tolwinski, B. (1985). Definition and properties of cooperative equilibria in a two-player game of infinite duration. *Journal of Optimization Theory and Applications*, 46(4):525–534.

Jørgensen, S. and Gromova, E. (2016). Sustaining cooperation in a differential game of advertising goodwill accumulation. *European Journal of Operational Research*, 254(1):294–303.

Jørgensen, S., Martín-Herrán, G., and Zaccour, G. (2003). Agreeability and time consistency in linear-state differential games. *Journal of Optimization Theory and Applications*, 119(1):49–63.

Jørgensen, S., Martín-Herrán, G., and Zaccour, G. (2005). Sustainability of cooperation overtime in linear-quadratic differential games. *International Game Theory Review*, 7(04):395–406.

Jørgensen, S. and Zaccour, G. (2001). Time consistent side payments in a dynamic game of down-stream pollution. *Journal of Economic Dynamics and Control*, 25(12):1973–1987.

Jørgensen, S. and Zaccour, G. (2003). Channel coordination over time: incentive equilibria and credibility. *Journal of Economic Dynamics and Control*, 27(5):801–822.

Kaitala, V. and Pohjola, M. (1988). Optimal recovery of a shared resource stock: A differential game model with efficient memory equilibria. *Natural Resource Modeling*, 3(1):91–119.

Kaitala, V. and Pohjola, M. (1990). Economic development and agreeable redistribution in capital-ism: Efficient game equilibria in a two-class neoclassical growth model. *International Economic Review*, 31(2):421–438.

Kuzyutin, D., Gromova, E., and Pankratova, Y. (2018). Sustainable cooperation in multicriteria multistage games. *Operations Research Letters*, 46(6):557–562.

Kuzyutin, D. and Nikitina, M. (2017). Time consistent cooperative solutions for multistage games with vector payoffs. *Operations Research Letters*, 45(3):269–274.

Kuzyutin, D., Smirnova, N., and Gromova, E. (2019). Long-term implementation of the cooperative solution in a multistage multicriteria game. *Operations Research Perspectives*, 6:100107.

Mailath, G. and Samuelson, L. (2006). *Repeated Games and Reputations: Long-Run Relationships*. Oxford scholarship online. Oxford University Press, USA.

Marín-Solano, J. and Shevkoplyas, E. V. (2011). Non-constant discounting and differential games with random time horizon. *Automatica*, 47(12):2626–2638.

Martín-Herrán, G. and Rincón-Zapatero, J. P. (2005). Efficient Markov perfect Nash equilibria: theory and application to dynamic fishery games. *Journal of Economic Dynamics and Control*, 29(6):1073–1096.

Martín-Herrán, G. and Zaccour, G. (2005). Credibility of incentive equilibrium strategies in linear-state differential games. *Journal of Optimization Theory and Applications*, 126(2):367–389.

Martín-Herrán, G. and Zaccour, G. (2009). Credible linear-incentive equilibrium strategies in linear-quadratic differential games. In Pourtallier, O., Gaitsgory, V., and Bernhard, P., editors, *Advances in Dynamic Games and Their Applications: Analytical and Numerical Developments*, pages 1–31, Boston. Birkhäuser Boston.

Mazalov, V. V. and Rettieva, A. N. (2010). Fish wars and cooperation maintenance. *Ecological Modelling*, 221(12):1545–1553.

Osborne, M. and Rubinstein, A. (1994). *A Course in Game Theory*. The MIT Press. MIT Press.

Petrosyan, L. (1977). Stable solution of differential game with many participants. *Viestnik of Leniversity University*, 19:46–52.

Petrosyan, L. (1991). The time consistency of the optimality principles in non-zero sum differential games. In Hämäläinen, R. P. and Ehtamo, H. K., editors, *Dynamic Games in Economic Analysis*, pages 299–311, Berlin, Heidelberg. Springer Berlin Heidelberg.

Petrosyan, L. (1993a). *Differential Games of Pursuit*. Series on Optimization. World Scientific Publishing.

Petrosyan, L. (1993b). Strongly time consistent differential optimality principles. *Vestnik Sankt-Peterburgskogo Universiteta. Ser 1. Matematika Mekhanika Astronomiya*, (4):35–40.

Petrosyan, L. (1997). Agreeable solutions in differential games. *Game Theory and Applications*, 3:165–177.

Petrosyan, L., Ayoshin, D., and Tanaka, T. (1998). Construction of a time consistent core in multi-choice multistage games. *Decision Theory and Its Related Topics*, 1043:198–206.

Petrosyan, L., Bulgakova, M., and Sedakov, A. (2021). Time-consistent solutions for two-stage network games with pairwise interactions. *Mobile Networks and Applications*, 26(2):491–500.

Petrosyan, L. and Danilov, N. N. (1979). Stability of solutions in non-zero sum differential games with transferable payoffs. *Viestnik of Leningrad Universtiy*, 1:52–59.

Petrosyan, L. and Sedakov, A. (2016). The subgame-consistent Shapley value for dynamic network games with shock. *Dynamic Games and Applications*, 6(4):520–537.

Petrosyan, L. and Shevkoplyas, E. (2000). Cooperative differential games with random duration. *Vestnik Sankt-Peterburgskogo Universiteta. Ser 1. Matematika Mekhanika Astronomiya*, (4):18–23.

Petrosyan, L. and Tanaka, T. (1996). Multistage games with vector playoffs. *Nova Journal of Mathematics, Game Theory and Algebra*, 6:97–102.

Petrosyan, L. and Zaccour, G. (2003). Time-consistent Shapley value allocation of pollution cost reduction. *Journal of Economic Dynamics and Control*, 27(3):381–398.

Petrosyan, L. and Zaccour, G. (2018). Cooperative differential games with transferable payoffs. In Basar, T. and Zaccour, G., editors, *Handbook of Dynamic Game Theory*, pages 1–38, Cham. Springer International Publishing.

Petrosyan, L. and Zenkevich, N. A. (2016). *Game theory: Second edition*.

Rettieva, A. (2018). Dynamic multicriteria games with finite horizon. *Mathematics*, 6(9).

Rettieva, A. N. (2020). Cooperation in dynamic multicriteria games with random horizons. *Journal of Global Optimization*, 76(3):455–470.

Rincón-Zapatero, J. P., Martín-Herrán, G., and Martínez, J. (2000). Identification of efficient subgame-perfect Nash equilibria in a class of differential games. *Journal of Optimization Theory and Applications*, 104(1):235–242.

Seierstad, A. (2014). Pareto improvements of Nash equilibria in differential games. *Dynamic Games and Applications*, 4(3):363–375.

Smala Fanokoa, P., Telahigue, I., and Zaccour, G. (2011). Buying cooperation in an asymmetric environmental differential game. *Journal of Economic Dynamics and Control*, 35(6):935–946.

Tolwinski, B., Haurie, A., and Leitmann, G. (1986). Cooperative equilibria in differential games. *Journal of Mathematical Analysis and Applications*, 119(1):182–202.

Tur, A. V. and Gromova, E. V. (2020). On optimal control of pollution emissions: An example of the largest industrial enterprises of irkutsk oblast. *Automation and Remote Control*, 81(3):548–565.

Wrzaczek, S., Shevkoplyas, E., and Kostyunin, S. (2014). A differential game of pollution control with overlapping generations. *International Game Theory Review*, 16(3).

Yeung, D. W. K. and Petrosyan, L. (2018). Nontransferable utility cooperative dynamic games. In Başar, T. and Zaccour, G., editors, *Handbook of Dynamic Game Theory*, pages 633–670, Cham. Springer International Publishing.

Zaccour, G. (2008). Time consistency in cooperative differential games: A tutorial. *INFOR: Information Systems and Operational Research*, 46(1):81–92.

Zakharov, V. V. and Shchegryaev, A. N. (2015). Stable cooperation in dynamic vehicle routing problems. *Automation and Remote Control*, 76(5):935–943.

Part III
Dynamic Games Played over Event Trees

Chapter 5
Noncooperative Dynamic Games Played over Event Trees

In this chapter, we introduce a class of dynamic games played over event trees (DGPETs). We define the elements of the game, in particular the S-adapted information structure, and the corresponding concepts. We state the existence and uniqueness of the equilibrium results and a maximum principle for this class of games. Also, we extend the formalism to DGPETs that can terminate at any intermediate node. An example and some additional readings are provided.

5.1 Motivation

Dynamic games played over event trees constitute a natural methodological choice whenever (i) the strategic interaction between players recurs over time, (ii) the decisions made at any period affect both current and future payoffs, and the evolution of the state variables, and (iii) the future parameter values are not known with certainty. There are many approaches to represent this randomness, with each having its pros and cons. An event tree, as a representation of the stochastic process, has at least two important advantages. First, it mimics how decision-makers deal with uncertain parameter values. Second, it leads to computable and implementable models, and therefore useful ones. We illustrate the DGPET setup with two examples, behind each of which there is an ever-developing literature.

Competition among the few: Any oligopoly with firms making intertemporal decisions, e.g., investment in production capacity, while facing some uncertainty about the parameter values, fits the DGPET model, for as long as this uncertainty can be described by an event tree. In most practical problems, decision-makers only consider future values for each parameter perceived as significantly different. For instance, to assess the next-period production cost, firms will consider a few potential inflation rates, e.g., low rate (say 2%), medium (3%), and high (say

5%). If the production decision is highly sensitive to this rate, then one could add values in between or expand the range.

Closed-loop supply chain: Collecting previously sold products when they reach their end of life is environmentally sound and can be profitable. Indeed, these *returns* can be recycled or refurbished, or some of their parts can be used in remanufacturing and put on the market again instead of ending in a landfill. Consumers typically receive a reward when returning their used products, be it monetary or symbolic (saving the environment). For manufacturers and retailers, returns are beneficial from an operations perspective (lower cost when producing with used parts) and from a marketing point of view as returns can lead to rebuys. The collection of these products can be done by either the manufacturer or the retailer. When demand is stochastic, which is the norm rather than the exception, the returns are also stochastic. To determine equilibrium strategies when the players act selfishly, or a cooperative solution when they agree to coordinate their strategies, the closed-loop supply chain can be represented by a DGPET, where the uncertainty in demand is described by an event tree, and the returns are modeled as a state variable whose evolution depends on past demand and some green activities undertaken by the supply chain (e.g., setting up the infrastructure for returns and running advertising campaigns to raise awareness about the recycling policy.

These two examples differ in terms of the type of strategic interactions between the players. In an oligopoly, the interaction is horizontal, that is, the firms compete in the same market. In the closed-loop supply chain example, the interaction is vertical, that is, the mode of play is typically sequential, with the manufacturer deciding on the wholesale price and the retailer following by choosing the price to consumers. Further, in imperfectly competitive markets, cooperation (collusion) is detrimental to consumers and social welfare. In supply chains, coordinating strategies leads to higher total profits, a higher consumer surplus, and therefore higher social welfare.

5.2   Event Tree as a Model of Uncertainty

Let $\mathbb{T} = \{0, 1, \ldots, T\}$ be the set of periods and denote by $\{\xi(t) : t \in \mathbb{T} \backslash \{T\}\}$ the exogenous stochastic process represented by an event tree.[1] The tree has a root node n_0 in period 0 and a finite set of nodes $\mathbf{n}_t$ in period $t \in \mathbb{T}$. Each node $n_t \in \mathbf{n}_t$ represents a possible sample value of the history of the $\xi(\cdot)$ process up to time t. The evolution of the stochastic process ξ is depicted in Fig. 5.1.

Remark 5.1 As there is typically more than one node at any given period, we should refer to a specific node at period t by $n_t^l \in \mathbf{n}_t$, where $l = \{1, \ldots, L\}$ and where L is

---

[1] Modeling uncertainty with scenario trees is a common practice in the stochastic programming literature.

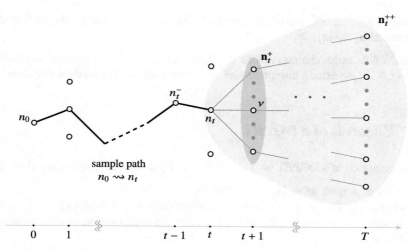

Fig. 5.1 Evolution of uncertainty described by an event tree

the number of elements in $\mathbf{n}_t$. To avoid further complicating the notation, we will omit the superscript l when no ambiguity may arise.

Denote by $n_t^- \in \mathbf{n}_{t-1}$ the unique predecessor of node $n_t \in \mathbf{n}_t$, and by ν a successor of node n_t. We denote by $\mathbf{n}_t^+ \subset \mathbf{n}_{t+1}$ the set of all possible direct successors of node n_t, and by $\mathbf{n}_t^{++}$ the set of all nodes of the event tree having n_t as their root node. Having reached node n_t at time t, $\mathbf{n}_t^{++}$ represents the residual uncertainty in the form of a subtree emanating from node n_t. Consequently, $\mathbf{n}_0^{++}$ describes the entire event tree. We illustrate a branch of the event tree in Fig. 5.1. A path from the root node n_0 to a terminal node n_T is called a *scenario*, with the probabilities of all scenarios summing up to 1. We denote by $n_0 \rightsquigarrow n_t$ the set of all nodes encountered along the unique sample path starting from n_0 and ending in n_t.

Let π_{n_t} be the probability of passing through node n_t, which corresponds to the sum of the probabilities of all scenarios containing this node. In particular, $\pi_{n_0} = 1$, and π_{n_T} is equal to the probability of the single scenario that terminates in node $n_T \in \mathbf{n}_T$. We notice that, for any node $n_t \in \mathbf{n}_t$, $t \in \mathbb{T} \backslash \{T\}$, the following equality holds:

$$\pi_{n_t} = \sum_{\nu \in \mathbf{n}_t^+} \pi_\nu.$$

We denote by $\pi_{n_t}^\nu := \frac{\pi_\nu}{\pi_{n_t}}$ the transition probability from node n_t to a particular node $\nu \in \mathbf{n}_t^+$, and by $\pi_{n_t}^{\mathbf{n}_t^+}$ the row vector of transition probabilities, that is,

$$\pi_{n_t}^{\mathbf{n}_t^+} = \left[\pi_{n_t}^{\nu^1} \ \pi_{n_t}^{\nu^2} \ \cdots \ \pi_{n_t}^{\nu^{|\mathbf{n}_t^+|}} \right],$$

where $\nu^1, \nu^2, \ldots, \nu^{|\mathbf{n}_t^+|}$ are the successors of node n_t. We enumerate the set of nodes at time t as $\mathbf{n}_t := \{n_t^1, n_t^2, \ldots, n_t^{|\mathbf{n}_t|}\}$.

Remark 5.2 In this chapter, we first consider a setup where the terminal date is given. In Sect. 5.10, we extend the framework to an event tree with a random duration.

5.3 Elements of a DGPET

The description of a DGPET on $\mathbb{T} = \{0, 1, \ldots, T\}$ involves the following elements:

1. A set of players $M = \{1, 2, \ldots, m\}$;
2. For each player $i \in M$, a vector of control (or decision) variables $u_{n_t}^i \in U_{n_t}^i \subseteq \mathbb{R}^{\ell_i}$ at node $n_t \in \mathbf{n}_t$, $t \in \mathbb{T}\backslash\{T\}$, where $U_{n_t}^i$ is the set of admissible control values for Player i;

   Denote by $\mathbf{U}^i$ the product of the sets of decision variables of Player i, i.e., $\prod_{\nu \in \mathbf{n}_0^{++}\backslash\mathbf{n}_T} U_\nu^i$ and by $U_{n_t} = \prod_{i \in M} U_{n_t}^i$ the joint decision set at node n_t;

3. A vector of state variables $x_{n_t} \in X \subset \mathbb{R}^q$ at node $n_t \in \mathbf{n}_t$, $t \in \mathbb{T}$, where X is the set of admissible states; where the evolution over time of the state is given by the transition function $f^{n_t}(\cdot, \cdot) : X \times U_{n_t} \mapsto X$ associated with each node n_t; and where the state equations are given by

$$x_{n_t} = f^{n_t^-}\left(x_{n_t^-}, u_{n_t^-}\right), \quad x_{n_0} = x_0, \tag{5.1}$$

   where $u_{n_t^-} \in U_{n_t^-}$, $n_t \in \mathbf{n}_t$, $t \in \mathbb{T} \backslash \{0\}$ and x_0 is the initial state at root node n_0;
4. A payoff functional for Player $i \in M$,

$$J_i(\mathbf{x}, \mathbf{u}) = \sum_{t=0}^{T-1} \rho^t \sum_{n_t \in \mathbf{n}_t} \pi_{n_t} \phi_i(n_t, x_{n_t}, u_{n_t}) + \rho^T \sum_{n_T \in \mathbf{n}_T} \pi_{n_T} \Phi_i(n_T, x_{n_T}), \tag{5.2}$$

   where $\rho \in (0, 1)$ is the discount factor; $\mathbf{x}$ and $\mathbf{u}$ are given by

$$\mathbf{x} = \{x_{n_\tau} : n_\tau \in \mathbf{n}_0^{++}\}, \tag{5.3}$$

$$\mathbf{u} = \{u_{n_\tau}^i \in U_{n_\tau}^i : n_\tau \in \mathbf{n}_0^{++} \backslash \mathbf{n}_T, i \in M\}; \tag{5.4}$$

   $\phi_i(n_t, x_{n_t}, u_{n_t})$ is the reward to Player i at node $n_t \in \mathbf{n}_t$, $t \in \mathbb{T} \backslash \{T\}$; and $\Phi_i(n_T, x_{n_T})$ is the reward to Player i at terminal node $n_T \in \mathbf{n}_T$;
5. An information structure that defines the information available to Player $i \in M$ when she selects her control vector $u_{n_t}^i$ at node $n_t \in \mathbf{n}_t$, $t \in \mathbb{T}\backslash\{T\}$;
6. A strategy σ_i for Player $i \in M$, which is an m_i-dimensional vector-decision rule that defines the control $u_{n_t}^i \in U_{n_t}^i$ as a function of the information available at node $n_t \in \mathbf{n}_t$, $t \in \mathbb{T}\backslash\{T\}$.

Remark 5.3 If at each t the number of nodes is exactly one, then we have a standard multistage game. For an introduction to this class of games, see Chap. 3.

We briefly comment on the elements involved in the description of a DGPET.

Control variables and control set: The control chosen by each player at any node affects her gain directly through the reward function $\phi_i(n_t, x_{n_t}, u_{n_t})$ and indirectly through the state equation (5.1). The choice of an action by Player i at node $n_t \in \mathbf{n}_t$, $t \in \mathbb{T}\setminus\{T\}$ must satisfy the feasibility constraint $u_{n_t}^i \in U_{n_t}^i$. Player i's control set $U_{n_t}^i$ is assumed to be independent of the state of the system, but could in general depend on the *position of the game*, i.e., (n_t, x_{n_t}). In such a case, the choice of a control must respect the constraint $u_{n_t}^i \in U_{n_t}^i(x_{n_t})$, where the point-to-set correspondence $(n_t, x_{n_t}) \rightarrow U_{n_t}^i(x_{n_t})$ is assumed to be upper-semicontinuous. More generally, we could also consider coupled constraints on players' controls. In this chapter, we start with the standard case where the control set of each player is independent of the control variables of the other players. In Sect. 5.8, we consider coupled control sets.

State set and dynamics: For simplicity, the set of admissible states X is assumed invariant over nodes and time periods. According to the state equation (5.1), at each node n_t, a profile of decisions u_{n_t} will determine, in association with the current state x_{n_t}, the same state x_ν for all descendent nodes $\nu \in \mathbf{n}_t^+$. In this state-equation formalism, once the decisions $u_{n_t}^i$, $i \in M$, are made by the players, the vector of state variables x_{n_t} is determined. The state variables are shared by all the players and enter their reward functions as shown in (5.2).

Information structure: By information structure, we mean the piece of information that players use when making their decisions. In deterministic multistage games, i.e., when the set of nodes $\mathbf{n}_t$ for all $t \in \mathbb{T}$ is a singleton, we distinguish between an open-loop information structure (OLIS) and a feedback information structure (FIS).[2] In an OLIS, the players use the current time and initial state to make their decisions, whereas in FIS, the decisions depend on the time and the current value of the state.

In a DGPET, the decision variables are indexed over the set of nonterminal nodes of the event tree. As there is a single path between the initial node and any other node of the event tree, we say that the decisions are *adapted* to the history of the stochastic process $\xi(\cdot)$. Further, at any period of time, we only consider a *sample* of possible events, e.g., the next-period demand for a product can be low, medium, or high. For these reasons, the generic information structure of a DGPET is referred to as an S-adapted information structure; see Haurie et al. (1990a) and Zaccour (1987).

Strategies: In an S-adapted open-loop information structure, each player designs a rule to select values of the decision variables that depend on the node of the event tree n_t and initial state x_0, that is, $u_{n_t}^i = u_{n_t}^i(x_0) \in U_{n_t}^i$. As the initial state is a given constant, we do not distinguish between the control $u_{n_t}^i$ and the decision

---

[2] Other information structures can be considered; see, e.g., Haurie et al. (2012).

rule $u_{n_t}^i(x_0)$. The strategy, or the collection of the decisions of Player $i \in M$ for all nodes is denoted by $\mathbf{u}^i = \{u_{n_\tau}^i : n_\tau \in \mathbf{n}_0^{++} \setminus \mathbf{n}_T\}$. The strategy profile of all players is denoted by $\mathbf{u} = \{\mathbf{u}^i, i \in M\}$. We notice that the players choose their strategies before the realization of uncertainty.

In an S-adapted feedback (or Markovian) information structure, each player designs a rule for choosing actions at each event-tree node n_t that depends on the observed state variable x_{n_t}. The feedback rule is a mapping from the state space into the action set, that is, $u_{n_t}^i = \gamma_{n_t}^i(x_{n_t}) \in U_{n_t}^i$. In the remainder of this book, we stick to an S-adapted open-loop information structure. (We shall from now on drop the term open-loop when no ambiguity can arise.) For a treatment of the linear-quadratic S-adapted feedback structure, see Reddy and Zaccour (2019).

Definition 5.1 (*S-adapted OL strategy*) An admissible S-adapted OL strategy of Player i is given by $\mathbf{u}^i = \{u_{n_\tau}^i : n_\tau \in \mathbf{n}_0^{++} \setminus \mathbf{n}_T\}$, that is, a plan of decisions adapted to the history of the random process $\xi(\cdot)$ represented by the event tree.

5.4 S-adapted Nash Equilibrium

5.4.1 A Normal-Form Representation

A DGPET can be formulated as a game in normal (or strategic) form, which allows for the use of established results for the existence and uniqueness of the Nash equilibrium.

Let $\mathbf{x}$ be obtained from $\mathbf{u}$ as the unique solution of the state equations (5.1) emanating from the initial state x_0. Then, the payoff function $J_i(\mathbf{x}, \mathbf{u})$, $i \in M$, can be rewritten as

$$
W_i(x_0, \mathbf{u}) = \sum_{t=0}^{T-1} \rho^t \sum_{n_t \in \mathbf{n}_t} \pi_{n_t} \phi_i \left(n_t, f^{n_t^-} \left(x_{n_t^-}, u_{n_t^-} \right), u_{n_t} \right)
$$
$$
+ \rho^T \sum_{n_T \in \mathbf{n}_T} \pi_{n_T} \Phi_i \left(n_T, f^{n_T^-} \left(x_{n_T^-}, u_{n_T^-} \right) \right),
$$

where $\mathbf{u} = \{\mathbf{u}^i : i \in M\}$ is the profile of the m players' S-adapted OL strategies.

Therefore, the game in normal form is defined by $(M, \{\mathbf{U}^i\}_{i \in M}, \{W_i\}_{i \in M})$.

Definition 5.2 An S-adapted Nash equilibrium is an admissible S-adapted strategy profile $\mathbf{u}^* = \{\mathbf{u}^{i*} \in \mathbf{U}^i : i \in M\}$ such that, for every player $i \in M$, the following condition holds:
$$
W_i(x_0, \mathbf{u}^*) \geq W_i(x_0, (\mathbf{u}^i, \mathbf{u}^{-i*})),
$$

where $(\mathbf{u}^i, \mathbf{u}^{-i*})$ is the S-adapted strategy profile when all players $j \neq i$, $j \in M$, use their Nash equilibrium strategies.

The normal-form game with payoff functions $W_i(x_0, \mathbf{u})$ is a concave game if, for any strategy profile $\mathbf{u}$, the payoff function $W_i(x_0, \mathbf{u})$ is continuous in $\mathbf{u}$ and is concave in $\mathbf{u}^i$ for each fixed value of $\mathbf{u}^{-i}$ (see Rosen 1965). We reformulate the existence and uniqueness theorems in Rosen (1965) for the class of DGPETs.

Theorem 5.1 *Let functions ϕ_i, $i \in M$, be concave in $(u_{n_t^-}, u_{n_t})$ for any node $n_t \in \mathbf{n}_t$, $t \in \mathbb{T} \setminus \{T\}$ and Φ_i be concave in $u_{n_T^-}$ for any terminal node $n_T \in \mathbf{n}_T$. Assume that the set of admissible S-adapted strategies is compact. Then, there exists an S-adapted Nash equilibrium.*

Proof Following Theorem 1 in Rosen (1965), which is based on Kakutani's fixed-point theorem (Kakutani 1941), we can easily prove the existence of an equilibrium for a dynamic game played over an event tree.   □

Also, we can adapt the uniqueness of the equilibrium result in Rosen (1965) to provide conditions under which the S-adapted Nash equilibrium is unique. It is convenient to introduce the notations $\tilde{u}_{n_t}^i = (u_{n_t^-}^i, u_{n_t}^i)$ and $\tilde{u}_{n_t} = (u_{n_t^-}, u_{n_t})$. Define the pseudo-gradients

$$G(n_t, \tilde{u}_{n_t}) = \left(\frac{\partial \phi_1(n_t, \tilde{u}_{n_t})}{\partial \tilde{u}_{n_t}^1}, \cdots, \frac{\partial \phi_i(n_t, \tilde{u}_{n_t})}{\partial \tilde{u}_{n_t}^i}, \cdots, \frac{\partial \phi_m(n_t, \tilde{u}_{n_t})}{\partial \tilde{u}_{n_t}^m} \right),$$

$$n_t \in \mathbf{n}_t, \ t \in \mathbb{T} \setminus \{T\},$$

$$G(n_T, u_{n_T^-}) = \left(\frac{\partial \Phi_1(n_T, u_{n_T^-})}{\partial u_{n_T^-}^1}, \cdots, \frac{\partial \Phi_i(n_T, u_{n_T^-})}{\partial u_{n_T^-}^i}, \cdots, \frac{\partial \Phi_m(n_T, u_{n_T^-})}{\partial u_{n_T^-}^m} \right),$$

$$n_T \in \mathbf{n}_T,$$

and the Jacobian matrices

$$\mathcal{J}(n_t, \tilde{u}_{n_t}) = \frac{\partial G(n_t, \tilde{u}_{n_t})}{\partial \tilde{u}_{n_t}}, \quad n_t \in \mathbf{n}_t, \ t \in \mathbb{T} \setminus \{T\},$$

$$\mathcal{J}(n_T, u_{n_T^-}) = \frac{\partial G(n_T, u_{n_T^-})}{\partial u_{n_T^-}}, \quad n_T \in \mathbf{n}_T.$$

Theorem 5.2 *If matrices $Q(n_t, \tilde{u}_{n_t}) = \frac{1}{2}[\mathcal{J}(n_t, \tilde{u}_{n_t}) + (\mathcal{J}(n_t, \tilde{u}_{n_t}))']$ and $Q(n_T, u_{n_T^-}) = \frac{1}{2}[\mathcal{J}(n_T, u_{n_T^-}) + (\mathcal{J}(n_T, u_{n_T^-}))']$ are negative definite for all nodes $n_t \in \mathbf{n}_t$, $t \in \mathbb{T} \setminus \{T\}$ and $n_T \in \mathbf{n}_T$, respectively, then the S-adapted Nash equilibrium is unique.*

Proof The result follows from Theorem 1.7 (we also refer the reader to Theorems 4–6 in Rosen 1965). The negative definiteness of the matrices $Q(n_t, \tilde{u}_{n_t})$ and $Q(n_T, u_{n_T^-})$ implies the strict diagonal concavity of the function $\sum_{i \in M} W_i(x_0, \mathbf{u})$, which then implies the uniqueness of the equilibrium.   □

Remark 5.4 In Rosen (1965), the coupled-constrained game is also considered, and the existence of the so-called normalized equilibria is proved. Coupled-constrained DGPETs are considered in Sect. 5.8 of this chapter.[3]

Remark 5.5 Nash equilibria for concave games can be obtained by solving a variational inequality. This well-known result extends readily to S-adapted equilibria. See Gürkan et al. (1999), Haurie and Moresino (2002) for details and for the design of efficient numerical methods.

5.5   The Maximum Principle

We derive a maximum principle (see Pontryagin et al. 1962) for the characterization of S-adapted Nash equilibria. We also establish a link between the theories of DGPET and of multistage games with an open-loop information structure.

We use a multistage game formalism and consider the game with a player set M, players' strategies $\mathbf{u}^i \in \mathbf{U}^i$, and payoff functions $J_i(\mathbf{x}, \mathbf{u})$ defined by (5.2), where $\mathbf{x}$ is given by (5.3) satisfying the state equation (5.1) with initial condition $x(n_0) = x_0$.

The Lagrangian of player $i \in M$ is given by

$$
\mathcal{L}_i(\lambda^i, \mathbf{x}, \mathbf{u}) = \phi_i(n_0, x_{n_0}, u_{n_0}) + \sum_{t=1}^{T-1} \rho^t \sum_{n_t \in \mathbf{n}_t} \pi_{n_t} \Big\{ \phi_i(n_t, x_{n_t}, u_{n_t}) \tag{5.5}
$$

$$
+ (\lambda_{n_t}^i)' \left(f^{n_t^-}\left(x_{n_t^-}, u_{n_t^-} \right) - x_{n_t} \right) \Big\}
$$

$$
+ \rho^T \sum_{n_T \in \mathbf{n}_T} \pi_{n_T} \Big\{ \Phi_i(n_T, x_{n_T})
$$

$$
+ (\lambda_{n_T}^i)' \left(f^{n_T^-}\left(x_{n_T^-}, u_{n_T^-} \right) - x_{n_T} \right) \Big\},
$$

where $\lambda^i = \left(\lambda_\nu^i : \nu \in \mathbf{n}_0^{++} \setminus \{n_0\} \right)$, and λ_ν^i is a costate (or adjoint) variable of Player i, defined over the set of nodes but excluding the root node. Then, for each player i and each node $n_t \in \mathbf{n}_t$, $t \in \mathbb{T} \setminus \{T\}$, we define the Hamiltonian function

$$
H_i(n_t, \lambda_{\mathbf{n}_t^+}^i, x_{n_t}, u_{n_t}) = \phi_i(n_t, x_{n_t}, u_{n_t}) + \rho \sum_{\nu \in \mathbf{n}_t^+} \pi_{n_t}^\nu (\lambda_\nu^i)' f^{n_t}\left(x_{n_t}, u_{n_t} \right), \tag{5.6}
$$

where $\lambda_{\mathbf{n}_t^+}^i$ stands for the collection of λ_ν^i with $\nu \in \mathbf{n}_t^+$.

Interpretation of the Hamiltonian: The right-hand side of (5.6) is the sum of two terms. The first term is the payoff at the current node n_t. Recalling that the

---

[3] We refer interested readers to Carlson and Haurie (2000), Haurie (1995), Haurie and Zaccour (1995) for a more detailed treatment of (deterministic) dynamic game models with coupled constraints.

adjoint variable λ_ν^i gives the marginal value of the corresponding state variable, the second term is the discounted sum of the expected state (or asset in the language of economics) values at the successor nodes.

Remark 5.6 The main difference between the usual Hamiltonian of a multistage game with an open-loop information structure and the Hamiltonian (5.6) is that the latter allows for an average "sensitivity vector," $\sum_{\nu \in \mathbf{n}_t^+} \pi_{n_t}^\nu (\lambda_\nu^i)'$. The interpretation of this average sensitivity becomes clear when we notice that $\pi_{n_t}^\nu = \frac{\pi_\nu}{\pi_{n_t}}$ is the transition probability from node n_t to node ν.

Theorem 5.3 *Let $\mathbf{u}^* = \{u_\nu^{i*} : \nu \in \mathbf{n}_0^{++} \setminus \mathbf{n}_T, i \in M\}$ be an S-adapted Nash equilibrium generating the state trajectory $\mathbf{x}^* = \{x_\nu^* : \nu \in \mathbf{n}_0^{++}, x_{n_0}^* = x_0\}$, over the event tree. Then, for any player $i \in M$, there exists a costate trajectory $\lambda_\nu^i, \nu \in \mathbf{n}_0^{++} \setminus \{n_0\}$ such that, for any player i, the following conditions hold:*

$$0 = \frac{\partial H_i(\nu, \lambda_{\nu^+}^i, x_\nu, u_\nu)}{\partial u_\nu}, \quad \nu \in \mathbf{n}_0^{++} \setminus \mathbf{n}_T, \tag{5.7}$$

$$\lambda_\nu^i = \left[\frac{\partial H_i(\nu, \lambda_{\nu^+}^i, x_\nu, u_\nu)}{\partial x_\nu} \right]', \quad \nu \in \mathbf{n}_0^{++} \setminus \{\mathbf{n}_T \cup n_0\}, \tag{5.8}$$

$$\lambda_\nu^i = \left[\frac{\partial \Phi_i(\nu, x_\nu)}{\partial x_\nu} \right]', \quad \nu \in \mathbf{n}_T, \tag{5.9}$$

where $u_\nu = u_\nu^*$, $x_\nu = x_\nu^*$.

Proof In the expression of the Lagrangian in (5.5), we group together the terms that contain x_ν to get

$$\mathcal{L}_i(\lambda^i, \mathbf{x}, \mathbf{u}) = \sum_{t=0}^{T-1} \rho^t \sum_{n_t \in \mathbf{n}_t} \pi_{n_t} \left\{ H_i(n_t, \lambda_{\mathbf{n}_t^+}^i, x_{n_t}, u_{n_t}) - (\lambda_{n_t}^i)' x_{n_t} \right\} \tag{5.10}$$

$$+ \rho^T \sum_{n_T \in \mathbf{n}_T} \pi_{n_T} \left\{ \Phi_i(n_T, x_{n_T}) - (\lambda_{n_T}^i)' x_{n_T} \right\}.$$

Expression (5.6) and Eqs. (5.7)–(5.9) are then obtained by equalling to zero the partial derivatives of the Lagrangian (5.5) with respect to x_ν and u_ν. □

5.5.1 Node Consistency

In a discrete-time dynamic game, the open-loop Nash equilibrium is *time consistent*. This property means that, if at any intermediate date $\tau > 0$, we recompute the equilibrium for the remaining horizon $T - \tau$, with initial state x_τ^*, i.e., the value reached by playing the equilibrium from 0 to τ, then we will obtain the same result (equilibrium control and state trajectories) computed at initial time 0.

In a DGPET, a similar property is *node consistency*. Let $\mathbf{u}^* = \{u_\nu^{i*}, \; \nu \in \mathbf{n}_0^{++} \setminus \mathbf{n}_T, \; i \in M\}$ be an S-adapted Nash equilibrium with an associated state trajectory $\mathbf{x}^* = \{x_\nu^*, \; \nu \in \mathbf{n}_0^{++}\}$ emanating from state x_0. Let $\mathbf{u}_{n_t}^* = \{u_v^{i*}, \; v \in \mathbf{n}_t^{++} \setminus \mathbf{n}_T, \; i \in M\}$ and $\mathbf{x}_{n_t}^* = \{x_v^*, \; v \in \mathbf{n}_t^{++}\}$ be the continuation of this equilibrium in the subgame starting at node n_t.

Definition 5.3 An S-adapted Nash equilibrium is node consistent if, for all $n_\tau \in \mathbf{n}_\tau$, $\forall \tau \in \mathbb{T} \setminus \{T\}$, the continuation of the same equilibrium will still be an admissible equilibrium for a new game that starts at position $(n_\tau, x_{n_\tau}^*)$.

We prove the following result about the time consistency of S-adapted Nash equilibria.

Lemma 5.1 *An S-adapted Nash equilibrium is node consistent.*

Proof Let $(\mathbf{u}^{i*}, \; i \in M)$ be the S-adapted Nash equilibrium and $\mathbf{x}^*$ be a corresponding state trajectory. To prove that it is node consistent, we need to show that, for any node $n_t \in \mathbf{n}_t$, $t \in \mathbb{T} \setminus \{T\}$, the restriction $\mathbf{u}^*[n_t, x_{n_t}^*]$ of the equilibrium strategies on the subtree $\mathbf{n}_t^{++}$ will be the S-adapted Nash equilibrium for the game starting at node n_t with state $x_{n_t}^*$.

Assume that this is not true at node $n_t \in \mathbf{n}_t$, $t \in \mathbb{T} \setminus \{T\}$, with the current state $x_{n_t}^*$. Then, there exists Player $i \in M$ and a strategy $\hat{\mathbf{u}}^i[n_t, x_{n_t}^*]$ that is a collection of strategies $(\hat{u}_\nu^i : \nu \in \mathbf{n}_t^{++})$, such that

$$W_i(x_{n_t}^*, (\hat{\mathbf{u}}^i[n_t, x_{n_t}^*], \mathbf{u}^{-i*}[n_t, x_{n_t}^*])) > W_i(x_{n_t}^*, (\mathbf{u}^{i*}[n_t, x_{n_t}^*], \mathbf{u}^{-i*}[n_t, x_{n_t}^*])),$$

where, in the left-hand side, there is a payoff for Player i when Player i uses strategy $\hat{\mathbf{u}}^i[n_t, x_{n_t}^*]$ but all other players $j \in M \setminus \{i\}$ use strategies $\mathbf{u}^{j*}[n_t, x_{n_t}^*]$.

Now construct a strategy for Player i in the game starting from n_0, such that, (i) for any node ν belonging to the subtree $\mathbf{n}_t^{++}$, the strategy of Player i will be $\hat{u}_\nu^i$; and (ii) for any node ν belonging to the set $\mathbf{n}_0^{++} \setminus \mathbf{n}_t^{++}$, the strategy of Player i will be u_ν^{i*}. Let this strategy of Player i be denoted by $\tilde{\mathbf{u}}^i$.

Obviously, we obtain an inequality

$$W_i(x_0, (\tilde{\mathbf{u}}^i, \mathbf{u}^{-i*})) > W_i(x_0, (\mathbf{u}^{i*}, \mathbf{u}^{-i*})),$$

which contradicts the assumption that $(\mathbf{u}^{i*}, \mathbf{u}^{-i*})$ is the S-adapted Nash equilibrium. $\qquad\qquad\square$

Another property sought in of an equilibrium is *subgame perfectness*, which says that, if we have defined an equilibrium strategy profile $\mathbf{u}^*$ and the players have not played accordingly for some time (meaning that some players use strategies different from $\mathbf{u}^{i*}$, $i \in M$) and if a position (n_t, x_{n_t}) is reached, then resuming the play according to $\mathbf{u}^*$ from the initial condition $(n_t, x_{n_t}^*)$ will still be an equilibrium. S-adapted Nash equilibria are not subgame perfect. The reason for this is that the strategies forming these equilibria are open loop. If players do not play correctly at a node n_t, choosing a strategy profile u_{n_t} instead of $u_{n_t}^*$, then the trajectory is perturbed and the initial S-adapted strategy profile is no longer an equilibrium for the subgames that would start out of the nodes $\nu \in \mathbf{n}_t^+$ with the state trajectory $x_\nu' = f^{n_t}(x_{n_t}^*, u_{n_t})$. Feedback-Nash equilibria are subgame perfect but much harder to compute, especially when the game does not belong to one of the tractable classes of dynamic games, that is, linear-quadratic and linear-state games; see the references in Sect. 5.11.

5.6   Karush-Kuhn-Tucker Conditions and Equilibrium Formulation in DGPETs as a Nonlinear Complementarity Problem

In this section, we extend the results of Sect. 5.5 and characterize the S-adapted Nash equilibrium with constraints on the players' control sets. We use a mathematical-programming approach with Karush-Kuhn-Tucker (KKT) conditions when Assumption 5.1 is satisfied.

Assumption 5.1 We make the following assumptions:

1. Functions $f^{n_t}(x_{n_t}, u_{n_t})$ and $\phi_i(n_t, x_{n_t}, u_{n_t})$ are continuously differentiable in state x_{n_t} and u_{n_t} for any node $n_t \in \mathbf{n}_t$, $t \in \mathbb{T} \setminus \{T\}$ and any player $i \in M$;
2. Function $\Phi_i(n_T, x_{n_T})$ is continuously differentiable in state x_{n_T} for any node $n_T \in \mathbf{n}_T$ and any player $i \in M$;
3. For each player $i \in M$, the control set $U_{n_t}^i$ is defined by inequalities $h_i(u_{n_t}^i) \leq 0$, where $h_i : \mathbb{R}^{\ell_i} \to \mathbb{R}^{c_i}$, $c_i < m_i$, are given differentiable mappings.

The pseudo-Lagrangian of Player $i \in M$ is defined as[4]

$$\bar{\mathcal{L}}_i(n_t, \lambda_{\mathbf{n}_t^+}^i, \mu_{n_t}^i, x_{n_t}, u_{n_t}) = \phi_i(n_t, x_{n_t}, u_{n_t}) \tag{5.11}$$
$$+ \rho \sum_{\nu \in \mathbf{n}_t^+} \pi_{n_t}^\nu (\lambda_\nu^i)' f^{n_t}(x_{n_t}, u_{n_t}) + (\mu_{n_t}^i)' h_i(u_{n_t}^i),$$

where $\lambda_{\mathbf{n}_t^+}^i$ stands for the collection of λ_ν^i with $\nu \in \mathbf{n}_t^+$, and $\mu_{n_t}^i \in \mathbb{R}^{c_i}$ is the vector of Lagrange multipliers.

Theorem 5.4 *Under Assumption 5.1, if $\mathbf{u}^*$ is the S-adapted Nash equilibrium generating state trajectory $\mathbf{x}^*$ with initial state x_0 for the DGPET and if the constraint-qualification KKT conditions hold, then there exist functions $\lambda_{n_t}^i \in \mathbb{R}^q$ and functions $\mu_{n_t}^i \in \mathbb{R}^{c_i}$ such that the following conditions hold true:*

---

[4] In optimal control problems involving constraints the pseudo-Lagrangian is obtained by adding the penalty for violating the constraints to the Hamiltonian; see Sethi and Thompson (2000).

$$0 = \frac{\partial \bar{\mathcal{L}}_i(\nu, \lambda^i_{\nu+}, \mu^i_\nu, x^*_\nu, u^{-i*}_\nu, u^{i*}_\nu)}{\partial u^i_\nu}, \quad \nu \in \mathbf{n}_0^{++} \setminus \mathbf{n}_T, \tag{5.12}$$

$$\lambda^i_\nu = \left[\frac{\partial \bar{\mathcal{L}}_i(\nu, \lambda^i_{\nu+}, \mu^i_\nu, x^*_\nu, u^*_\nu)}{\partial x_\nu} \right]', \quad \nu \in \mathbf{n}_0^{++} \setminus \{\mathbf{n}_T \cup n_0\}, \tag{5.13}$$

$$\lambda^i_\nu = \left[\frac{\partial \Phi_i(\nu, x^*_\nu)}{\partial x_\nu} \right]', \quad \nu \in \mathbf{n}_T, \tag{5.14}$$

$$0 = (\mu^i_\nu)' h_i(u^{i*}_\nu), \quad \nu \in \mathbf{n}_0^{++} \setminus \mathbf{n}_T, \tag{5.15}$$

$$0 \le \mu^i_\nu, \quad \nu \in \mathbf{n}_0^{++} \setminus \mathbf{n}_T. \tag{5.16}$$

Proof The proof is straightforward; it is based on the fact that any player i's best reply to other players' equilibrium controls is an optimal control problem. In the theorem, we write the necessary optimality conditions when Assumption 5.1 is satisfied. $\square$

5.7 An Example of a DGPET

Consider a two-player, three-period dynamic game played over an event tree. To distinguish explicitly between nodes belonging to the same period, we denote a node by $n^l_t \in \mathbf{n}_t, l \in \{1, \ldots, L_t\}, t = 0, 1, 2$. The players are producers of a homogeneous commodity sold in a competitive market. Denote by $q^i_{n^l_t}$ the output of Player i at node $n^l_t \in \mathbf{n}_t, t = 0, 1, 2$. The price of the product is given by the following stochastic inverse-demand law:

$$p_{n^l_t} = f(n^l_t, Q_{n^l_t}),$$

where $Q_{n^l_t} = q^1_{n^l_t} + q^2_{n^l_t}$ is the total quantity available on the market. We assume that the price $p_{n^l_t}$ is decreasing in $Q_{n^l_t}$.

Denote by $n^{l-}_t \in \mathbf{n}_{t-1}, t = 1, 2$ the unique predecessor of $n^l_t \in \mathbf{n}_t, t = 1, 2$. Let $\mathbf{n}^{l+}_t \subset \mathbf{n}_{t+1}$ be the set of all possible direct successors of node $n^l_t, t = 0, 1$. Player i, $i = 1, 2$, is described by the following data:

- A production capacity $K^i_{n^l_t}, n^l_t \in \mathbf{n}_t, t = 0, 1, 2$, which accumulates over time according to the following difference equation:

$$K^i_{n^l_t} = K^i_{n^{l-}_t} + I^i_{n^{l-}_t}, \quad K^i_{n_0} \text{ given}, \tag{5.17}$$

where $I^i_{n^l_t}$ is the physical investment in the production capacity at node $n^l_t \in \mathbf{n}_t$, $t = 0, 1$;
- A production cost function $C_i(q^i)$, assumed to be strictly convex, increasing, and twice continuously differentiable;

- An investment cost function $F_i(I^i)$, assumed to be strictly convex, increasing, and twice continuously differentiable;
- A payoff function

$$J_i(\cdot) = \sum_{t=0}^{1} \rho^t \sum_{l=1}^{L_t} \sum_{n_t^l \in \mathbf{n}_t} \pi_{n_t^l} \left(p_{n_t^l} q_{n_t^l}^i - C_i(q_{n_t^l}^i) - F_i(I_{n_t^l}^i) \right) \qquad (5.18)$$

$$+ \rho^2 \sum_{l=1}^{L_2} \sum_{n_2^l \in \mathbf{n}_2} \pi_{n_2^l} \left(p_{n_2^l} q_{n_2^l}^i - C_i(q_{n_2^l}^i) \right),$$

where $\pi_{n_t^l}$ is the probability of passing through node n_t^l, and $\rho \in (0,1)$ is the discount factor;
- A capacity constraint $q_{n_t^l}^i \le K_{n_t^l}^i$, $n_t^l \in \mathbf{n}_t$, $t = 0, 1, 2$.

From (5.17) we see that it takes one period for an investment to become productive. Given the absence of a salvage value at terminal date $T = 2$, there is no reason to invest in capacity at that period.

Introduce the following notation:

$$K_{n_t^l} = \left(K_{n_t^l}^1, K_{n_t^l}^2 \right), \quad \mathbf{K}^i = (K_{n_t^l}^i : n_t^l \in \mathbf{n}_t,\ t = 0, 1, 2), \quad \mathbf{K} = (\mathbf{K}^1, \mathbf{K}^2),$$

$$q_{n_t^l} = \left(q_{n_t^l}^1, q_{n_t^l}^2 \right), \quad \mathbf{q}^i = (q_{n_t^l}^i : n_t^l \in \mathbf{n}_t,\ t = 0, 1, 2), \quad \mathbf{q} = (\mathbf{q}^1, \mathbf{q}^2),$$

$$I_{n_t^l} = \left(I_{n_t^l}^1, I_{n_t^l}^2 \right), \quad \mathbf{I}^i = (I_{n_t^l}^i : n_t^l \in \mathbf{n}_t,\ t = 0, 1), \quad \mathbf{I} = (\mathbf{I}^1, \mathbf{I}^2).$$

The Hamiltonian of Player i is defined by

$$H_i\left(n_t^l, \lambda_{\mathbf{n}_t^+}^i, K_{n_t^l}^i, q_{n_t^l}^i, I_{n_t^l}^i \right) = p_{n_t^l} q_{n_t^l}^i - C_i(q_{n_t^l}^i) - F_i(I_{n_t^l}^i) + \rho \sum_{\nu \in \mathbf{n}_t^{l+}} \pi_{n_t}^\nu \lambda_\nu^i \left(K_{n_t^l}^i + I_{n_t^l}^i \right)$$

for any node $n_t^l \in \mathbf{n}_t$, $t = 0, 1$.

The Lagrangian of Player i is given by

$$\mathcal{L}_i(\lambda^i, \mathbf{K}, \mathbf{q}, \mathbf{I}) = p_{n_0} q_{n_0}^i - C_i(q_{n_0}^i) - F_i(I_{n_0}^i)$$

$$+ \rho \sum_{n_1^l \in \mathbf{n}_1} \pi_{n_1^l} \left\{ p_{n_1^l} q_{n_1^l}^i - C_i(q_{n_1^l}^i) - F_i(I_{n_1^l}^i) \right.$$

$$\left. + \lambda_{n_1^l}^i \left(K_{n_1^l}^i - K_{n_0}^i - I_{n_0}^i \right) \right\} + \rho^2 \sum_{n_2^l \in \mathbf{n}_2} \pi_{n_2^l} \left\{ p_{n_2^l} q_{n_2^l}^i - C_i(q_{n_2^l}^i) \right.$$

$$\left. + \lambda_{n_2^l}^i \left(K_{n_2^l}^i - K_{n_2^{l-}}^i - I_{n_2^{l-}}^i \right) \right\} + \mu_{n_0}^i \left(K_{n_0}^i - q_{n_0}^i \right)$$

$$+ \mu_{n_1}^i \left(K_{n_1}^i - q_{n_1}^i \right) + \mu_{n_1}^2 \left(K_{n_1}^i - q_{n_1}^2 \right)$$

$$+ \mu_{n_2^1} \left(K_{n_2^1}^i - q_{n_2^1}^i \right) + \mu_{n_2^2} \left(K_{n_2^2}^i - q_{n_2^2}^i \right)$$
$$+ \mu_{n_2^3} \left(K_{n_2^3}^i - q_{n_2^3}^i \right) + \mu_{n_2^4} \left(K_{n_2^4}^i - q_{n_2^4}^i \right),$$

where $\mu_{n_t^l} = (\mu_{n_t^l}^1, \mu_{n_t^l}^2), n_t^l \in \mathbf{n}_t, t = 0, 1, 2$, is the Lagrange multiplier associated to the capacity constraint $q_{n_t^l}^i \leq K_{n_t^l}^i$.

Remark 5.7 The quantity produced $q_{n_t^l}^i, n_t^l \in \mathbf{n}_t, t = 0, 1, 2$, and the investment in production capacity $I_{n_t^l}^i, n_t^l \in \mathbf{n}_t, t = 0, 1$, must be nonnegative. We shall verify that these constraints are satisfied in equilibrium.

Assumptions: To simplify the exposition and provide a numerical illustration, we assume the following functional forms for the inverse-demand and cost functions:

$$p_{n_t^l} = \alpha_{n_t^l} - \beta_{n_t^l} Q_{n_t^l},$$
$$C_1(q^1) = 2(q^1)^2, \quad C_2(q^2) = 4(q^2)^2,$$
$$F_1(I_1) = 10I_1^2, \quad F_2(I_2) = 6I_2^2,$$

where $\alpha_{n_t^l}$ and $\beta_{n_t^l}$ are strictly positive parameters for any node n_t^l. Their values and the probability transitions are shown in Fig. 5.2.

The initial production capacities are

$$K_{n_0}^1 = 4, \quad K_{n_0}^2 = 2.$$

Existence of an S-adapted equilibrium: The payoff at any node n_t^l depends only on the decision variables at that node and indirectly on the state through the capacity constraint. Therefore, to show the existence of at least one S-adapted equilibrium, we need to show (i) that at any nonterminal node $n_t^l \in \mathbf{n}_t, t = 0, 1$, the profit

$$p_{n_t^l} q_{n_t^l}^i - C_i\left(q_{n_t^l}^i\right) - F_i\left(I_{n_t^l}^i\right)$$

is jointly concave in $q_{n_t^l}^i$ and $I_{n_t^l}^i$; and (ii) that at any terminal node $n_2^l \in \mathbf{n}_2$, the profit is concave in the control variable $q_{n_2^l}^i$.

The Hessian matrices at node $n_t^l \in \mathbf{n}_t, t = 0, 1$ of Players 1 and 2 are given by

$$H_1\left(q_{n_t^l}^1, I_{n_t^l}^1\right) = \begin{pmatrix} -2\beta_{n_t^l} - 4 & 0 \\ 0 & -20 \end{pmatrix},$$
$$H_2\left(q_{n_t^l}^2, I_{n_t^l}^2\right) = \begin{pmatrix} -2\beta_{n_t^l} - 8 & 0 \\ 0 & -12 \end{pmatrix}.$$

Clearly, both matrices are negative definite, and therefore, the profit functions of both players are strictly concave.

Further, the set of admissible strategies

$$\left\{ 0 \le q_{n_t^i}^i \le K_{n_t^i}^i, \; n_t^i \in \mathbf{n}_t, \; t = 0, 1, 2; \; I_{n_t^i}^i \ge 0, \; n_t^i \in \mathbf{n}_t, \; t = 0, 1 \right\}$$

is compact (and also convex). Therefore, by Theorem 5.1, there exists at least one S-adapted Nash equilibrium.

Uniqueness of the S-adapted equilibrium: The uniqueness of the equilibrium can be verified using Theorem 5.2. Let us compute the pseudo-gradients

$$\mathcal{G}(q_{n_t^l}) = \begin{pmatrix} \alpha_{n_t^l} - 2\beta_{n_t^l} q_{n_t^l}^1 - \beta_{n_t^l} q_{n_t^l}^2 - 4q_{n_t^l}^1 \\ \alpha_{n_t^l} - 2\beta_{n_t^l} q_{n_t^l}^2 - \beta_{n_t^l} q_{n_t^l}^1 - 8q_{n_t^l}^2 \end{pmatrix}, \; n_t^l \in \mathbf{n}_t, \; t = 0, 1,$$

$$\mathcal{G}(I_{n_t^l}) = \begin{pmatrix} -20 I_{n_t^l}^1 \\ -12 I_{n_t^l}^2 \end{pmatrix}, \; n_t^l \in \mathbf{n}_t, \; t = 0, 1,$$

$$\left(\mathcal{G}\left(q_{n_T^l} \right) \right)' = \begin{pmatrix} \alpha_{n_T^l} - 2\beta_{n_T^l} q_{n_T^l}^1 - \beta_{n_T^l} q_{n_T^l}^2 - 4q_{n_T^l}^1 \\ \alpha_{n_T^l} - 2\beta_{n_T^l} q_{n_T^l}^2 - \beta_{n_T^l} q_{n_T^l}^1 - 8q_{n_T^l}^2 \end{pmatrix}, \; n_T^l \in \mathbf{n}_2,$$

and the Jacobian matrices

$$\mathcal{J}\left(q_{n_t^l} \right) = \begin{pmatrix} -2\beta_{n_t^l} - 4 & -\beta_{n_t^l} \\ -\beta_{n_t^l} & -2\beta_{n_t^l} - 8 \end{pmatrix}, \; n_t^l \in \mathbf{n}_t, \; t = 0, 1$$

$$\mathcal{J}\left(I_{n_t^l} \right) = \begin{pmatrix} -20 & 0 \\ 0 & -12 \end{pmatrix}, \; n_t^l \in \mathbf{n}_t, \; t = 0, 1,$$

$$\mathcal{J}\left(q_{n_T^l} \right) = \begin{pmatrix} -2\beta_{n_T^l} - 4 & -\beta_{n_T^l} \\ -\beta_{n_T^l} & -2\beta_{n_T^l} - 8 \end{pmatrix}, \; n_T^l \in \mathbf{n}_2.$$

The matrices

$$Q\left(n_t, q_{n_t^l} \right) = \frac{1}{2} \left[\mathcal{J}\left(q_{n_t^l} \right) + \mathcal{J}\left(q_{n_t^l} \right)' \right],$$

$$Q\left(n_t, I_{n_t^l} \right) = \frac{1}{2} \left[\mathcal{J}\left(I_{n_t^l} \right) + \mathcal{J}\left(I_{n_t^l} \right)' \right],$$

$$Q\left(n_T, q_{n_T^l} \right) = \frac{1}{2} \left[\mathcal{J}\left(q_{n_T^l} \right) + \mathcal{J}\left(q_{n_T^l} \right)' \right]$$

are given by

$$Q\left(n_t, q_{n_t^l}\right) = \begin{pmatrix} -2\beta_{n_t^l} - 4 & -\beta_{n_t^l} \\ -\beta_{n_t} & -2\beta_{n_t^l} - 8 \end{pmatrix},$$

$$Q\left(n_t, I_{n_t^l}\right) = \begin{pmatrix} -20 & 0 \\ 0 & -12 \end{pmatrix},$$

$$Q\left(n_T, q_{n_T^l}\right) = \begin{pmatrix} -2\beta_{n_T^l} - 4 & -\beta_{n_T^l} \\ -\beta_{n_T^l} & -2\beta_{n_T^l} - 8 \end{pmatrix}.$$

All matrices are negative definite for all nodes $n_t \in \mathbf{n}_t$, $t \in \mathbb{T} \setminus \{T\}$ (the first and the second ones) and $n_T \in \mathbf{n}_T$, respectively. Indeed, the term $-2\beta_{n_t^l} - 4$ is negative, and the determinant $\left|Q(n_t, \tilde{u}_{n_t})\right| = 3\beta_{n_t^l}^2 + 24\beta_{n_t^l} + 32$ is positive. Therefore, from Theorem 5.2, the equilibrium is unique.

Numerical results: In Fig. 5.2, we report the investment and production decisions as well as the values of the production capacities at each node.

In all nodes, the output and investment variables are nonnegative. At initial node n_0, both players produce at the maximum possible levels and invest in their production capacities. In period 1, the capacity constraint is binding only in node n_1^3, that is, at the node where the market potential is the highest (120 at that node, compared to 80 and 100 at the other two nodes). Also, both players invest at node n_1^3, and their production capacities are binding in node n_2^5. Investment decisions seem to be driven by the so-called *good news* principle, that is, they are made under the assumption that high demand will follow. Finally, we note that we have a corner solution. Indeed, at each node, either the investment is zero or the capacity constraint is binding.

5.8 DGPETs with Coupled Constraints

In this section, we assume that the players also face a coupled constraint at each node. The basic definition of a DGPET, given in Sect. 5.3, is extended by presenting a coupled constraint. We define the normalized equilibrium à la (Rosen 1965), providing existence and uniqueness conditions for this equilibrium.

We consider a dynamic game played over an event tree defined in Sect. 5.3 with the coupled constraint given by the $k-$dimensional vector,

$$h_{n_t}\left(u_{n_t}^1, \ldots, u_{n_t}^i, \ldots, u_{n_t}^m\right) \geq 0 \tag{5.19}$$

for any $n_t \in \mathbf{n}_t, t \in \mathbb{T} \setminus \{T\}$. This constraint implies that each player's strategy space may depend on other players' strategies. Let function h_{n_t} be a concave function for any $n_t \in \mathbf{n}_t, t \in \mathbb{T} \setminus \{T\}$.

As before, the state equations are

$$x_{n_t} = f^{n_t^-}\left(x_{n_t^-}, u_{n_t^-}\right), \quad x_{n_0} = x_0, \tag{5.20}$$

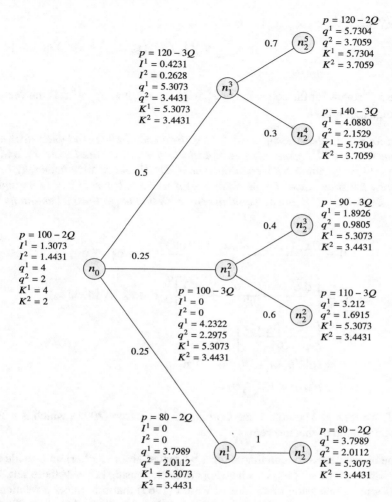

Fig. 5.2 Investment and production decisions over an event tree

where $u_{n_t} \in U_{n_t}$, $n_t \in \mathbf{n}_t$, $t \in \mathbb{T} \setminus \{0\}$ and x_0 is the initial state at root node n_0.

The necessary conditions of the constrained Nash equilibrium in the game when any player i maximizes the functional (5.2) are given in Proposition 5.4.

We use a mathematical-programming approach with KKT conditions when Assumption 5.1 is satisfied and taking into account that function h_{n_t} is concave for any $n_t \in \mathbf{n}_t$.

Let us consider the auxiliary game when Player i maximizes the weighted payoff function $r_i J_i$, where $r_i > 0$, with the state dynamics and constraints defined above. For this game the pseudo-Lagrangian of Player i is defined as follows:

$$\bar{\mathcal{L}}_i^{r_i}(n_t, \lambda_{\mathbf{n}_t^+}^i, \mu_{n_t}^i, x_{n_t}, u_{n_t}) = r_i \phi_i(n_t, x_{n_t}, u_{n_t})$$
$$+ \rho \sum_{\nu \in \mathbf{n}_t^+} \pi_{n_t}^\nu (\lambda_\nu^i)' f^{n_t}(x_{n_t}, u_{n_t}) + (\mu_{n_t}^i)' h_{n_t}(u_{n_t}),$$

where $\lambda_{\mathbf{n}_t^+}^i$ stands for the collection of λ_ν^i with $\nu \in \mathbf{n}_t^+$, and $\mu_{n_t}^i \in \mathbb{R}^k$ is the vector of Lagrange multipliers.

Theorem 5.5 *Under Assumption 5.1, let $\mathbf{u}^*$ be an admissible S-adapted Nash equilibrium in a DGPET generating state trajectory $\mathbf{x}^*$ with initial state x_0, and the constraint-qualification KKT conditions hold. Let $\mathbf{U}$ and X have nonempty interiors; then for each player $i \in M$ there exist a real number $r_i \geq 0$, and multipliers $\lambda_{n_t}^i \in \mathbb{R}^q$ and $\mu_{n_t}^i \in \mathbb{R}^k$ not all equal to zero, such that the following conditions hold true:*

$$0 = \frac{\partial \bar{\mathcal{L}}_i^{r_i}(\nu, \lambda_{\nu^+}^i, \mu_\nu^i, x_\nu^*, u_\nu^{-i*}, u_\nu^{i*})}{\partial u_\nu^i}, \quad \nu \in \mathbf{n}_0^{++} \setminus \mathbf{n}_T,$$

$$\lambda_\nu^i = \left[\frac{\partial \bar{\mathcal{L}}_i^{r_i}(\nu, \lambda_{\nu^+}^i, \mu_\nu^i, x_\nu^*, u_\nu^*)}{\partial x_\nu} \right]', \quad \nu \in \mathbf{n}_0^{++} \setminus \{\mathbf{n}_T \cup n_0\},$$

$$\lambda_\nu^i = \left[\frac{r_i \partial \Phi_i(\nu, x_\nu^*)}{\partial x_\nu} \right]', \quad \nu \in \mathbf{n}_T,$$

$$0 = (\mu_\nu^i)' h_\nu(u_\nu^*), \quad \nu \in \mathbf{n}_0^{++} \setminus \mathbf{n}_T,$$

$$0 \leq \mu_\nu^i, \quad \nu \in \mathbf{n}_0^{++} \setminus \mathbf{n}_T.$$

Proof We refer to Theorem 1 and Corollary 1 in Carlson (2002), which is a deterministic version of this theorem.   □

The above necessary conditions for a S-adapted Nash equilibrium provide a set of multipliers for each player with no apparent relationship between these sets. Now we present an extension of an idea of Rosen (1965) that introduces a relationship between these sets of multipliers. This notion, called a normalized Nash equilibrium, has been used in Carlson and Haurie (1996) and Carlson et al. (2018).

We define a S-adapted normalized equilibrium for DGPETs and characterize the necessary conditions for this equilibrium.

Definition 5.4 We say that the S-adapted Nash equilibrium is normalized if the Lagrange multipliers μ_ν^i, $\nu \in \mathbf{n}_0^{++} \setminus \mathbf{n}_T$ from Theorem 5.5 are colinear with a common vector μ_ν^0 for any $i \in M$, that is,

$$\mu_\nu^i = \frac{1}{r_i} \mu_\nu^0, \tag{5.21}$$

where $r_i > 0, i \in M$, is a weight assigned to Player i's payoff in the joint optimization problem defined below.

Consider the auxiliary optimization problem:

$$\max_{\mathbf{u} \in \mathbf{U}} J_r(\mathbf{x}, \mathbf{u}^*, \mathbf{u}), \tag{5.22}$$

where

$$J_r(\mathbf{x}, \mathbf{u}^*, \mathbf{u}) = \sum_{t=0}^{T-1} \rho^t \sum_{n_t \in \mathbf{n}_t} \pi_{n_t} \phi_r(n_t, x_{n_t}, u^*_{n_t}, u_{n_t}) + \rho^T \sum_{n_T \in \mathbf{n}_T} \pi_{n_T} \Phi_r(n_T, x_{n_T}),$$

where $r = (r_i : i \in M)$, $r_i > 0$ is a weight assigned to Player i's payoff; and

$$\phi_r(n_t, x_{n_t}, u^*_{n_t}, u_{n_t}) = \sum_{i \in M} r_i \phi_i(n_t, x_{n_t}, (u^{-i*}_{n_t}, u^i_{n_t})),$$

$$\Phi_r(n_T, x_{n_T}) = \sum_{i \in M} r_i \Phi_i(n_T, x_{n_T}),$$

subject to (5.19) and (5.20).

We use a mathematical-programming approach with KKT conditions when Assumption 5.1 is satisfied and taking into account that function h_{n_t} is concave for any $n_t \in \mathbf{n}_t$.

The pseudo-Lagrangian of the problem is defined as follows

$$\begin{aligned} \bar{\mathcal{L}}_r(n_t, \lambda_{\mathbf{n}_t^+}, \mu^0_{n_t}, x_{n_t}, u^*_{n_t}, u_{n_t}) = {} & \phi_r(n_t, x_{n_t}, u^*_{n_t}, u_{n_t}) \\ & + \rho \sum_{\nu \in \mathbf{n}_t^+} \pi^\nu_{n_t} \lambda'_\nu f^{n_t}(x_{n_t}, u_{n_t}) + (\mu^0_{n_t})' h_{n_t}(u_{n_t}), \quad (5.23) \end{aligned}$$

where $\lambda_{\mathbf{n}_t^+}$ stands for the collection of λ_ν with $\nu \in \mathbf{n}_t^+$ and $\mu^0_{n_t} \in \mathbb{R}^k$ is the vector of Lagrange multipliers.

Theorem 5.6 *Under Assumption 5.1, if $\mathbf{u}^*$ is an admissible S-adapted Nash equilibrium DGPET generating a state trajectory $\mathbf{x}^*$ with an initial state x_0 if the constraint-qualification KKT conditions hold. Let $\mathbf{U}$ and X have nonempty interiors; then there exists a real number $\ell \geq 0$, functions $\lambda_{n_t} \in \mathbb{R}^q$, and $\mu^0_{n_t} \in \mathbb{R}^k$ not all equal to zero, such that the following conditions hold true:*

$$0 = \frac{\partial \bar{\mathcal{L}}^\ell_r(\nu, \lambda_{\nu^+}, \mu^0_\nu, x^*_\nu, u^{-i*}_\nu, u^i_\nu)}{\partial u^i_\nu}, \quad \nu \in \mathbf{n}_0^{++} \setminus \mathbf{n}_T, \ i \in M, \tag{5.24}$$

$$\lambda_\nu = \left[\frac{\partial \bar{\mathcal{L}}^\ell_r(\nu, \lambda_{\nu^+}, \mu^0_\nu, x_\nu, u^*_\nu)}{\partial x_\nu} \right]', \quad \nu \in \mathbf{n}_0^{++} \setminus \{\mathbf{n}_T \cup n_0\}, \tag{5.25}$$

$$\lambda_\nu = \left[\frac{\ell \partial \Phi_r(\nu, x_\nu)}{\partial x_\nu} \right]', \ \nu \in \mathbf{n}_T, \tag{5.26}$$

$$0 = (\mu_\nu^0)' h_\nu(u_\nu^*), \ \nu \in \mathbf{n}_0^{++} \setminus \mathbf{n}_T, \tag{5.27}$$

$$0 \leq \mu_\nu^0, \ \nu \in \mathbf{n}_0^{++} \setminus \mathbf{n}_T, \tag{5.28}$$

where $\bar{\mathcal{L}}_r^\ell(\nu, \lambda_{\nu^+}, \mu_\nu^0, x_\nu^*, u_\nu^{-i*}, u_\nu^i) = \ell \phi_r(\nu, x_\nu, u_\nu^*, u_\nu) + \rho \sum_{n \in \nu^+} \pi_\nu^n \lambda_n f^\nu(x_\nu, u_\nu)$
$+ (\mu_\nu^0)' h_\nu(u_\nu)$.

Proof See Theorem 3 in Carlson (2002).   □

The relationship between the KKT multipliers, $\mu_{n_t}^i = \frac{1}{r_i} \mu_{n_t}^0$, shows that $\left(x_{n_t}^*, u_{n_t}^* \right)$ defined in Theorem 5.6 is a normalized equilibrium. Once we obtain the value of the multiplier $\mu_{n_t}^0$ associated with the coupled constraint at node $n_t \in \mathbf{n}_t, t = 0, \ldots, T - 1$, the determination of an S-adapted Rosen equilibrium can be achieved by solving the decoupled game with the players' payoffs defined as follows:

$$J_i(\mathbf{x}, \mathbf{u}) = \sum_{t=0}^{T-1} \rho^t \sum_{n_t \in \mathbf{n}_t} \pi_{n_t} \left\{ \phi_i(n_t, x_{n_t}, u_{n_t}) - \frac{1}{r_i} (\mu_{n_t}^0)' h_{n_t}(u_{n_t}) \right\}$$
$$+ \rho^T \sum_{n_T \in \mathbf{n}_T} \pi_{n_T} \Phi_i(n_T, x_{n_T})$$

for any $i \in M$, subject to (5.19) and (5.20). We should notice that the additional term in Player i's objective function, namely, $\frac{1}{r_i}(\mu_{n_t}^0)' h_{n_t}(u_{n_t})$ for any nonterminal node plays the role of a penalty term. Thus, the normalized equilibrium of the game in which the players satisfy the coupled constraints can be implemented by a taxation scheme. This is done by setting the taxes at the appropriate levels, i.e., by choosing them equal to the common KKT multipliers while adjusted for each player i by her weighting factor $1/r_i$. To wrap up, solving for a normalized equilibrium requires the determination of the solution of the combined optimization problem, and next solving, for a given vector of weights $r = (r_1, \ldots, r_m)$, the above uncoupled m-player game.

Remark 5.8 A similar approach (enforcement through taxation) has been widely used in the literature. In Krawczyk (2005), the regulator calculates a Rosen coupled-constraint equilibrium (normalized Nash), and then uses the coupled-constraint Lagrange multiplier to formulate a threat under which the agents will play a decoupled Nash game. Carlson and Haurie (2000) also use the equilibrium of the relaxed game as a penalizing term for the coupled constraint in the main game. See Haurie (1995), Haurie and Krawczyk (1997), Haurie and Zaccour (1995), Krawczyk (2005, 2007), and Krawczyk and Uryasev (2000) for more applications of a normalized equilibrium to define a penalty tax in environmental games. Tidball and Zaccour (2005) and Tidball and Zaccour (2009) study a game with environmental constraints using the concept of a normalized equilibrium in a static and dynamic context, respectively. Conceptually similar studies of equilibria with coupled constraints have been explored in Contreras et al. (2004), Contreras et al. (2007), Pang and Fukushima (2005), and Drouet et al. (2008, 2011) for the games in which the competitive agents maximize their utility functions subject to coupled constraints, which define their joint strategy space.

5.9 An Example of a DGPET with Coupled Constraints

We illustrate the theory with an environmental management example. Consider an industry with a finite number of competing firms. The regulatory authority requires that the industry not exceed a certain level of cumulative pollution (that is, the total cumulative pollution is capped), at a given date. This problem involves a coupled constraint set in the combined strategy space of all agents. To implement a normalized equilibrium, we suppose that the regulator selects a vector of weights $r = (r_1, \ldots, r_m)$ assigned for the players in the joint optimization problem[5] given by (5.22).

Consider an oligopoly with three firms or players competing à la Cournot in a three-period game, $\mathbb{T} = \{0, 1, 2\}$. We determine the equilibrium in two scenarios, that is, a benchmark scenario with no environmental constraint and a coupled-constraint scenario where players must jointly satisfy an environmental standard.

Denote by $q_{n_t}^i$ Player i's production at node $n_t \in \mathbf{n}_t$, $t \in \mathbb{T}$ and by $I_{n_t}^i$ her investment in production capacity at $n_t \in \mathbf{n}_t$, $t \in \mathbb{T} \setminus \{T\}$, that is, no investment can take place at a terminal node. Denote by $q_{n_t} = \sum_{i \in M} q_{n_t}^i$ the total quantity produced at node n_t, and by $e_{n_t}^i$ the pollution emitted by Player i in that node. To keep it simple, we suppose that each unit of production generates a unit of pollution emissions, that is, $q_{n_t}^i = e_{n_t}^i$.

The consumer's demand is stochastic, with the random process being described by the event tree in Fig. 5.3. The probabilities of passing through the nodes are as follows: $\pi_{n_1^1} = 0.30$, $\pi_{n_1^2} = 0.70$ for $t = 1$, and $\pi_{n_2^1} = 0.20$, $\pi_{n_2^2} = 0.10$, $\pi_{n_2^3} = 0.20$, $\pi_{n_2^4} = 0.50$ for $t = 2$.

Fig. 5.3 Event tree representing the uncertain demand, $T = 2$

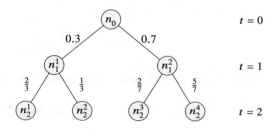

---

[5] In international pollution control, such an authority does not exist and the players (countries) need to agree on a particular vector of weights. Therefore, the game with coupled constraints cannot be considered as a purely noncooperative game.

The inverse demand is given by the following linear function:

$$P(n_t, q_{n_t}) = A_{n_t} - b q_{n_t}, \quad n_t \in \mathbf{n}_t, \quad t \in \mathbb{T}, \tag{5.29}$$

where A_{n_t} and b are positive parameters. Note that the randomness does not affect the slope of the demand, only the choke price (or maximum willingness-to-pay). The values of A_{n_t} are given by

Node	n_0	n_1^1	n_1^2	n_2^1	n_2^2	n_2^3	n_2^4
A_{n_t}	120	140	90	160	120	115	85

Let $K_{n_t}^i$ be the available production capacity node n_t. A vector of production capacities $(K_{n_t}^i : i \in M, \ n_t \in \mathbf{n}_t, \ t \in \mathbb{T})$ represents the state variable. Assuming a one-period lag before an investment becomes productive, the evolution of the production capacity of Player i is described by the following state equation:

$$K_{n_t}^i = (1 - \delta) K_{n_t^-}^i + I_{n_t^-}^i, \quad K_{n_0}^i = k_0^i, \tag{5.30}$$

where k_0^i denotes the given initial capacity of Player i, and δ is the depreciation rate of capacity, $0 < \delta < 1$. The quantity produced is subject to the available capacity, i.e.,

$$q_{n_t}^i \leqslant K_{n_t}^i, \quad n_t \in \mathbf{n}_t, t \in \mathbb{T}. \tag{5.31}$$

The strategy of Player i at node $n_t \in \mathbf{n}_t, t \in \mathbb{T}$, is a vector $(q_{n_t}^i, I_{n_t}^i)$ satisfying (5.31), and $I_{n_T}^i \equiv 0$ for any $n_T \in \mathbf{n}_T$.

The production and investment costs are defined by the quadratic functions:

$$C_i(q_{n_t}^i) = \frac{c_i}{2} (q_{n_t}^i)^2, \quad c_i > 0, \tag{5.32}$$

$$D_i(I_{n_t}^i) = \frac{d_i}{2} (I_{n_t}^i)^2, \quad d_i > 0. \tag{5.33}$$

Denote by $S_i(K_{n_T}^i)$ the salvage value of the production capacity of Player i at terminal nodes $n_T \in \mathbf{n}_T$ in time $T = 2$, given by

$$S_i(K_{n_T}^i) = \frac{v_i}{2} (K_{n_T}^i)^2, \quad v_i > 0. \tag{5.34}$$

Player i's payoff function in a nonterminal node n_t is defined by

$$\phi_i(n_t, q_{n_t}^i, q_{n_t}, I_{n_t}^i) = q_{n_t}^i P(n_t, q_{n_t}) - C_i(q_{n_t}^i) - D_i(I_{n_t}^i),$$

and at a terminal node n_T by

$$\Phi_i(n_T, q_{n_T}^i, q_{n_T}, K_{n_T}^i) = q_{n_T}^i P(n_T, q_{n_T}) - C_i(q_{n_T}^i) + S_i(K_{n_T}^i),$$

where the linear-inverse demand $P(n_t, q_{n_t})$ is defined by (5.29), production costs $C_i(q_{n_t}^i)$, investment costs $D_i(I_{n_t}^i)$, and the salvage value $S_i(K_{n_T}^i)$ are given in (5.32), (5.33), and (5.34), respectively.

Player $i \in M$ aims to maximize

$$J_i = \sum_{t=0}^{T-1} \rho^t \sum_{n_t \in \mathbf{n}_t} \pi_{n_t} \phi_i(n_t, q_{n_t}^i, q_{n_t}, I_{n_t}^i) + \rho^T \sum_{n_T \in \mathbf{n}_T} \pi_{n_T} \Phi_i(n_T, q_{n_T}^i, q_{n_T}, K_{n_T}^i)$$

$$= \sum_{t=0}^{1} \rho^t \sum_{n_t \in \mathbf{n}_t} \pi_{n_t} \big(q_{n_t}^i P(n_t, q_{n_t}) - C_i(q_{n_t}^i) - D_i(I_{n_t}^i)\big)$$

$$+ \rho^2 \sum_{n_2 \in \mathbf{n}_2} \pi_{n_2} \Big(q_{n_2}^i P(n_2, q_{n_2}) - C_i(q_{n_2}^i) + S_i(K_{n_2}^i)\Big),$$

s.t. (5.30), (5.31), and nonnegativity constraints:

$$I_{n_t}^i \geq 0, \ q_{n_t}^i \geq 0, \ n_t \in \mathbf{n}_t, \ t \in \mathbb{T}.$$

In the coupled-constraint scenario, the regulator imposes a cap on cumulative emissions. Denote by E_{n_t} the pollution stock at node $n_t \in \mathbf{n}_t$, $t \in \mathbb{T}$, whose evolution is given by

$$E_{n_0} = \sum_{i \in M} e_{n_0}^i, \tag{5.35}$$

$$E_{n_t} = (1 - \mu) E_{n_t^-} + \sum_{i \in M} e_{n_t}^i, \quad \forall n_t \in \mathbf{n}_0^{++} \setminus \{n_0\}, \tag{5.36}$$

where μ is nature's absorption rate of pollution. The joint pollution accumulation constraint is as follows:

$$E_{n_t} \leq \bar{E}, \quad \forall n_t \in \mathbf{n}_0^{++}, \tag{5.37}$$

where $\bar{E}$ is the cap set by the regulator.

We retain the following parameters in the numerical simulation:

$$\delta = 0.15, \ \rho = 0.90, \ \mu = 0.4, \ b = 5, \ \bar{E} = 15,$$
$$c_1 = 2, \ c_2 = 3.5, \ c_3 = 6,$$
$$d_1 = 3, \ d_2 = 2.5, \ d_3 = 2,$$
$$v_1 = 0.10, \ v_2 = 0.15, \ v_3 = 0.12,$$
$$K_0^1 = 6, \ K_0^2 = 4, \ K_0^3 = 2.$$

Nash equilibrium: The players choose production and investment levels in the absence of any environmental constraint. It is straightforward to verify that the equilibrium exists and is unique. The equilibrium control and state values are given in Table 5.1. The players' profits are shown in Table 5.2. The results show that

Table 5.1 Nash equilibrium strategy and capacity trajectories

	n_0	n_1^1	n_1^2	n_2^1	n_2^2	n_2^3	n_2^4
$K_{n_t}^1$	6	6.2508	6.2508	6.9704	6.9704	5.5526	5.5526
$q_{n_t}^1$	6	6.2508	4.4124	6.9704	5.8832	5.5526	4.1673
$I_{n_t}^1$	1.1508	1.6572	0.2394				4.6799
$K_{n_t}^2$	4	4.6799	4.6799	5.5506	5.5506	4.2449	4.2449
$q_{n_t}^2$	4	4.6799	3.3093	5.5506	4.4124	4.2449	3.1255
$I_{n_t}^2$	1.2799	1.5727	0.2670				
$K_{n_t}^3$	2	2.9635	2.9635	4.0361	4.0361	2.8679	2.8679
$q_{n_t}^3$	2	2.9635	2.3360	4.0361	3.1146	2.8679	2.2062
$I_{n_t}^3$	1.2635	1.5172	0.3490				
E_{n_t}	12	21.0942	17.2577	29.2136	26.0667	23.0200	19.8535

Table 5.2 Nash equilibrium profits along each path in the event tree

	$n_0 \rightsquigarrow n_2^1$	$n_0 \rightsquigarrow n_2^2$	$n_0 \rightsquigarrow n_2^3$	$n_0 \rightsquigarrow n_2^4$
Player 1	964.2394	803.2444	591.4841	507.4920
Player 2	665.9674	540.2022	392.2990	332.9681
Player 3	404.2633	317.4354	227.5067	190.8109

Table 5.3 Nash equilibrium production along each path in the event tree

	$n_0 \rightsquigarrow n_2^1$	$n_0 \rightsquigarrow n_2^2$	$n_0 \rightsquigarrow n_2^3$	$n_0 \rightsquigarrow n_2^4$
Player 1	19.2212	18.1340	15.9650	14.5797
Player 2	14.2305	13.0923	11.5542	10.4348
Player 3	8.9996	8.0781	7.2039	6.5422

Player 1 produces (and pollutes) the most, followed by Player 2, and finally Player 3. The profits follow the same order. Table 5.3 illustrates the Nash equilibrium production along each path in the event tree.

Coupled-constraint equilibrium: In this scenario, the regulator imposes the joint constraint in (5.37). Although a coupled-constraint Nash equilibrium can be computed for any vector of weights, only one will be implemented in practice. In the theoretical developments of the previous section, we had a coupled constraint at each node, whereas here only terminal nodes are concerned. To select a weight vector, let us suppose that the regulator adopts the following approach:

1. Compute the cumulative emissions by each player in the Nash equilibrium benchmark to obtain $(\hat{E}_1, \hat{E}_2, \hat{E}_3) = (16.1405, 11.6835, 7.3196)$, and $\hat{E}_1 + \hat{E}_2 + \hat{E}_3 = 35.1436$. Denote by $s = (s_1, s_2, s_3)$ the vector of players' shares in this total, that is, $s = (0.4593, 0.3325, 0.2083)$.

Table 5.4 Couple-constrained equilibrium strategy and capacity trajectories

	n_0	n_1^1	n_1^2	n_2^1	n_2^2	n_2^3	n_2^4
$K_{n_t}^1$	6	5.2047	5.2047	4.5608	4.5608	4.5608	4.5608
$q_{n_t}^1$	4.8586	2.2661	2.8444	0	0.5809	1.1238	1.9292
$I_{n_t}^1$	0.1047	0.1368	0.1368				
$K_{n_t}^2$	4	3.5311	3.5311	3.1727	3.1727	3.1727	3.1727
$q_{n_t}^2$	4	3.4756	2.7102	2.4554	2.3864	2.5234	2.2703
$I_{n_t}^2$	0.1311	0.1713	0.1713				
$K_{n_t}^3$	2	2.7431	2.7431	3.5446	3.5446	2.7239	2.7239
$q_{n_t}^3$	2	2.7431	2.3118	3.5446	3.0327	2.7239	2.1716
$I_{n_t}^3$	1.0431	1.2130	0.3923				
E_{n_t}	10.8586	15	14.3815	15	15	15	15

Table 5.5 Coupled-constraint profits along each path in the event tree

	$n_0 \rightsquigarrow n_2^1$	$n_0 \rightsquigarrow n_2^2$	$n_0 \rightsquigarrow n_2^3$	$n_0 \rightsquigarrow n_2^4$
Player 1	463.3986	505.1992	462.4198	465.7968
Player 2	716.5730	632.9344	460.2825	391.5093
Player 3	616.3066	480.5079	329.6610	252.8356

2. Recall that the individual and common Lagrange multipliers are related by $\mu_{n_T}^i = \frac{1}{r_i}\mu_{n_T}^0$. Let the vector of weights be defined by

$$\left(\frac{1}{r_1}, \frac{1}{r_2}, \frac{1}{r_3}\right) = (0.4593, \ 0.3325, \ 0.2083),$$

which means that the responsibility of each player in satisfying the common constraint is proportional to her expected cumulative emissions in the benchmark scenario.

With this vector of weights, we obtain the coupled-constraint equilibrium strategies and production capacities reported in Table 5.4, and the corresponding profits in Table 5.5. Note that Player 1's profits in all scenarios are considerably lower than what she would get if there was no environmental constraint. On the other hand, Players 2 and 3 obtain a higher profit in the coupled-constraint equilibrium than in the benchmark equilibrium. Table 5.6 illustrates the coupled-constraint production along each path in the event tree.

Table 5.6 Coupled-constraint production along each path in the event tree

	$n_0 \rightsquigarrow n_2^1$	$n_0 \rightsquigarrow n_2^2$	$n_0 \rightsquigarrow n_2^3$	$n_0 \rightsquigarrow n_2^4$
Player 1	7.1248	7.7057	8.8268	9.6322
Player 2	9.9310	9.8620	9.2335	8.9804
Player 3	8.2877	7.7758	7.0357	6.4834

5.10 DGPETs with Random Termination

So far, the assumption has been that the terminal date of the DGPET is known with certainty. In this section, we extend the framework to the case where the game may terminate at any intermediate node of the event tree. Consequently, the stochastic process is defined by two probabilities for any node $n_t \in \mathbf{n}_t$, $t \in \mathbb{T}$, namely, a termination probability $\tau_{n_t} \in [0, 1]$, with $\tau_{n_0} = 0$, and as before, a probability π_{n_t} of passing through node n_t, which corresponds to the sum of the probabilities of all scenarios containing this node.[6] If the stochastic process does not terminate at node n_t, it transits to one of the successor nodes in $\mathbf{n}_t^+$. In particular, $\pi_{n_0} = 1$, and π_{n_T} is equal to the probability of the single scenario that terminates in (leaf) node $n_T \in \mathbf{n}_T$. Notice that $\tau_{n_T} = \pi_{n_T}$ for any $n_T \in \mathbf{n}_T$; therefore, it is certain that the process will terminate in the nodes from $\mathbf{n}_T$. The given probabilities satisfy the following condition:

$$\pi_{n_t} = \tau_{n_t} + \sum_{n_{t+1} \in \mathbf{n}_t^+} \pi_{n_{t+1}}, \tag{5.38}$$

that is, the probability of passing through node n_t is equal to the sum of the probability of the game terminating at this node and the sum of the probabilities of passing through the direct successors of this node.

The control and state variables and the sets are as defined in Sect. 5.3. At a node $n_t \in \mathbf{n}_t$, $t = 0, \ldots, T - 1$, the reward to Player i is $\phi_i(n_t, x_{n_t}, u_{n_t})$ if the game does not terminate at this node; otherwise, it is $\Phi_i(n_t, x_{n_t})$. In particular, at $t = T$, the reward to Player i is $\Phi_i(n_T, x_{n_T})$. The information that a node n_t is terminal is revealed before the players choose their controls at that node, but not at n_t^-. Observe that, in any nonterminal node, the reward to a player depends on the state and the controls of all players.

The reward function $J_i(\mathbf{x}, \mathbf{u})$ to Player $i \in M$ is

$$J_i(\mathbf{x}, \mathbf{u}) = \sum_{t=0}^{T-1} \rho^t \sum_{n_t \in \mathbf{n}_t} (\pi_{n_t} - \tau_{n_t}) \phi_i(n_t, x_{n_t}, u_{n_t})$$

$$+ \sum_{t=1}^{T} \rho^t \sum_{n_t \in \mathbf{n}_t} \tau_{n_t} \Phi_i(n_t, x_{n_t}). \tag{5.39}$$

---

[6] By setting probabilities τ_{n_t} for any $n_t \in \mathbf{n}_t$ for all $t = 0, \ldots, T$ we determine the probability distribution of the process termination over the set of nodes. For the distribution, it holds that $\sum_{\nu \in \mathbf{n}_0^{++}} \tau_\nu = 1$.

The game in normal form, with the set of players M, is defined by the payoff functions $W_i(x_0, \mathbf{u}) = J_i(\mathbf{x}, \mathbf{u})$, $i \in M$, where $\mathbf{x}$ is obtained from $\mathbf{u}$ as the unique solution of the state equations (5.1) emanating from the initial state x_0.

5.10.1   Necessary Conditions

To formulate the necessary conditions for the S-adapted equilibrium in the form of Pontryagin's maximum principle, we rewrite the payoff of Player $i \in M$ in the following form:

$$
\begin{aligned}
J_i(\mathbf{x}, \mathbf{u}) = {} & \phi_i(n_0, x_0, u_{n_0}) \\
& + \sum_{t=1}^{T-1} \rho^t \sum_{n_t \in \mathbf{n}_t} \left\{ (\pi_{n_t} - \tau_{n_t}) \phi_i(n_t, x_{n_t}, u_{n_t}) + \tau_{n_t} \Phi_i(n_t, x_{n_t}) \right\} \\
& + \rho^T \sum_{n_T \in \mathbf{n}_T} \tau_{n_T} \Phi_i(n_T, x_{n_T}),
\end{aligned}
$$

where $\tau_{n_T} = \pi_{n_T}$ for any $n_T \in \mathbf{n}_T$.

For each player $i \in M$, we form the Lagrangian

$$
\begin{aligned}
\mathcal{L}_i = {} & \phi_i(n_0, x_0, u_{n_0}) + \sum_{t=1}^{T-1} \rho^t \sum_{n_t \in \mathbf{n}_t} \left\{ (\pi_{n_t} - \tau_{n_t}) \phi_i(n_t, x_{n_t}, u_{n_t}) + \tau_{n_t} \Phi_i(n_t, x_{n_t}) \right. \\
& \hspace{8cm} (5.40) \\
& \left. + \pi_{n_t} (\lambda^i_{n_t})' \left(f^{n_t^-}(x_{n_t^-}, u_{n_t^-}) - x_{n_t} \right) \right\} \\
& + \rho^T \sum_{n_T \in \mathbf{n}_T} \tau_{n_T} \left\{ \Phi_i(n_T, x_{n_T}) + (\lambda^i_{n_T})' \left(f^{n_T^-}(x_{n_T^-}, u_{n_T^-}) - x_{n_T} \right) \right\} \\
= {} & \phi_i(n_0, x_0, u_{n_0}) + \sum_{t=1}^{T-1} \rho^t \sum_{n_t \in \mathbf{n}_t} \pi_{n_t} \left\{ \frac{\pi_{n_t} - \tau_{n_t}}{\pi_{n_t}} \phi_i(n_t, x_{n_t}, u_{n_t}) + \frac{\tau_{n_t}}{\pi_{n_t}} \Phi_i(n_t, x_{n_t}) \right. \\
& \left. + (\lambda^i_{n_t})' \left(f^{n_t^-}(x_{n_t^-}, u_{n_t^-}) - x_{n_t} \right) \right\} \\
& + \rho^T \sum_{n_T \in \mathbf{n}_T} \pi_{n_T} \left\{ \Phi_i(n_T, x_{n_T}) + (\lambda^i_{n_T})' \left(f^{n_T^-}(x_{n_T^-}, u_{n_T^-}) - x_{n_T} \right) \right\},
\end{aligned}
$$

where $\pi_{n_T} = \tau_{n_T}$ for any $n_T \in \mathbf{n}_T$. For each player $i \in M$, we introduce a costate variable $\lambda^i_{n_t}$ indexed over the set of nodes of period $t = 1, \ldots, T$. The costate variable $\lambda^i_{n_t}$ has the same dimension as x_{n_t}. Define the Hamiltonian function for any player $i \in M$ and any node $n_t \in \mathbf{n}_t$, $t = 0, \ldots, T - 1$ by

$$H_i\left(n_t, \lambda^i_{\mathbf{n}^+_t}, x_{n_t}, u_{n_t}\right) = \frac{\pi_{n_t} - \tau_{n_t}}{\pi_{n_t}} \phi_i(n_t, x_{n_t}, u_{n_t}) + \frac{\tau_{n_t}}{\pi_{n_t}} \Phi_i(n_t, x_{n_t}) \qquad (5.41)$$

$$+ \rho \sum_{\nu \in \mathbf{n}^+_t} \pi^\nu_{n_t} \lambda^i_\nu f^{n_t}(x_{n_t}, u_{n_t}),$$

where $\lambda^i_{\mathbf{n}^+_t}$ is the collection of $\lambda_i(\nu)$ with $\nu \in \mathbf{n}^+_t$.

Theorem 5.7 *Let $\tilde{\mathbf{u}}$ be an S-adapted Nash equilibrium in the game played over an event tree with a random termination time and with an initial state x_0, generating the state trajectory $\tilde{\mathbf{x}} = \{\tilde{x}_{n_t} : n_t \in \mathbf{n}_t, t = 1, \ldots, T\}$. Then, for each player $i \in M$ there exist costate variables $\lambda^i_{n_t}, n_t \in \mathbf{n}_t, t = 1, \ldots, T$ such that the following conditions hold for any $i \in M$, and $u^i_{n_t} = \tilde{u}^i_{n_t}, x_{n_t} = \tilde{x}_{n_t}$:*

$$0 = \frac{\partial H_i\left(n_t, \lambda^i_{\mathbf{n}^+_t}, x_{n_t}, u_{n_t}\right)}{\partial u^i_{n_t}}, \quad n_t \in \mathbf{n}_t, \ t = 0, \ldots, T-1, \qquad (5.42)$$

$$\lambda^i_{n_t} = \frac{\partial H_i\left(n_t, \lambda^i_{\mathbf{n}^+_t}, x_{n_t}, u_{n_t}\right)}{\partial x_{n_t}}, \quad n_t \in \mathbf{n}_t, \ t = 1, \ldots, T-1, \qquad (5.43)$$

$$\lambda^i_{n_T} = \frac{\partial \Phi_i(n_T, x_{n_T})}{\partial x_{n_T}}, \quad x_{n_T} \in \mathbf{n}_T. \qquad (5.44)$$

Proof Consider the expression of Lagrangian (5.40) and rewrite it in the following form:

$$\mathcal{L}_i = \sum_{t=0}^{T-1} \rho^t \sum_{n_t \in \mathbf{n}_t} \pi_{n_t} \left\{ H_i\left(n_t, \lambda^i_{\mathbf{n}^+_t}, x_{n_t}, u_{n_t}\right) - (\lambda^i_{n_t})' x_{n_t} \right\}$$

$$+ \rho^T \sum_{n_T \in \mathbf{n}_T} \pi_{n_T} \left\{ \Phi_i(n_T, x_{n_T}) - (\lambda^i_{n_T})' x_{n_T} \right\}.$$

Taking into account (5.41), equations (5.42)–(5.44) are obtained by equating to zero the partial derivatives of the Lagrangian with respect to x_{n_t} and u_{n_t}. □

Remark 5.9 Identifying the set of sufficient conditions for the S-adapted equilibrium for a general class of DGPETs is in itself an interesting problem. Of course, Pontryagin's maximum principle may also give sufficient conditions for some classes of payoff functions and state equations. Here, we focus on the explicit solution of the equations that are necessary conditions for the S-adapted equilibrium, without taking into account the issue of sufficient conditions. The sufficient conditions to solve optimal control problems with or without constraints on the state set are given in Seierstad and Sydsæter (1987). This topic is also discussed by Dockner et al. (2000), with applications to management problems.

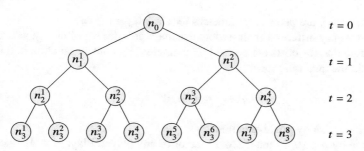

Fig. 5.4 Event tree graph

5.10.2 An Example of a DGPET with Random Termination

We illustrate our results with an example[7] of transboundary pollution control.[8] Denote by $M = \{1, 2, 3\}$ the set of players representing countries, and by $\mathbb{T} = \{0, \ldots, T\}$ the set of time periods. Production activities in each country generate revenues and, as a by-product, pollutant emissions, e.g., CO_2. Denote by $u_{n_t} = (u^1_{n_t}, u^2_{n_t}, u^3_{n_t})$ the profile of countries' emissions in node $n_t \in \mathbf{n}_t$ in time period t, and by x_{n_t} the stock of pollution at this node. The evolution of this stock is governed by the following difference equation:

$$x_{n_t} = (1 - \delta_{n_t^-}) \, x_{n_t^-} + \sum_{j \in M} u^j_{n_t^-}, \ x_{n_0} = x_0, \tag{5.45}$$

where the initial state x_0 is given, node n_t^- is a predecessor of node n_t, and $\delta_{n_t^-} \in (0, 1)$ is a rate of pollution absorption by Mother Nature at node n_t^-. We suppose that δ_{n_t} can take two possible values, that is, $\delta_{n_t} \in \{\underline{\delta}, \overline{\delta}\}$, with $\underline{\delta} < \overline{\delta}$.

The event tree is depicted in Fig. 5.4. The event tree is defined by two sets of probabilities: $\{\pi_{n_t} : n_t \in \mathbf{n}_t, t \in \mathbb{T}\}$ and $\{\tau_{n_t} : n_t \in \mathbf{n}_t, t \in \mathbb{T}\}$, such that $\pi_{n_0} = 1$, $\tau_{n_T} = \pi_{n_T}$ for any $n_T \in \mathbf{n}_T$, and $\pi_{n_t} = \tau_{n_t} + \sum_{n_{t+1} \in \mathbf{n}_t^+} \pi_{n_{t+1}}$. Let nodes n_0, n_1^2, n_2^2, $n_2^4, n_3^2, n_3^4, n_3^6, n_3^8$ correspond to the low level of pollution reduction $\underline{\delta}$, and nodes n_1^1, $n_2^1, n_2^3, n_3^1, n_3^3, n_3^5, n_3^7$ corresponds to the high level of pollution reduction $\overline{\delta}$.

Denote by $Q^i_{n_t}$ Player i's production of goods and services at node $n_t \in \mathbf{n}_t$. The resulting emissions $u^i_{n_t}$ at node n_t are an increasing function of production, i.e., $u^i_{n_t} = h_i \left(Q^i_{n_t}\right)$ satisfying $h_i(0) = 0$. Assuming a monotone increasing relationship between production and revenues, we can express the revenues, denoted by R_i, directly as a function of emissions. We adopt the following specification:

$$R_i \left(u^i_{n_t}\right) = \alpha_i u^i_{n_t} - \frac{1}{2} \beta_i \left(u^i_{n_t}\right)^2, \tag{5.46}$$

---

[7] The cooperative version of this game is examined in Sect. 6.3.2.

[8] For a background on this class of games, see the survey in Jørgensen et al. (2010).

where α_i and β_i are positive parameters for any player $i \in M$.

Each country suffers an environmental damage cost due to pollution accumulation. We assume that the costs are defined by an increasing convex function in this stock and retain the quadratic form

$$D_i(x_{n_t}) = c_i(x_{n_t})^2, \ i \in M,$$

where c_i is a strictly positive parameter.

Denote by $\rho \in (0, 1)$ the discount factor of any player. Player $i \in M$ maximizes the following objective functional:

$$J_i(\mathbf{x}, \mathbf{u}) = \sum_{t=0}^{T-1} \rho^t \sum_{n_t \in \mathbf{n}_t} (\pi_{n_t} - \tau_{n_t}) \left(R_i\left(u_{n_t}^i\right) - D_i(x_{n_t})\right) + \sum_{t=1}^{T} \rho^t \sum_{n_t \in \mathbf{n}_t} \tau_{n_t} \Phi_i(x_{n_t})$$

$$(5.47)$$

subject to (5.45), and $u_{n_t}^i \geq 0$ for all $i \in M$ and any node $n_t \in \mathbf{n}_t$, $t \in \mathbb{T} \setminus \{T\}$, where $\mathbf{x} = \{x_{n_t} : n_t \in \mathbf{n}_t, \ t \in \mathbb{T}\}$ is a state trajectory.

Let the payoff function of Player i at the terminal node $n_T \in \mathbf{n}_T$ be

$$\Phi_i(x_{n_T}) = -d_i x_{n_T},$$

where $d_i > 0$ for any $i \in M$.

Substituting $\Phi_i(x_{n_T})$, $R_i\left(u_{n_t}\right)$, and $D_i(x_{n_t})$ by their values, we get

$$J_i(\mathbf{x}, \mathbf{u}) = \sum_{t=0}^{T-1} \rho^t \sum_{n_t \in \mathbf{n}_t} (\pi_{n_t} - \tau_{n_t}) \left(\alpha_i u_{n_t}^i - \frac{1}{2}\beta_i\left(u_{n_t}^i\right)^2 - c_i(x_{n_t})^2\right)$$

$$- \sum_{t=1}^{T} \rho^t \sum_{n_t \in \mathbf{n}_t} \tau_{n_t} d_i x_{n_T}.$$

Using Theorem 5.7, we find the Nash equilibrium in the game with random termination. The Hamiltonian function for Player i is as follows:

$$H_i\left(n_t, \lambda_{\mathbf{n}_t^+}^i, x_{n_t}, u_{n_t}\right) = \frac{\pi_{n_t} - \tau_{n_t}}{\pi_{n_t}} \left(\alpha_i u_{n_t}^i - \frac{1}{2}\beta_i\left(u_{n_t}^i\right)^2 - c_i(x_{n_t})^2\right) - \frac{\tau_{n_t}}{\pi_{n_t}} d_i x_{n_t}$$

$$+ \rho \sum_{\nu \in \mathbf{n}_t^+} \frac{\pi_\nu}{\pi_{n_t}} \lambda_\nu^i \left[(1 - \delta_{n_t})x_{n_t} + \sum_{j \in M} u_{n_t}^j\right].$$

The necessary conditions are given by the system

$$0 = \frac{\partial H_i\left(n_t, \lambda^i_{\mathbf{n}_t^+}, x_{n_t}, u_{n_t}\right)}{\partial u^i_{n_t}} = \frac{\pi_{n_t} - \tau_{n_t}}{\pi_{n_t}} \left(\alpha_i - \beta_i u^i_{n_t}\right) + \rho \sum_{\nu \in \mathbf{n}_t^+} \pi^\nu_{n_t} \lambda^i_\nu,$$

$$n_t \in \mathbf{n}_t,\ t = 0, 1, 2,$$

$$\lambda^i_{n_t} = \frac{\partial H_i\left(n_t, \lambda^i_{\mathbf{n}_t^+}, x_{n_t}, u_{n_t}\right)}{\partial x_{n_t}} = \frac{\pi_{n_t} - \tau_{n_t}}{\pi_{n_t}} \left(-2c_i x_{n_t}\right) - \frac{\tau_{n_t}}{\pi_{n_t}} d_i$$

$$+ \rho(1 - \delta_{n_t}) \sum_{\nu \in \mathbf{n}_t^+} \pi^\nu_{n_t} \lambda^i_\nu,\ n_t \in \mathbf{n}_t,\ t = 1, 2,$$

$$\lambda^i_{n_T} = \frac{\partial \Phi_i(n_T, x_{n_T})}{\partial x_{n_T}} = -d_i,\ x_{n_T} \in \mathbf{n}_T.$$

We use the following parameter values in the numerical illustration:

$$x_0 = 0,\ \rho = 0.9,\ \underline{\delta} = 0.4,\ \overline{\delta} = 0.8,$$
$$\alpha_1 = 30,\ \alpha_2 = 29,\ \alpha_3 = 28,$$
$$\beta_1 = 3,\ \beta_2 = 4,\ \beta_3 = 4.5,$$
$$c_1 = 0.15,\ c_2 = 0.10,\ c_3 = 0.05,$$
$$d_1 = 0.25,\ d_2 = 0.20,\ d_3 = 0.15.$$

The values of probabilities π_{n_t} and τ_{n_t} are given in Table 5.7.

The Nash equilibrium control trajectories and the corresponding state values are given in Table 5.8.

The players' payoffs in the Nash equilibrium are $J_1^{nc} = 212.292$, $J_2^{nc} = 156.609$, $J_3^{nc} = 157.157$.

Table 5.7 Probabilities π_{n_t} and τ_{n_t} defined on the event tree represented in Fig. 5.4

Time period	$t = 0$	$t = 1$		$t = 2$			
Node	n_0	n_1^1	n_1^2	n_2^1	n_2^2	n_2^3	n_2^4
π_{n_t}	1.00	0.30	0.70	0.20	0.05	0.30	0.30
τ_{n_t}	0.00	0.05	0.10	0.05	0.01	0.10	0.05
$t = 3$							
Node	n_3^1	n_3^2	n_3^3	n_3^4	n_3^5	n_3^6	n_3^7	n_3^8
π_{n_t}	0.05	0.10	0.01	0.03	0.10	0.10	0.15	0.10
τ_{n_t}	0.05	0.10	0.01	0.03	0.10	0.10	0.15	0.10

Table 5.8 Nash equilibrium control trajectories $u_{n_t}^{i,nc}$, $i \in M$, $n_t \in \mathbf{n}_t$, $t \in \mathbb{T} \setminus \{T\}$, and Nash equilibrium state trajectory $x_{n_t}^{nc}$, $n_t \in \mathbf{n}_t$, $t \in \mathbb{T}$

Time period	$t = 0$	$t = 1$		$t = 2$			
Node	n_0	n_1^1	n_1^2	n_2^1	n_2^2	n_2^3	n_2^4
$u_{n_t}^{1,nc}$	7.7169	8.2961	7.8367	9.9250	9.9250	9.9250	9.9250
$u_{n_t}^{2,nc}$	6.1059	6.3948	6.1643	7.2050	7.2050	7.2050	7.2050
$u_{n_t}^{3,nc}$	5.7103	5.8378	5.7344	6.1922	6.1922	6.1922	6.1922
$x_{n_t}^{nc}$	0.0000	19.533	19.533	24.4352	24.4352	31.4551	31.4551

$t = 3$							
Node	n_3^1	n_3^2	n_3^3	n_3^4	n_3^5	n_3^6	n_3^7	n_3^8
$x_{n_t}^{nc}$	28.2093	28.2093	37.9833	37.9833	29.6132	29.6132	42.1953	42.1953

5.11 Additional Readings

DGPETs generalize the class of multistage games to a stochastic setup, where the randomness is described by an event tree. For an introduction (and also some advanced material) on multistage games, see, e.g., Başar and Olsder (1999), Haurie et al. (2012), and Krawczyk and Petkov (2018).

The DGPET class was introduced in Zaccour (1987) and Haurie et al. (1990b), and further developed in Haurie and Zaccour (2005). The initial motivation was an analysis of the European natural gas market, and more specifically, the forecasting of long-term gas deliveries from four producing countries to nine consuming European regions.

The DGPET formalism has been used to study electricity markets in, e.g., Pineau and Murto (2003), Genc et al. (2007), Genc and Sen (2008), and Pineau et al. (2011). In these papers, the main objective is to predict equilibrium investments in various generation technologies in deregulated electricity markets.

Characterization of equilibria and computational issues are considered in Haurie and Moresino (2002) and Reddy and Zaccour (2019). In Haurie and Moresino (2002), the equilibrium is characterized as a solution of a variational inequality, and an approximation method is proposed, based on the use of a random sampling of scenarios in the event tree. Reddy and Zaccour (2019) characterize open-loop and feedback-Nash equilibria in constrained linear-quadratic DGPETs.

5.12 Exercises

Exercise 5.1 Consider an industry made up of m firms competing in a two-period Cournot game. At period 0, the demand is certain. At period 1, the demand can be either high (H) with probability p, or low (L) with probability $1 - p$.

Each firm is endowed with an initial production capacity K_0^i, $i = 1, \ldots, m$. At period 1, the capacity is given by

$$K_1^i = K_0^i + I_0^i,$$

where $I_0^i \geq 0$ is the investment made by firm i at period 0. This means that it takes one period before an investment becomes productive. The investment cost of firm $i = 1, \ldots, m$ is given by the convex function

$$c_i(I_i) = \frac{1}{2}\alpha_i I_i^2,$$

where α_i is a positive parameter.

Denote by q_t^i the quantity produced by firm i at time t, and by Q_t the total quantity produced by m firms at period $t = 0, 1$. The production costs of firm $i = 1, \ldots, m$ are assumed to be quadratic and given by

$$g_i(q^i) = \frac{1}{2}\beta_i(q^i)^2,$$

where β_i is a positive parameter.

Denote by P_0 the price at period 0 and by P_1^H and P_1^L the prices at period 1 when the demand is high and low, respectively. The inverse-demand laws at the different nodes are given by

$$
\begin{aligned}
P_0 &= \max\{a_0 - b_0 Q_0; 0\}, \\
P_1^H &= \max\{A_H - B Q_0; 0\}, \\
P_1^L &= \max\{A_L - B Q_0; 0\},
\end{aligned}
$$

where a_0, b_0, A_H, A_L, B are strictly positive parameters, with $A_H > A_L$. Let each player maximizes the sum of her profits over the two periods.

1. Show the existence and uniqueness of an S-adapted Nash equilibrium for this game.
2. Find equilibrium investments and quantities for each firm when $m = 3$, $p = 0.2$, $\alpha_1 = \alpha_2 = 1.5, \alpha_3 = 0.8, \beta_1 = 0.5, \beta_2 = \beta_3 = 1$, and $a_0 = 4.5, b_0 = 0.4, A_H = 3, A_L = 2.5, B = 0.6$.
3. Is the S-adapted Nash equilibrium subgame perfect?
4. Suppose that the firms form a cartel and optimize their joint profit.

   a. Determine the optimal quantities, investments, and joint profit.
   b. Compare this solution to the S-adapted Nash equilibrium and discuss the results (in particular the impact of the cartel's formation on consumers).

Exercise 5.2 Suppose that the m firms in the above example compete à la Cournot in two different markets.

1. Extend the model to account for the additional market.
2. Write down the necessary conditions for an S-adapted Nash equilibrium.

Exercise 5.3 Consider a three-player stochastic version of the deterministic model of pollution control in Germain et al. (2003). Let $M = \{1, 2, 3\}$ be the set of players representing countries, and $\mathbb{T} = \{0, 1, 2, 3\}$ be the set of periods. Countries produce emissions of some pollutant at any time period except the terminal one, that is $u_{n_t} = (u_{n_t}^1, u_{n_t}^2, u_{n_t}^3)$, $n_t \in \mathbf{n}_t$, $t = 0, 1, 2$. The stock of pollution in node n_t at time period t is x_{n_t}. Suppose that the evolution of this is governed by the following difference equation:

$$x_{n_t} = (1 - \delta_{n_t^-})x_{n_t^-} + \sum_{i \in M} u_{n_t^-}^i, \tag{5.48}$$

$$x_{n_0} = x_0, \tag{5.49}$$

with the initial stock x_0 at root node n_0 being given, and where $\delta_{n_t} \in (0, 1)$ is the stochastic rate of pollution absorption by nature in node n_t. We suppose that, at any time period $t = 1, 2, 3$, the parameter δ_{n_t} can take two possible values, that is, $\delta_{n_t} \in \{\underline{\delta}, \overline{\delta}\}$, with $\underline{\delta} < \overline{\delta}$. At node n_0, $\delta_{n_0} = \overline{\delta}$.

The damage cost is an increasing convex function in the pollution stock and has the quadratic form $D_i(x_{n_t}) = \alpha_i (x_{n_t})^2$, $i \in M$, where α_i is a strictly positive parameter. The cost of emissions is also given by a quadratic function $C_i(u_{n_t}^i) = \frac{\gamma_i}{2}(u_{n_t}^i - e)^2$, where e and γ_i are strictly positive constants.

1. Let the event tree be depicted in Fig. 5.5, where nodes n_1^2, n_2^3, n_2^4 correspond to the low level of pollution reduction $\underline{\delta}$, and nodes n_0, n_1^1, n_2^1, n_2^3 correspond to the high level of pollution reduction $\overline{\delta}$. Write the payoff function of each player and the necessary conditions of the S-adapted Nash equilibrium from Theorem 5.3.
2. Find the S-adapted Nash equilibrium using the following parameters:

$$\alpha_1 = 0.1, \ \alpha_2 = 0.2, \ \alpha_3 = 0.3,$$
$$\gamma_1 = 0.9, \ \gamma_2 = 0.8, \ \gamma_3 = 0.7,$$
$$\underline{\delta} = 0.45, \ \overline{\delta} = 0.8, \ e = 30, \ \rho = 0.9,$$
$$\pi_{n_1^1} = 0.6, \ \pi_{n_1^2} = 0.4,$$
$$\pi_{n_2^1} = 0.3, \ \pi_{n_2^2} = 0.3, \ \pi_{n_2^3} = 0.3, \ \pi_{n_2^4} = 0.1,$$
$$\pi_{n_3^1} = 0.1, \ \pi_{n_3^2} = 0.2, \ \pi_{n_3^3} = 0.1, \ \pi_{n_3^4} = 0.2,$$
$$\pi_{n_3^5} = 0.05, \ \pi_{n_3^6} = 0.25, \ \pi_{n_3^7} = 0.05, \ \pi_{n_3^8} = 0.05,$$

where ρ is the common discount factor.
3. Is the S-adapted Nash equilibrium unique?
4. Does the equilibrium satisfy the properties of time consistency and subgame perfectness?

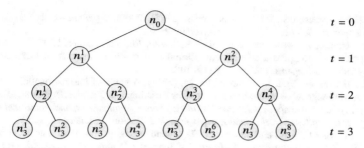

Fig. 5.5 Event tree for $T = 3$.

Exercise 5.4 Consider example in Sect. 5.9.

1. Compute the Nash equilibrium and the coupled-constraint equilibrium when the players also invest in their production capacities in period 2.
2. In the given solution of the coupled-constraint scenario, the regulator assigns different weights to the three players. How would the results change if the regulator treats in the same way the players, i.e., $r_1 = r_2 = r_3$? Would it be fair to proceed in this manner?

Exercise 5.5 Redo example in Sect. 5.7 assuming that there is a one-in-three probability of the game terminating in each of the nodes in period 1.

References

Başar, T. and Olsder, G. (1999). *Dynamic Noncooperative Game Theory: Second Edition*. Classics in Applied Mathematics. Society for Industrial and Applied Mathematics.

Carlson, D. (2002). Uniqueness of normalized Nash equilibrium for a class of games with strategies in banach spaces. In Zaccour, G., editor, *Decision & Control in Management Science: Essays in Honor of Alain Haurie*, pages 333–348, Boston, MA. Springer US.

Carlson, D. and Haurie, A. (1996). A turnpike theory for infinite-horizon open-loop competitive processes. *SIAM Journal on Control and Optimization*, 34(4):1405–1419.

Carlson, D., Haurie, A., and Zaccour, G. (2018). Infinite horizon concave games with coupled constraints. In Başar, T. and Zaccour, G., editors, *Handbook of Dynamic Game Theory*, pages 111–155, Cham. In: Başar T., Zaccour G. (eds) Handbook of Dynamic Game Theory. Springer International Publishing.

Carlson, D. and Haurie, A. B. (2000). Infinite horizon dynamic games with coupled state constraints. In Filar, J. A., Gaitsgory, V., and Mizukami, K., editors, *Advances in Dynamic Games and Applications*, pages 195–212, Boston, MA. Birkhäuser Boston.

Contreras, J., Klusch, M., and Krawczyk, J. (2004). Numerical solutions to Nash-Cournot equilibria in coupled constraint electricity markets. *IEEE Transactions on Power Systems*, 19(1):195–206.

Contreras, J., Krawczyk, J. B., and Zuccollo, J. (2007). Electricity market games with constraints on transmission capacity and emissions. In *30th Conference of the International Association for Energy Economics*.

Dockner, E. J., Jørgensen, S., Van Long, N., and Sorger, G. (2000). *Differential Games in Economics and Management Science*. Cambridge University Press.

Drouet, L., Haurie, A., Moresino, F., Vial, J.-P., Vielle, M., and Viguier, L. (2008). An oracle based method to compute a coupled equilibrium in a model of international climate policy. *Computational Management Science*, 5:119–140.

Drouet, L., Haurie, A., Vial, J.-P., and Vielle, M. (2011). A game of international climate policy solved by a homogeneous oracle-based method for variational inequalities. In Breton, M. and Szajowski, K., editors, *Advances in Dynamic Games: Theory, Applications, and Numerical Methods for Differential and Stochastic Games*, pages 469–488, Boston. Birkhäuser Boston.

Genc, T. S., Reynolds, S. S., and Sen, S. (2007). Dynamic oligopolistic games under uncertainty: A stochastic programming approach. *Journal of Economic Dynamics and Control*, 31(1):55–80.

Genc, T. S. and Sen, S. (2008). An analysis of capacity and price trajectories for the ontario electricity market using dynamic Nash equilibrium under uncertainty. *Energy Economics*, 30(1):173–191.

Germain, M., Toint, P., Tulkens, H., and de Zeeuw, A. (2003). Transfers to sustain dynamic core-theoretic cooperation in international stock pollutant control. *Journal of Economic Dynamics and Control*, 28(1):79–99.

Gürkan, G., Özge, A. Y., and Robinson, S. M. (1999). Sample-path solution of stochastic variational inequalities. *Mathematical Programming*, 84(2):313–333.

Haurie, A. (1995). Environmental coordination in dynamic oligopolistic markets. *Group Decision and Negotiation*, 4(1):39–57.

Haurie, A. and Krawczyk, J. B. (1997). Optimal charges on river effluent from lumped and distributed sources. *Environmental Modeling & Assessment*, 2(3):177–189.

Haurie, A., Krawczyk, J. B., and Zaccour, G. (2012). *Games and Dynamic Games*. World Scientific, Singapore.

Haurie, A. and Moresino, F. (2002). S-adapted oligopoly equilibria and approximations in stochastic variational inequalities. *Annals of Operations Research*, 114(1):183–201.

Haurie, A. and Zaccour, G. (1995). Differential game models of global environmental management. In Carraro, C. and Filar, J. A., editors, *Control and Game-Theoretic Models of the Environment*, pages 3–23, Boston, MA. Birkhäuser Boston.

Haurie, A. and Zaccour, G. (2005). *S-Adapted Equilibria in Games Played over Event Trees: An Overview*, pages 417–444. Birkhäuser Boston, Boston, MA.

Haurie, A., Zaccour, G., and Smeers, Y. (1990a). Stochastic equilibrium programming for dynamic oligopolistic markets. *Journal of Optimization Theory and Applications*, 66(2):243–253.

Haurie, A., Zaccour, G., and Smeers, Y. (1990b). Stochastic equilibrium programming for dynamic oligopolistic markets. *Journal of Optimization Theory and applications*, 66(2):243–253.

Jørgensen, S., Martín-Herrán, G., and Zaccour, G. (2010). Dynamic games in the economics and management of pollution. *Environmental Modeling & Assessment*, 15:433–467.

Kakutani, S. (1941). A generalization of Brouwer's fixed point theorem. *Duke Mathematical Journal*, 8(3):457 – 459.

Krawczyk, J. (2007). Numerical solutions to coupled-constraint (or generalised Nash) equilibrium problems. *Computational Management Science*, 4(2):183–204.

Krawczyk, J. B. (2005). Coupled constraint Nash equilibria in environmental games. *Resource and Energy Economics*, 27(2):157–181.

Krawczyk, J. B. and Petkov, V. (2018). *Multistage Games*, pages 157–213. In: Başar T., Zaccour G. (eds) Handbook of Dynamic Game Theory. Springer International Publishing, Cham.

Krawczyk, J. B. and Uryasev, S. (2000). Relaxation algorithms to find Nash equilibria with economic applications. *Environmental Modeling & Assessment*, 5(1):63–73.

Pang, J.-S. and Fukushima, M. (2005). Quasi-variational inequalities, generalized Nash equilibria, and multi-leader-follower games. *Computational Management Science*, 2(1):21–56.

Pineau, P.-O. and Murto, P. (2003). An Oligopolistic Investment Model of the Finnish Electricity Market. *Annals of Operations Research*, 121(1):123–148.

Pineau, P.-O., Rasata, H., and Zaccour, G. (2011). Impact of some parameters on investments in oligopolistic electricity markets. *European Journal of Operational Research*, 213(1):180–195.

Pontryagin, L. S., Boltyanskii, V. G., Gamkrelidze, R. V., and Mishechenko, E. F. (1962). *The Mathematical Theory of Optimal Processes*. Interscience publishers. Interscience Publishers.

Reddy, P. V. and Zaccour, G. (2019). Open-loop and feedback Nash equilibria in constrained linear–quadratic dynamic games played over event trees. *Automatica*, 107:162–174.

Rosen, J. B. (1965). Existence and uniqueness of equilibrium points for concave n-person games. *Econometrica*, 33(3):520–534.

Seierstad, A. and Sydsæter, K. (1987). *Optimal Control Theory with Economic Applications*. Advanced Textbooks in Economics. Elsevier Science.

Sethi, S. and Thompson, G. (2000). *Optimal Control Theory: Applications to Management Science and Economics*. Springer Nature Book Archives Millennium. Springer.

Tidball, M. and Zaccour, G. (2005). An environmental game with coupling constraints. *Environmental Modeling & Assessment*, 10(2):153–158.

Tidball, M. and Zaccour, G. (2009). A differential environmental game with coupling constraints. *Optimal Control Applications and Methods*, 30(2):197–207.

Zaccour, G. (1987). *Théorie des jeux et marchés énergétiques: marché européen du gaz naturel et échanges d'électricité*. PhD thesis, École des hautes études commerciales, Montréal, Canada.

Chapter 6
Node-Consistent Single-Valued Solutions in DGPETs

In this chapter, we examine the problem of the sustainability of cooperative solutions in dynamic games played over event trees (DGPETs). We introduce a cooperative version of the noncooperative DGPET defined in Chap. 5. We describe the problem of node inconsistency of cooperative solutions, which is a key issue in sustainable cooperation. We propose a procedure to define a node-consistent Shapley value and a Nash bargaining solution in DGPETs with a prescribed duration and with a random terminal time.

6.1 Cooperative DGPETs

We recall the ingredients of a DGPET, as introduced in Chap. 5. The set of players is $M = \{1, \ldots, m\}$, and each player $i \in M$ chooses a control $u_{n_t}^i \in U_{n_t}^i$ at node $n_t \in \mathbf{n}_t$, $t \in \mathbb{T} \setminus \{T\}$. The state equations are given as

$$x_{n_t} = f^{n_t^-}\left(x_{n_t^-}, u_{n_t^-}\right), \quad x_{n_0} = x_0, \tag{6.1}$$

where $u_{n_t^-} \in U_{n_t^-}$, $n_t \in \mathbf{n}_t$, $t \in \mathbb{T} \setminus \{0\}$, and x_0 is the initial state at root node n_0.

A payoff functional for Player $i \in M$ is

$$J_i(\mathbf{x}, \mathbf{u}) = \sum_{t=0}^{T-1} \rho^t \sum_{n_t \in \mathbf{n}_t} \pi_{n_t} \phi_i(n_t, x_{n_t}, u_{n_t}) + \rho^T \sum_{n_T \in \mathbf{n}_T} \pi_{n_T} \Phi_i(n_T, x_{n_T}). \tag{6.2}$$

Profile $\mathbf{u} = (\mathbf{u}^i : i \in M)$ is the S-adapted strategy vector of m players. The game in normal form is defined by the payoffs $W_i(x_0, \mathbf{u}) = J_i(\mathbf{x}, \mathbf{u})$, $i \in M$, where $\mathbf{x}$ is obtained from $\mathbf{u}$ as the unique solution of the state equations (6.1) that emanate from the initial state x_0. Profile $\mathbf{u}^{nc} = (\mathbf{u}^{i,nc} : i \in M)$ is the game's Nash equilibrium in open-loop S-adapted strategies.

Suppose that the players agree to cooperate by jointly choosing their strategies in any node to maximize their total discounted payoffs over the entire horizon. Then, they solve the following optimization problem:

$$\max_{\mathbf{u}^i, i \in M} \sum_{i \in M} W_i(x_0, \mathbf{u}). \tag{6.3}$$

Denote by $\mathbf{u}^*$ the resulting vector of cooperative controls, i.e.,

$$\mathbf{u}^* = \arg \max_{\mathbf{u}^i, i \in M} \sum_{i \in M} W_i(x_0, \mathbf{u}). \tag{6.4}$$

We denote by $\mathbf{x}^* = \{x_{n_\tau}^* : n_\tau \in \mathbf{n}_0^{++}\}$ the cooperative state trajectory generated by $\mathbf{u}^*$. In the sequel, we distinguish between the cooperative and noncooperative strategies and outcomes realized in different subgames along the cooperative state trajectory $\mathbf{x}^*$.

Remark 6.1 In the computations to follow, we will suppose that the joint-optimization solution and the Nash equilibrium in any subgame are unique. The uniqueness of the S-adapted open-loop Nash equilibrium is discussed in Chap. 4. If the joint-optimization solution is not unique, we assume that the players will select an optimal solution and a Nash equilibrium.[1]

In Chap. 2, we introduced the concept of the characteristic function (CF) and the different approaches to computing its values. In this chapter, we use the γ CF approach,[2] which states that the value of a coalition K corresponds to its Nash equilibrium outcome in the noncooperative game between coalition K (acting as a single player) and the remaining players acting individually. In a DGPET, the equilibrium solution is the open-loop S-adapted Nash equilibrium, and the characteristic-function value must be computed for all coalitions in each node, assuming that cooperation has prevailed till that node.

Denote by $\Gamma(M, v, x_{n_t}^*)$ a cooperative game starting from node n_t and state $x_{n_t}^*$ where

1. $M = \{1, \ldots, m\}$ is the set of players;
2. $v(K; x_{n_t}^*)$ is the *characteristic function* (CF) defined by

$$v(\cdot; x_{n_t}^*) : 2^M \to \mathbb{R}, \tag{6.5}$$

with $v(\varnothing; x_{n_t}^*) = 0$ for any $n_t \in \mathbf{n}_0^{++}$.

The solution of a cooperative game with transferable utility is an imputation or a subset of the imputation set, which determines the payoff to any player according

---

[1] Equilibrium selection falls outside the scope of this book.

[2] All the results in this chapter can be easily adapted to another choice of characteristic function.

to some rule. The imputation set $Y(x_{n_t}^*)$ in the subgame starting at node n_t and state $x_{n_t}^*$ is defined as follows:

$$Y(x_{n_t}^*) = \left\{ (y_1, \ldots, y_m) \mid y_i \geq v\left(\{i\}; x_{n_t}^*\right), \forall i \in M, \text{ and } \sum_{i \in M} y_i = v\left(M; x_{n_t}^*\right) \right\}.$$

6.2   Node Consistency of a Single-Valued Cooperative Solution

In this section, we consider the case when the cooperative solution of the DGPET is a unique point in the imputation set. The Shapley value, the nucleolus, and the egalitarian solution are examples of such a solution. For the sake of generality, we do not specify (at least for the moment) which single-valued solution is chosen by the players, and denote it by $\xi(x_0) = (\xi_i(x_0) : i \in M) \in Y(x_0)$. Note that this solution is determined at initial state x_0. The choice of $\xi(x_0)$ gives information about the reward that each player will get in the entire game but not its distribution over nodes. The decomposition of a cooperative solution over time and nodes is called an imputation distribution procedure (IDP).

Definition 6.1 An imputation distribution procedure of the cooperative solution $\xi(x_0) = (\xi_i(x_0) : i \in M)$ is the collection of payment functions

$$\left\{ \beta_i\left(x_{n_t}^*\right) : n_t \in \mathbf{n}_0^{++}, i \in M \right\},$$

where $\beta_i\left(x_{n_t}^*\right)$ is the payment to Player i at node n_t in state $x_{n_t}^*$, such that, for any player $i \in M$, it holds that

$$\xi_i(x_0) = \sum_{t=0}^{T} \rho^t \sum_{n_t \in \mathbf{n}_t} \pi_{n_t} \beta_i\left(x_{n_t}^*\right). \tag{6.6}$$

In general, the IDP is not unique, that is, there often exists many functions $\beta_i\left(x_{n_t}^*\right)$ satisfying (6.6). Consequently, one needs to implement a refinement procedure to make it unique; otherwise, the sharing problem would not be completely solved. In Chap. 4, we introduced the concept of a time-consistent IDP. In a DGPET, where subgames start at nodes rather than at periods, the corresponding idea is node consistency.

Suppose that the players agree on an imputation $\xi(x_0)$, and let its decomposition over nodes be given by payment functions that satisfy (6.6). Consider the cooperative subgame that starts at node n_t, with a state value $x_{n_t}^*$. Using the IDP, at this node, Player $i \in M$ would have thus far collected $\sum_{\theta=0}^{t-1} \sum_{n_\theta \in \mathbf{n}_\theta} \rho^\theta \pi_{n_\theta} \beta_i(x_{n_\theta}^*) \triangleq X_i$. Now, it makes sense that, in the subgame starting at n_t, the players implement the same solution as the one chosen initially, that is, at n_0. Consequently, they select the imputation $\xi\left(x_{n_t}^*\right) = \left(\xi_i(x_{n_t}^*) : i \in M\right)$. The total discounted expected payoff of Player i from n_t till the end of the game is then given by $\rho^t \sum_{n_t \in \mathbf{n}_t} \pi_{n_t} \xi_i\left(x_{n_t}^*\right) \triangleq Y_i$. If $X_i + Y_i$ is equal to the total reward that Player i is entitled to in the entire game, that is, $\xi_i\left(x_0\right)$, then the IDP is node consistent.

Definition 6.2 The cooperative solution $\xi\left(x_0\right)$ and its imputation distribution procedure $\left\{\beta_i\left(x_{n_t}^*\right) : n_t \in \mathbf{n}_0^{++}, i \in M\right\}$ are called node consistent at initial state x_0 if, at any node $n_t \in \mathbf{n}_t, t = 0, \ldots, T$, in state $x_{n_t}^*$, it holds that

$$\sum_{\theta=0}^{t-1} \sum_{n_\theta \in \mathbf{n}_\theta} \rho^\theta \pi_{n_\theta} \beta_i(x_{n_\theta}^*) + \rho^t \sum_{n_t \in \mathbf{n}_t} \pi_{n_t} \xi_i\left(x_{n_t}^*\right) = \xi_i\left(x_0\right), \qquad (6.7)$$

for any player $i \in M$.

The following theorem defines a specific node-consistent IDP and, by the same token, shows that such an IDP always exists.

Theorem 6.1 *Let*

$$\beta_i\left(x_{n_t}^*\right) = \xi_i\left(x_{n_t}^*\right) - \rho \sum_{n_{t+1} \in \mathbf{n}_t^+} \pi_{n_t}^{n_{t+1}} \xi_i\left(x_{n_{t+1}}^*\right), \ n_t \in \mathbf{n}_t, \ t = 0, \ldots, T-1, \quad (6.8)$$

$$\beta_i\left(x_{n_T}^*\right) = \xi_i\left(x_{n_T}^*\right), \ n_T \in \mathbf{n}_T. \qquad (6.9)$$

Then, the imputation distribution procedure $\left\{\beta_i\left(x_{n_t}^*\right) : n_t \in \mathbf{n}_0^{++}, i \in M\right\}$ *satisfies* (6.7).

Proof To prove property (6.7), we first compute the weighted sum of the imputations $\beta_i\left(x_{n_t}^*\right)$ over the set of all nodes $n_t \in \mathbf{n}_t, t = 0, \ldots, T-1$:

$$\rho^t \sum_{n_t \in \mathbf{n}_t} \pi_{n_t} \beta_i\left(x_{n_t}^*\right) = \rho^t \sum_{n_t \in \mathbf{n}_t} \pi_{n_t} \xi_i\left(x_{n_t}^*\right)$$

$$- \rho^{t+1} \sum_{n_t \in \mathbf{n}_t} \pi_{n_t} \left(\sum_{n_{t+1} \in \mathbf{n}_t^+} \pi_{n_t}^{n_{t+1}} \xi_i\left(x_{n_{t+1}}^*\right)\right)$$

$$= \rho^t \sum_{n_t \in \mathbf{n}_t} \pi_{n_t} \xi_i\left(x_{n_t}^*\right) - \rho^{t+1} \sum_{n_t \in \mathbf{n}_t} \left(\sum_{n_{t+1} \in \mathbf{n}_t^+} \pi_{n_{t+1}} \xi_i\left(x_{n_{t+1}}^*\right)\right)$$

$$= \rho^t \sum_{n_t \in \mathbf{n}_t} \pi_{n_t} \xi_i\left(x_{n_t}^*\right) - \rho^{t+1} \sum_{n_{t+1} \in \mathbf{n}_{t+1}} \pi(n_{t+1}) \xi_i\left(x_{n_{t+1}}^*\right).$$

Next, summing over $\theta = 0, \ldots, t-1$, we get

$$\sum_{\theta=0}^{t-1} \sum_{n_\theta \in \mathbf{n}_\theta} \rho^\theta \pi_{n_\theta} \beta(x_{n_\theta}^*) = \pi_{n_0} \xi_i(x_0) - \rho^t \sum_{n_t \in \mathbf{n}_t} \pi_{n_t} \xi_i\left(x_{n_t}^*\right).$$

Taking into account that $\pi_{n_0} = 1$, we get (6.7). $\qquad\square$

Equations (6.8)–(6.9) uniquely define the IDP for any single-valued cooperative solution of the DGPET. Moreover, if the payments to the players are defined by (6.8)–(6.9), the cooperative solution $\xi(x_0)$ is node consistent.

The next corollary shows that, at each node, the IDP defined in Theorem 6.1 is budget balanced, that is, the total distributed reward is equal to the total payoff realized by the players.

Proposition 6.1 *Consider the IDP defined by (6.8) and (6.9). Then, for any nonterminal node $n_t \in \mathbf{n}_0^{++} \setminus \mathbf{n}_T$, we have*

$$\sum_{i \in M} \beta_i\left(x_{n_t}^*\right) = \sum_{i \in M} \phi_i(n_t, x_{n_t}^*, u_{n_t}^*),$$

and for any terminal node $n_T \in \mathbf{n}_T$,

$$\sum_{i \in M} \beta_i\left(x_{n_T}^*\right) = \sum_{i \in M} \Phi_i(n_T, x_{n_T}^*).$$

Proof For a terminal node $n_T \in \mathbf{n}_T$, in the IDP defined by (6.9), we have

$$\sum_{i \in M} \beta_i\left(x_{n_T}^*\right) = \sum_{i \in M} \xi_i\left(x_{n_T}^*\right) = v(M; x_{n_T}^*) = \sum_{i \in M} \Phi_i(n_T, x_{n_T}^*).$$

First, we consider a nonterminal node $n_t \in n_T^-$ from which the terminal node is reached at the next time period. We have

$$\sum_{i \in M} \beta_i\left(x_{n_t}^*\right) = \sum_{i \in M} \xi_i\left(x_{n_t}^*\right) - \rho \sum_{n_{t+1} \in \mathbf{n}_t^+} \pi_{n_t}^{n_{t+1}} \sum_{i \in M} \xi_i(x_{n_{t+1}}^*)$$

$$= v(M; x_{n_t}^*) - \rho \sum_{n_{t+1} \in \mathbf{n}_t^+} \pi_{n_t}^{n_{t+1}} \sum_{i \in M} \Phi_i(n_{t+1}, x_{n_{t+1}}^*)$$

$$= \sum_{i \in M} \phi_i(n_t, x_{n_t}^*, u_{n_t}^*).$$

Then, we proceed backward in the same way until the root node is reached. $\qquad\square$

We can express the IDP in terms of transfers, or side payments, made by the players at any node. Denote by $\zeta_i(n_t)$ the transfer made by Player $i \in M$ at node $n_t \in \mathbf{n}_0^{++}$. Using the definition of a node-consistent distribution procedure of the cooperative solution $\xi(x_0)$, we determine transfer $\zeta_i(n_t)$ in the following way:

1. If $n_T \in \mathbf{n}_T$ is a terminal node, then

$$\zeta_i(n_T) = 0.$$

2. If $n_t \in \mathbf{n}_0^{++} \setminus \mathbf{n}_T$ is a nonterminal node, then

$$\zeta_i(n_t) = \beta_i(x_{n_t}^*) - \phi_i(n_t, x_{n_t}^*, u_{n_t}^*). \tag{6.10}$$

At a nonterminal node, the player's transfer may be negative (the player pays), positive (the player is paid), or zero where the component of the distribution procedure coincides with the value of the payoff function. At the terminal nodes, the players always have zero transfers because they do not make any decisions at these nodes.

All the results discussed in this section are valid for any single-valued cooperative solution. We illustrate them for two of these, namely, the Nash bargaining procedure and the Shapley value.

6.2.1 Node-Consistent Nash Bargaining Solution

Consider a two-player game and suppose that the Nash bargaining solution (NBS) is chosen as a solution of the cooperative game (see Chap. 2 for a discussion of the NBS).

To obtain the optimal collective controls in (6.4), the players solve the maximization problem in (6.3), subject to the state dynamics (6.1), with Player i's payoff function defined by (6.2). The resulting joint cooperative payoff is $\sum_{j \in \{1,2\}} W_j(x_0, \mathbf{u}^*)$. (Note that this joint payoff is equal to the value of characteristic function $v(\{1, 2\}; x_0)$.)

Let the status quo in the NBS be given by the S-adapted Nash equilibrium outcomes, that is, the pair $(W_1(x_0, \mathbf{u}^{nc}), W_2(x_0, \mathbf{u}^{nc}))$. The Nash bargaining solution of the game starting from node n_0 with state x_0 is a vector $\xi^{NB}(x_0) = (\xi_1^{NB}(x_0), \xi_2^{NB}(x_0))$, such that

$$\xi_i^{NB}(x_0) = W_i(x_0, \mathbf{u}^{nc}) + \frac{\sum_{j \in \{1,2\}} W_j(x_0, \mathbf{u}^*) - \sum_{j \in \{1,2\}} W_j(x_0, \mathbf{u}^{nc})}{2}, \quad i = 1, 2. \tag{6.11}$$

Recall that the NBS gives each player her status quo payoff plus half the dividend of cooperation, defined by the numerator of last term in the above equation.

The next step is decomposing the total individual NBS outcomes over nodes. Assuming that the players want to implement a node-consistent Nash bargaining solution, then a straightforward application of Definition 6.2 yields

$$\sum_{\theta=0}^{t-1} \sum_{n_\theta \in \mathbf{n}_\theta} \rho^\theta \pi_{n_\theta} \beta_i(x_{n_\theta}^*) + \rho^t \sum_{n_t \in \mathbf{n}_t} \pi_{n_t} \xi_i^{NB}(x_{n_t}^*) = \xi_i^{NB}(x_0), \qquad (6.12)$$

for $i = 1, 2$, where $\xi_i^{NB}(x_{n_t}^*)$ is Player i's component of the NBS in the subgame starting at node n_t, with a cooperative state $x_{n_t}^*$, and given by

$$\xi_i^{NB}(x_{n_t}^*) = W_i(x_{n_t}^*, \mathbf{u}^{nc}(n_t)) + \frac{\sum_{j \in \{1,2\}} W_j(x_{n_t}^*, \mathbf{u}^*(n_t)) - \sum_{j \in \{1,2\}} W_j(x_{n_t}^*, \mathbf{u}^{nc}(n_t))}{2}.$$

Remark 6.2 Similarly to the general case, determining the Nash bargaining solution for a subgame starting from n_t requires the recomputation of the Nash equilibrium in that subgame, with the cooperative state value $x_{n_t}^*$ as the initial state.

The next proposition explicitly defines the function $\beta_i(x_{n_t}^*)$ for $n_t \in \mathbf{n}_0^{++}$ and $i \in \{1, 2\}$.

Proposition 6.2 *Let*

$$\beta_i(x_{n_t}^*) = \xi_i^{NB}(x_{n_t}^*) - \rho \sum_{n_{t+1} \in \mathbf{n}_t^+} \pi_{n_t}^{n_{t+1}} \xi_i^{NB}(x_{n_{t+1}}^*), \quad n_t \in \mathbf{n}_t, \ t = 0, \ldots, T-1,$$

$$(6.13)$$

$$\beta_i(x_{n_T}^*) = \xi_i^{NB}(x_{n_T}^*), \quad n_T \in \mathbf{n}_T. \qquad (6.14)$$

Then, the imputation distribution procedure $\{\beta_i(x_{n_t}^*) : n_t \in \mathbf{n}_0^{++}, i \in \{1, 2\}\}$ *and the corresponding Nash bargaining solution* $\xi^{NB}(x_0)$ *are node consistent.*

Proof Similar to the proof of Theorem 6.1. □

Finally, the transfer payment $\zeta_i(n_t)$ of Player $i = 1, 2$ is defined as follows:

1. If $n_T \in \mathbf{n}_T$ is a terminal node, then

$$\zeta_i(n_T) = 0.$$

2. If $n_t \in \mathbf{n}_0^{++} \setminus \mathbf{n}_T$ is a nonterminal node, then

$$\zeta_i(n_t) = \beta_i(x_{n_t}^*) - \phi_i(n_t, x_{n_t}^*, u_{n_t}^*),$$

where $\beta_i(x_{n_t}^*)$ satisfies Eqs. (6.13)–(6.14).

Remark 6.3 If the game is fully symmetric, then the NBS outcomes are equal, i.e., $\xi_1^{NB}(x_0) = \xi_2^{NB}(x_0)$. In this case, the NBS is inherently node consistent, and there is no need to define transfer payments between players.

Remark 6.4 The results of this section can be generalized to more than two players. It is an excellent exercise to rewrite this subsection for $m > 2$.

6.2.2 Node-Consistent Shapley Value

Suppose that the Shapley value $Sh(x_0) = (Sh_i(x_0) : i \in M)$ is chosen as a cooperative solution in the DGPET. The i's component of the Shapley value in the entire game starting at node n_0 in state x_0 is given by

$$Sh_i(x_0) = \sum_{\substack{K \subset M \\ K \ni i}} \frac{(m-k)!(k-1)!}{m!} [v(K; x_0) - v(K \setminus \{i\}; x_0)], \qquad (6.15)$$

where k is the number of players in coalition K, and

$$\sum_{i \in M} Sh_i(x_0) = v(M; x_0).$$

Similarly, the Shapley value in a subgame starting at any node n_t with cooperative state value $x_{n_t}^*$ is given by

$$Sh_i(x_{n_t}^*) = \sum_{\substack{K \subset M \\ K \ni i}} \frac{(m-k)!(k-1)!}{m!} [v(K; x_{n_t}^*) - v(K \setminus \{i\}; x_{n_t}^*)], \qquad (6.16)$$

$$\sum_{i \in M} Sh_i(x_{n_t}^*) = v(M; x_{n_t}^*).$$

If the players agree at the initial date to play cooperatively for the whole duration of the game and to allocate the grand-coalition payoff according to the Shapley value, then the *global* sharing problem is solved by using (6.15). To define a node-consistent allocation of the Shapley value, introduce the payment functions $\beta_i(x_{n_t}^*)$, $i \in M, n_t \in \mathbf{n}_0^{++}$, satisfying

$$Sh_i(x_0) = \sum_{t=0}^{T} \rho^t \sum_{n_t \in \mathbf{n}_t} \pi_{n_t} \beta_i(x_{n_t}^*).$$

Rewriting Definition 6.2 in terms of the Shapley value, we have

Definition 6.3 The Shapley value $Sh(x_0)$ and its imputation distribution procedure $\{\beta_i(x_{n_t}^*) : n_t \in \mathbf{n}_0^{++}, i \in M\}$ are node consistent if, at any node $n_t \in \mathbf{n}_t, t = 0, \ldots, T$, in state $x_{n_t}^*$, it holds that

$$\sum_{\theta=0}^{t-1} \sum_{n_\theta \in \mathbf{n}_\theta} \rho^\theta \pi_{n_\theta} \beta_i(x_{n_\theta}^*) + \rho^t \sum_{n_t \in \mathbf{n}_t} \pi_{n_t} Sh_i(x_{n_t}^*) = Sh_i(x_0)$$

for any player $i \in M$.

The next proposition explicitly defines the function $\beta_i\left(x_{n_t}^*\right)$ for $n_t \in \mathbf{n}_0^{++}$ and $i \in M$.

Proposition 6.3 *Let*

$$\beta_i\left(x_{n_t}^*\right) = Sh_i\left(x_{n_t}^*\right) - \rho \sum_{n_{t+1} \in \mathbf{n}_t^+} \pi_{n_t}^{n_{t+1}} Sh_i\left(x_{n_{t+1}}^*\right), \quad n_t \in \mathbf{n}_t, \ t = 0, \dots, T-1,$$

(6.17)

$$\beta_i\left(x_{n_T}^*\right) = Sh_i\left(x_{n_T}^*\right), \quad n_T \in \mathbf{n}_T.$$

(6.18)

Then, the Shapley value $Sh(x_0)$ and its distribution procedure $\left\{\beta_i\left(x_{n_t}^*\right) : n_t \in \mathbf{n}_0^{++}, i \in M\right\}$ are node consistent.

Proof Similar to the proof of Theorem 6.1. □

Finally, to define the transfer payment $\zeta_i(n_t)$ of Player $i \in M$ at any nonterminal node $n_t \in \mathbf{n}_0^{++} \setminus \mathbf{n}_T$, it suffices to substitute $\beta_i(x_{n_t}^*)$ from (6.17) into

$$\zeta_i(n_t) = \beta_i(x_{n_t}^*) - \phi_i(n_t, x_{n_t}^*, u_{n_t}^*),$$

to get the result. As before, $\zeta_i(n_T) = 0$ at any terminal node $n_T \in \mathbf{n}_T$.

6.3 Node-Consistent Shapley Value in DGPETs with Random Termination

In Chap. 5, we introduced the class of DGPETs with a random terminal time. More specifically, we assumed that the game can terminate at any intermediate node of the event tree. If it does not terminate at one of these nodes, then it definitely terminates at one of the tree's terminal nodes. We recall that the analysis of such games requires that we define two probabilities at any node $n_t \in \mathbf{n}_t$, $t \in \mathbb{T}$, namely, a probability of the game terminating at this node and a probability of passing through this node.

Denote by $\tau_{n_t} \in [0, 1]$ the probability that the game terminates at node n_t.[3] In particular, $\tau_{n_0} = 0$. If the stochastic process does not terminate at node n_t, it transits to one of the successor nodes in $\mathbf{n}_t^+$. As before, we denote by π_{n_t} the probability of passing through node n_t, which corresponds to the sum of the probabilities of all scenarios that contain this node. In particular, $\pi_{n_0} = 1$ and π_{n_T} is equal to the probability of the single scenario that terminates at (leaf) node $n_T \in \mathbf{n}_T$. Notice that $\tau_{n_T} = \pi_{n_T}$ for any $n_T \in \mathbf{n}_T$; therefore, the process will terminate in the nodes from $\mathbf{n}_T$ for sure. The given probabilities satisfy the following condition:

---

[3] By setting probabilities τ_{n_t} for any $n_t \in \mathbf{n}_t$ for all $t = 0, \dots, T$ we determine the probability distribution of the process terminating over the set of nodes. For the distribution it is true that $\sum_{n \in \mathbf{n}_0^{++}} \tau_n = 1$.

$$\pi_{n_t} = \tau_{n_t} + \sum_{n_{t+1} \in \mathbf{n}_t^+} \pi_{n_{t+1}}, \tag{6.19}$$

that is, the probability of passing through node n_t is equal to the sum of the probability that the game terminates at this node and the sum of the probabilities of passing through to the direct successors of this node.

The control sets and the state dynamics are defined in the same way as above, and $u_{n_t}^i \in U_{n_t}^i$ for any Player $i \in M$ at any node $n_t \in \mathbf{n}_t$, $t \in \mathbb{T} \setminus \{T\}$, and the state equations by (6.1).

At a node $n_t \in \mathbf{n}_t$, $t = 0, \ldots, T-1$, the reward to Player i is given by $\phi_i(n_t, x_{n_t}, u_{n_t})$ if the game does not terminate at this node; otherwise, it is given by $\Phi_i(n_t, x_{n_t})$. In particular, at $t = T$, the reward to Player i is $\Phi_i(n_T, x_{n_T})$. The information that a node n_t is terminal is revealed before the players choose their controls at that node, but not at n_t^-.

The reward function $J_i(\mathbf{x}, \mathbf{u})$ to Player $i \in M$ is

$$J_i(\mathbf{x}, \mathbf{u}) = \sum_{t=0}^{T-1} \rho^t \sum_{n_t \in \mathbf{n}^t} (\pi_{n_t} - \tau_{n_t}) \phi_i(n_t, x_{n_t}, u_{n_t})$$

$$+ \sum_{t=1}^{T} \rho^t \sum_{n_t \in \mathbf{n}^t} \tau_{n_t} \Phi_i(n_t, x_{n_t}). \tag{6.20}$$

The game in normal form with the set of players M is defined by payoff functions $W_i(x_0, \mathbf{u}) = J_i(\mathbf{x}, \mathbf{u})$, $i \in M$, where $\mathbf{x}$ is obtained from $\mathbf{u}$ as the unique solution of the state equations (6.1) that emanate from the initial state x_0. Let $\mathbf{u}^{nc} = (\mathbf{u}^{i,nc} : i \in M)$ be the Nash equilibrium in open-loop S-adapted strategies in this game, and $\mathbf{u}^*$ be the vector of cooperative controls, i.e.,

$$\mathbf{u}^* = \arg \max_{\mathbf{u}^i, i \in M} \sum_{i \in M} W_i(x_0, \mathbf{u}). \tag{6.21}$$

The cooperative state trajectory generated by $\mathbf{u}^*$ is $\mathbf{x}^* = \{x_n^* : n \in \mathbf{n}_0^{++}\}$.

For later use, we also need to determine the subgame starting at node $n_t \in \mathbf{n}_t$, $t = 1, \ldots, T-1$, with state $x_{n_t}^*$. This subgame takes place on a tree subgraph with the root node n_t. The payoff to Player $i \in M$ in this subgame is given by

$$W_i\left(x_{n_t}^*, \mathbf{u}[n_t, x_{n_t}^*]\right) = \sum_{\theta=t}^{T-1} \rho^{\theta-t} \sum_{n_\theta \in \mathbf{n}_\theta \cap \mathbf{n}_t^+} (\pi_{n_t}^{n_\theta} - \tau_{n_t}^{n_\theta}) \phi_i(n_\theta, x_{n_\theta}, u_{n_\theta}) \tag{6.22}$$

$$+ \sum_{\theta=t+1}^{T} \rho^{\theta-t} \sum_{n_\theta \in \mathbf{n}_\theta \cap \mathbf{n}_t^+} \tau_{n_t}^{n_\theta} \Phi_i(n_\theta, x_{n_\theta}),$$

where $\mathbf{u}[n_t, x_{n_t}^*] = (\mathbf{u}^i[n_t, x_{n_t}^*] : i \in M)$ is an S-adapted strategy profile and $\mathbf{u}^i[n_t, x_{n_t}^*] = (u_{n_\theta}^i : n_\theta \in \mathbf{n}_t^{++})$ is an admissible S-adapted strategy of Player i in the subgame starting from node n_t, with initial state $x_{n_t}^*$. The term $\tau_{n_t}^{n_\theta}$ is the conditional probability of terminating at node n_θ when the subgame starts from node n_t.

The conditional probabilities $\pi_{n_t}^{n_\theta}$ and $\tau_{n_t}^{n_\theta}$ are computed as follows:

$$\pi_{n_t}^{n_\theta} = \frac{\pi_{n_\theta}}{\pi_{n_t} - \tau_{n_t}}, \quad \tau_{n_t}^{n_\theta} = \frac{\tau_{n_\theta}}{\pi_{n_t} - \tau_{n_t}},$$

if $\pi_{n_t} \neq \tau_{n_t}$; otherwise, the subgame starting from node n_t cannot materialize.

If the players act noncooperatively in the subgame beginning at node n_t with state $x_{n_t}^*$, they will implement the S-adapted Nash equilibrium strategies $\mathbf{u}^{nc}[n_t, x_{n_t}^*] = (\mathbf{u}^{i,nc}[n_t, x_{n_t}^*] : i \in M)$. The corresponding payoff of Player i is given by $W_i(x_{n_t}^*, \mathbf{u}^{nc}[n_t, x_{n_t}^*])$.

If the players cooperate in the subgame starting from node n_t and in state $x_{n_t}^*$, then they maximize the expected sum of their total discounted payoffs. The resulting cooperative controls in the subgame are as follows:

$$\mathbf{u}^*[n_t, x_{n_t}^*] = \arg \max_{\mathbf{u}^i[n_t, x_{n_t}^*]:i \in M} \sum_{i \in M} W_i\left(x_{n_t}^*, \mathbf{u}[n_t, x_{n_t}^*]\right). \tag{6.23}$$

Therefore, the payoff of Player i in the cooperative subgame starting from node $n_t \in \mathbf{n}_t$, with initial state $x_{n_t}^*$, is equal to $W_i\left(x_{n_t}^*, \mathbf{u}^*[n_t, x_{n_t}^*]\right)$, $t = 1, \ldots, T$.

6.3.1 Node-Consistent Shapley Value

Suppose that the players cooperate during the whole game and agree to use the Shapley value as an allocation mechanism for their joint payoff $\sum_{i \in M} W_i(x_0, \mathbf{u}^*)$. Let $v(K; x^*(n_t))$ be the γ characteristic function value of coalition K in the subgame starting at node $n_t \in \mathbf{n}_t$, $t = 0, \ldots, T$, in state $x^*(n_t)$.

Remark 6.5 We define the characteristic function $v(K; x_{n_t}^*)$ for any node $n_t \in \mathbf{n}_t$ based on the initially given noncooperative game, assuming that the players implement the cooperative controls $\mathbf{u}^*$, and moving along the cooperative trajectory $\mathbf{x}^*$. The values of the characteristic function $v(K; x_{n_t}^*)$ also depend on the initial node n_t and on the cooperative controls, but we omit them to keep the notation simple.

Let $Y(x_{n_t}^*)$ be the set of imputations for the subgame starting at node n_t with a state $x_{n_t}^*$. Let $Sh(x_{n_t}^*) = (Sh_1(x_{n_t}^*), \ldots, Sh_m(x_{n_t}^*))$ be the Shapley value for the subgame starting at node n_t with state $x_{n_t}^*$.

Now, we consider the Shapley value decomposition over time and nodes in the DGPET with a random terminal time. To start, we define an imputation distribution procedure in this setup.

Definition 6.4 The collection of functions $\beta_i(x_n^*)$, $n \in \mathbf{n}_0^{++}$, $i \in M$, where[4]

$$
\beta_i(x_n^*) = \begin{cases}
\beta_i^c(x_0), & \text{if } n = n_0, \\
\beta_i^c(x_n^*), & \text{if the process does not terminate at node } n \in \mathbf{n}_t, \\
& t = 0, \ldots, T-1 \\
\beta_i^\tau(x_n^*), & \text{if the process terminates at node } n \in \mathbf{n}_t, \\
& t = 1, \ldots, T-1, \\
\beta_i^\tau(x_{n_t}^*), & \text{if } n_t \in \mathbf{n}_T,
\end{cases}
\tag{6.24}
$$

is an IDP of the Shapley value $Sh(x_0) = (Sh_1(x_0), \ldots, Sh_m(x_0))$, if

$$
Sh_i(x_0) = \sum_{t=0}^{T-1} \rho^t \sum_{n_t \in \mathbf{n}^t} (\pi_{n_t} - \tau_{n_t}) \beta_i^c(x_{n_t}^*) + \sum_{t=1}^{T} \rho^t \sum_{n_t \in \mathbf{n}_t} \tau_{n_t} \beta_i^\tau(x_{n_t}^*), \text{ for all } i \in M.
\tag{6.25}
$$

The difference between the distribution procedures in Definitions 6.1 and 6.4 is the following. In the root node n_0 in which the stochastic process does not terminate and in nodes $n_T \in \mathbf{n}_T$ in which the process terminates, the components of the IDP are determined in a unique way because it is known when the game continues or terminates for sure. For an intermediate node $n_t \in \mathbf{n}_t$, $t = 1, \ldots, T-1$, unlike in Definition 6.1, we determine two possible IDP values for Player i, corresponding to the possibilities of the game continuing or terminating, namely, (i) $\beta_i^c(x_{n_t}^*)$ if the game does not terminate at node n_t, and (ii) $\beta_i^\tau(x_{n_t}^*)$ if it does. Equation (6.25) means that the component of the Shapley value corresponding to Player $i \in M$ is equal to the discounted payments made to this player in the whole game.

Definition 6.5 The Shapley value $Sh(x_0)$ and corresponding imputation distribution procedure $\{\beta_i(x_n^*) : n \in \mathbf{n}_0^{++}, i \in M\}$ determined by (6.24) are called node consistent in the game with a random termination if, for any $t = 0, \ldots, T$, the following condition is satisfied:

$$
Sh_i(x_0) = \sum_{\theta=0}^{t-1} \rho^\theta \sum_{n_\theta \in \mathbf{n}_\theta} \left\{ (\pi_{n_\theta} - \tau_{n_\theta}) \beta_i^c(x_{n_\theta}^*) + \tau_{n_\theta} \beta_i^\tau(x_{n_\theta}^*) \right\}
\tag{6.26}
$$
$$
+ \rho^t \sum_{n_t \in \mathbf{n}_t} \left\{ (\pi_{n_t} - \tau_{n_t}) Sh_i(x_{n_t}^*) + \tau_{n_t} Sh_i^\tau(x_{n_t}^*) \right\},
$$

where $Sh_i(x_{n_t}^*)$ is Player i's Shapley value in the subgame starting from node n_t in state $x_{n_t}^*$ when the stochastic process does not terminate at node n_t, and $Sh_i^\tau(x_{n_t}^*)$ is Player i's Shapley value in the one-period game played at n_t with state $x_{n_t}^*$ when the stochastic process terminates at this node.

---

[4] The superscripts c and τ for β_i in (6.24) correspond to continue and terminate, respectively.

Notice that if the game terminates at node n_t, the players do not choose any controls at this node, and the components of the Shapley value $Sh_i^\tau(x_{n_t}^*)$ are equal to the corresponding values of their payoff functions $\Phi_i(n_t, x_{n_t}^*)$.

If the payoffs in the nodes are allocated according to an IDP, then the node consistency of the Shapley value $Sh(x_0)$ means that one can define a feasible distribution procedure under which the continuation value at every node is also the Shapley value of the continuation game.

Theorem 6.2 *The Shapley value $Sh(x_0)$ is node consistent if the corresponding imputation distribution procedure determined by (6.24) satisfies the following conditions for $t = 0, \ldots, T-1$:*

$$\beta_i^c(x_{n_t}^*) = Sh_i(x_{n_t}^*) - \rho \sum_{n_{t+1} \in \mathbf{n}_t^+} \left\{ \left(\pi_{n_t}^{n_{t+1}} - \tau_{n_t}^{n_{t+1}} \right) Sh_i(x_{n_{t+1}}^*) \right. \tag{6.27}$$
$$\left. + \tau_{n_t}^{n_{t+1}} \Phi_i(n_{t+1}, x_{n_{t+1}}^*) \right\},$$

and for $t = 1, \ldots, T$:

$$\beta_i^\tau(x_{n_t}^*) = \Phi_i(n_t, x_{n_t}^*). \tag{6.28}$$

Proof First, we prove that $\beta_i(x_{n_t}^*)$, determined by (6.27) and (6.28), is an IDP of the Shapley value $Sh(x_0)$, i.e., we show that condition (6.25) is satisfied. Substituting the expressions of $\beta_i^c(x_{n_t}^*)$ and $\beta_i^\tau(x_{n_t}^*)$, given by (6.27) and (6.28), into the right-hand-side term of Eq. (6.25) and taking into account that

$$\pi_{n_t}^{n_{t+1}} - \tau_{n_t}^{n_{t+1}} = \frac{\pi_{n_{t+1}} - \tau_{n_{t+1}}}{\pi_{n_t} - \tau_{n_t}}, \tag{6.29}$$

we obtain

$$\sum_{t=0}^{T-1} \rho^t \sum_{n_t \in \mathbf{n}_t} (\pi_{n_t} - \tau_{n_t}) \left[Sh_i(x_{n_t}^*) \right.$$
$$\left. - \rho \sum_{n_{t+1} \in \mathbf{n}_t^+} \left\{ \left(\pi_{n_t}^{n_{t+1}} - \tau_{n_t}^{n_{t+1}} \right) Sh_i(x_{n_{t+1}}^*) + \tau_{n_t}^{n_{t+1}} \Phi_i(n_{t+1}, x_{n_{t+1}}^*) \right\} \right]$$
$$+ \sum_{t=1}^{T} \rho^t \sum_{n_t \in \mathbf{n}_t} \tau_{n_t} \Phi_i(n_t, x_{n_t}^*)$$
$$= (\pi(n_0) - \tau(n_0)) Sh_i(x_0) - \rho(\pi(n_0) - \tau(n_0)) \sum_{n_1 \in \mathbf{n}_0^+} \left\{ \left(\pi_{n_0}^{n_1} - \tau_{n_0}^{n_1} \right) Sh_i(x_{n_1}^*) \right.$$
$$\left. + \tau_{n_0}^{n_1} \Phi_i(n_1, x_{n_1}^*) \right\} + \rho \sum_{n_1 \in \mathbf{n}_1} \tau_{n_1} \Phi_i(n_1, x_{n_1}^*) + \ldots$$
$$+ \rho^{T-1} \sum_{n_{T-1} \in \mathbf{n}_{T-1}} (\pi_{n_{T-1}} - \tau_{n_{T-1}}) \left[Sh_i(x_{n_{T-1}}^*) \right.$$

$$- \rho \sum_{n_T \in \mathbf{n}_{T-1}^+} \left\{ \left(\pi_{n_{T-1}}^{n_T} - \tau_{n_{T-1}}^{n_T} \right) Sh_i(x_{n_T}^*) + \tau_{n_{T-1}}^{n_T} \Phi_i(n_T, x_{n_T}^*) \right\} \right]$$

$$+ \rho^T \sum_{n_T \in \mathbf{n}_T} \tau_{n_T} \Phi_i(n_T, x_{n_T}^*) = Sh_i(x_0).$$

We also use the condition that $Sh_i(x_{n_T}^*) = \Phi_i(n_T, x_{n_T}^*)$ for any $n_T \in \mathbf{n}_T$.

Now, we prove that the Shapley value and the corresponding IDP are node consistent, i.e., that they satisfy Eq. (6.26). To obtain this result, we substitute $\beta_i^c(x_{n_\theta}^*)$ from (6.27) and $\beta_i^\tau(x_{n_\theta}^*)$ from (6.28) into (6.26) and take into account that $\pi_{n_0} = 1$, $\tau_{n_0} = 0$ and Eq. (6.29) for any $n_{t+1} \in \mathbf{n}_t^+$, $t = 0, \ldots, T-1$. Indeed, we have

$$\sum_{\theta=0}^{t-1} \rho^\theta \sum_{n_\theta \in \mathbf{n}_\theta} \left\{ (\pi_{n_\theta} - \tau_{n_\theta}) Sh_i(x_{n_\theta}^*) - \rho \sum_{n_{\theta+1} \in \mathbf{n}_\theta^+} \left[(\pi_{n_{\theta+1}} - \tau_{n_{\theta+1}}) Sh_i(x_{n_{\theta+1}}^*) \right.\right.$$

$$\left.\left. + \tau_{n_{\theta+1}} \Phi_i(n_{\theta+1}, x_{n_{\theta+1}}^*) \right] \right\} + \rho^t \sum_{n_t \in \mathbf{n}_t} \left\{ \left(\pi_{n_t} - \tau_{n_t} \right) Sh_i(x_{n_t}^*) + \tau_{n_t} \Phi_i(n_t, x_{n_t}^*) \right\}$$

$$= Sh_i(x_0)$$

This proves the node consistency of the Shapley value when the corresponding IDP is defined by Eqs. (6.27) and (6.28).   □

The next corollary shows that, at any node, the IDP is budget balanced, that is, the total allocated amount is equal to the sum of rewards realized by the players.

Corollary 6.1 *Let the IPD be defined by* (6.27) *and* (6.28). *If the process does not terminate at node* $n_t \in \mathbf{n}_t$, $t = 0, \ldots, T-1$, *then*

$$\sum_{i \in M} \beta_i^c(x_{n_t}^*) = \sum_{i \in M} \phi_i\left(n_t, x_{n_t}^*, u_{n_t}^*\right). \tag{6.30}$$

If the process terminates at node $n_t \in \mathbf{n}_t$, $t = 1, \ldots, T$, *then*

$$\sum_{i \in M} \beta_i^\tau(x_{n_t}^*) = \sum_{i \in M} \Phi_i(n_t, x_{n_t}^*). \tag{6.31}$$

Proof We need to show that conditions (6.30) and (6.31) are satisfied when the IDP is defined by Eqs. (6.27) and (6.28). For terminal nodes, condition (6.31) is satisfied because the Shapley value is an imputation and satisfies the property of efficiency. Further, we verify that condition (6.30) is true for the root node and any intermediate node $n_t \in \mathbf{n}_t$, $t = 0, \ldots, T-1$:

$$\sum_{i \in M} \beta_i^c \left(x_{n_t}^*\right) = \sum_{i \in M} Sh_i(x_{n_t}^*)$$

$$- \sum_{i \in M} \rho \sum_{n_{t+1} \in \mathbf{n}_t^+} \left\{ \left(\pi_{n_t}^{n_{t+1}} - \tau_{n_t}^{n_{t+1}}\right) Sh_i(x_{n_{T+1}}^*) + \tau_{n_t}^{n_{t+1}} \Phi_i(n_{t+1}, x_{n_{t+1}}^*) \right\}$$

$$= v(M; x_{n_t}^*) - \left\{ \sum_{i \in M} W_i(x_{n_t}^*, \mathbf{u}^*[n_t, x_{n_t}^*]) - \sum_{i \in M} \phi_i(n_t, x_{n_t}^*, u_{n_t}^*) \right\},$$

and given that

$$v(M; x_{n_t}^*) = \sum_{i \in M} W_i(x_{n_t}^*, \mathbf{u}^*[n_t, x_{n_t}^*]),$$

and from the construction of the characteristic function, we obtain (6.30). □

The implementation of the IDP $\{\beta_i(x_n^*) : n \in \mathbf{n}_0^{++}, i \in M\}$ requires transfers, or side payments, between the players. They are defined as follows:

- If the game does not terminate at node $n_t \in \mathbf{n}_t$, $t = 0, \ldots, T - 1$, then

$$\zeta_i(n_t, x_{n_t}^*) = \beta_i^c(x_{n_t}^*) - \phi_i(n_t, x_{n_t}^*, u_{n_t}^*). \tag{6.32}$$

- If the game terminates at node $n_t \in \mathbf{n}_t$, $t = 1, \ldots, T$, then

$$\zeta_i(n_t, x_{n_t}^*) = 0,$$

where $\zeta_i(n_t, x_{n_t}^*)$ is the transfer payment that Player i makes at node n_t over the cooperative state trajectory $x_{n_t}^*$, such that $\sum_{i \in M} \zeta_i(n_t, x_{n_t}^*) = 0$ for any node n_t over $x_{n_t}^*$. Note that $\zeta_i(n_t, x_{n_t}^*)$ in (6.32) can assume any sign.

6.3.2 An Example

We consider a cooperative version of an example of transboundary pollution control represented in Example 5.10.2.

Denote by $M = \{1, \ldots, m\}$ the set of players representing countries, and by $\mathbb{T} = \{0, \ldots, T\}$ the set of time periods. Production activities in each country generate revenues and, as a by-product, pollutant emissions, e.g., CO_2. Denote by $u_{n_t} = (u_{n_t}^1, u_{n_t}^2, u_{n_t}^3)$ the profile of countries' emissions in node $n_t \in \mathbf{n}_t$ in time period t, and by x_{n_t} the stock of pollution at this node. The evolution of this stock is governed by the following difference equation:

$$x_{n_t} = (1 - \delta_{n_t^-})x_{n_t^-} + \sum_{j \in M} u_{n_t^-}^j, \quad x_{n_0} = x_0, \tag{6.33}$$

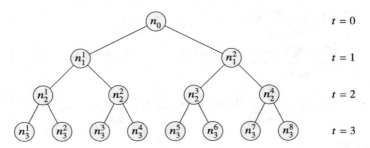

Fig. 6.1 Event tree graph

where the initial state x_0 is given, node n_t^- is a predecessor of node n_t, and $\delta_{n_t^-} \in (0, 1)$ is a rate of pollution absorption by Mother Nature at node n_t^-. Let δ_{n_t} $(0 < \delta_{n_t} < 1)$ be the stochastic rate of pollution absorption by nature at node n_t. We suppose that δ_{n_t} can take two possible values, namely, $\delta_{n_t} \in \{\underline{\delta}, \overline{\delta}\}$, with $\underline{\delta} < \overline{\delta}$.

The event tree is depicted in Fig. 6.1. The event tree is defined by two sets of probabilities: $\{\pi_{n_t} : n_t \in \mathbf{n}_t, t \in \mathbb{T}\}$ and $\{\tau_{n_t} : n_t \in \mathbf{n}_t, t \in \mathbb{T}\}$, such that $\pi_{n_0} = 1$, $\tau_{n_T} = \pi_{n_T}$ for any $n_T \in \mathbf{n}_T$, and $\pi_{n_t} = \tau_{n_t} + \sum_{n_{t+1} \in n_t^+} \pi_{n_{t+1}}$. Let nodes n_0, n_1^2, n_2^2, n_2^4, n_3^2, n_3^4, n_3^6, n_3^8 correspond to the low level of pollution reduction $\underline{\delta}$, and nodes n_1^1, n_2^1, n_2^3, n_3^1, n_3^3, n_3^5, n_3^7 corresponds to the high level of pollution reduction $\overline{\delta}$.

Denote by $q_{n_t}^i$, Player i's production of goods and services at node $n_t \in \mathbf{n}_t$. The resulting emissions $u_{n_t}^i$ at node $u_{n_t}^i$ are an increasing function of production, i.e., $u_{n_t}^i = h_i(q_{n_t}^i)$ satisfying $h_i(0) = 0$. Assuming a monotone increasing relationship between production and revenues, we can express the revenues, denoted by R_i, directly as a function of emissions. We adopt the following specification:

$$R_i(u_{n_t}^i) = \alpha_i u_{n_t}^i - \frac{1}{2}\beta_i \left(u_{n_t}^i\right)^2, \tag{6.34}$$

where α_i is a positive parameter for any player $i \in M$.

Each country suffers an environmental damage cost due to pollution accumulation. We assume that the costs are defined by an increasing convex function in this stock and retain the quadratic form

$$D_i(x_{n_t}) = c_i(x_{n_t})^2, \quad i \in M,$$

where c_i is a strictly positive parameter. Denote by $\rho \in (0, 1)$ the discount factor of any player. Assume that Player $i \in M$ maximizes the following objective functional:

$$J_i(\mathbf{x}, \mathbf{u}) = \sum_{t=0}^{T-1} \rho^t \sum_{n_t \in \mathbf{n}_t} (\pi_{n_t} - \tau_{n_t}) \left(R_i(u_{n_t}^i) - D_i(x_{n_t})\right) + \sum_{t=1}^{T} \rho^t \sum_{n_t \in \mathbf{n}_t} \tau_{n_t} \Phi_i(n_t, x_{n_t})$$

$$\tag{6.35}$$

subject to (6.33), and $u_{n_t}^i \geq 0$ for all $i \in M$ and any node $n_t \in \mathbf{n}_t$, $t \in \mathbb{T} \setminus \{T\}$, where $\mathbf{x} = \{x_{n_t} : n_t \in \mathbf{n}_t, \ t \in \mathbb{T}\}$ is a state trajectory.

Let the payoff function of Player i at the terminal node $n_T \in \mathbf{n}_T$ be

$$\Phi_i(x_{n_T}) = -d_i x_{n_T},$$

where $d_i > 0$ for any $i \in M$. Substituting $\Phi_i(x_{n_T})$, $R_i(u_{n_t}^i)$ and $D_i(x_{n_t})$ by their values, we get

$$J_i(\mathbf{x}, \mathbf{u}) = \sum_{t=0}^{T-1} \rho^t \sum_{n_t \in \mathbf{n}_t} (\pi_{n_t} - \tau_{n_t}) \left(\alpha_i u_{n_t}^i - \frac{1}{2} \beta_i \left(u_{n_t}^i \right)^2 - c_i(x_{n_t})^2 \right)$$
$$- \sum_{t=1}^{T} \rho^t \sum_{n_t \in \mathbf{n}_t} \tau_{n_t} d_i x_{n_T}.$$

We use the following parameter values in the numerical illustration:

$$x_0 = 0, \ \rho = 0.9, \ \underline{\delta} = 0.4, \ \overline{\delta} = 0.8,$$
$$\alpha_1 = 30, \ \alpha_2 = 29, \ \alpha_3 = 28,$$
$$\beta_1 = 3, \ \beta_2 = 4, \ \beta_3 = 4.5,$$
$$c_1 = 0.15, \ c_2 = 0.10, \ c_3 = 0.05,$$
$$d_1 = 0.25, \ d_2 = 0.20, \ d_3 = 0.15.$$

The values of probabilities π_{n_t} and τ_{n_t} are given in Table 6.1. Using (6.21) and (6.23), we compute the cooperative controls for each possible subgame and for the whole game with random termination. The cooperative control trajectories and the corresponding state values are given in Table 6.2.

The γ characteristic function values are given in Table 6.3. It is easy to verify that the γ characteristic function is superadditive for any subgame. The Shapley values

Table 6.1 Probabilities π_{n_t} and τ_{n_t} defined on the event tree represented in Fig. 6.1

Time period	$t = 0$	$t = 1$		$t = 2$			
Node	n_0	n_1^1	n_1^2	n_2^1	n_2^2	n_2^3	n_2^4
π_{n_t}	1.00	0.30	0.70	0.20	0.05	0.30	0.30
τ_{n_t}	0.00	0.05	0.10	0.05	0.01	0.10	0.05

$t = 3$							
Node	n_3^1	n_3^2	n_3^3	n_3^4	n_3^5	n_3^6	n_3^7	n_3^8
π_{n_t}	0.05	0.10	0.01	0.03	0.10	0.10	0.15	0.10
τ_{n_t}	0.05	0.10	0.01	0.03	0.10	0.10	0.15	0.10

Table 6.2 Cooperative control trajectories $u_{n_t}^{i*}$, $i \in M$, $n_t \in \mathbf{n}_t$, $t \in \mathbb{T} \setminus \{T\}$, and cooperative state trajectory $x_{n_t}^*$, $n_t \in \mathbf{n}_t$, $t \in \mathbb{T}$

Time period	$t = 0$	$t = 1$		$t = 2$			
Node	n_0	n_1^1	n_1^2	n_2^1	n_2^2	n_2^3	n_2^4
$u_{n_t}^{1,nc}$	6.4690	7.2209	6.6257	9.8200	9.8200	9.8200	9.8200
$u_{n_t}^{2,nc}$	4.6018	5.1657	4.7193	7.1150	7.1150	7.1150	7.1150
$u_{n_t}^{3,nc}$	3.8682	4.3695	3.9727	6.1022	6.1022	6.1022	6.1022
$x_{n_t}^{nc}$	0.0000	14.9390	14.9390	19.7438	19.7438	24.2812	24.2812

Time period				$t = 3$			
Node	n_3^1	n_3^2	n_3^3	n_3^4	n_3^5	n_3^6	n_3^7	n_3^8
$x_{n_t}^{nc}$	26.9860	26.9860	34.8835	34.8835	27.8935	27.8935	37.6059	37.6059

Table 6.3 Characteristic functions for the whole game and subgames

Time period	$t = 0$	$t = 1$		$t = 2$			
Node	n_0	n_1^1	n_1^2	n_2^1	n_2^2	n_2^3	n_2^4
$v(\{1\})$	212.292	151.193	119.067	83.738	82.372	52.180	51.816
$v(\{2\})$	156.609	109.888	88.9201	59.914	58.822	38.663	38.372
$v(\{3\})$	157.157	112.168	101.346	62.950	62.130	52.005	51.787
$v(\{1, 2\})$	379.500	264.966	213.704	143.664	141.206	90.855	90.199
$v(\{1, 3\})$	376.473	265.926	224.186	146.696	144.511	104.194	103.611
$v(\{2, 3\})$	316.567	223.179	191.920	122.870	120.958	90.674	90.164
$v(\{1, 2, 3\})$	561.443	386.372	328.646	206.653	203.375	142.899	142.025

Time period				$t = 3$			
Node	n_3^1	n_3^2	n_3^3	n_3^4	n_3^5	n_3^6	n_3^7	n_3^8
$v(\{1\})$	−6.747	−6.747	−8.721	−8.721	−6.973	−6.973	−9.402	−9.402
$v(\{2\})$	−5.397	−5.397	−6.977	−6.977	−5.579	−5.579	−7.522	−7.522
$v(\{3\})$	−4.048	−4.048	−5.233	−5.233	−4.184	−4.184	−5.641	−5.641
$v(\{1, 2\})$	−12.144	−12.144	−15.698	−15.698	−12.552	−12.552	−16.923	−16.923
$v(\{1, 3\})$	−10.794	−10.794	−13.953	−13.953	−11.157	−11.157	−15.042	−15.042
$v(\{2, 3\})$	−9.445	−9.445	−12.209	−12.209	−9.763	−9.763	−13.162	−13.162
$v(\{1, 2, 3\})$	−16.192	−16.192	−20.930	−20.930	−16.736	−16.736	−22.564	−22.564

of the whole game and all subgames, which are computed using formulas (6.15) and (6.16), are provided in Table 6.4.

Finally, based on the Shapley value, we determine the node-consistent IDP using Eqs. (6.27) and (6.28) in Theorem 6.2. The results, that is, the $\beta_j^c(x_{n_t}^*)$ and $\beta_j^\tau(x_{n_t}^*)$, $j \in M$, are shown in Fig. 6.2. We should mention that in the root node n_0, only β^c is determined (the game process never terminates at the root node) and in the nodes from $\mathbf{n}_T$, only β^τ is determined (the game process definitely terminates at the terminal nodes), while in all "intermediate" nodes, both β^c and β^τ are determined by (6.27) and (6.28) from Theorem 6.2. The payments along the cooperative state

Table 6.4 The Shapley values for the whole game and all subgames

Time period	$t = 0$	$t = 1$		$t = 2$			
Node	n^0	n_1^1	n_2^1	n_1^2	n_2^2	n_3^2	n_4^2
Sh_1	226.090	156.268	126.535	83.756	82.391	52.198	51.834
Sh_2	168.296	114.242	95.329	59.931	58.839	38.680	38.389
Sh_3	167.056	115.862	106.783	62.965	62.146	52.021	51.802

$t = 3$							
Node	n_1^3	n_2^3	n_3^3	n_4^3	n_5^3	n_6^3	n_7^3	n_8^3
Sh_1	−6.747	−6.747	−8.721	−8.721	−6.973	−6.973	−9.402	−9.402
Sh_2	−5.397	−5.397	−6.978	−6.978	−5.579	−5.579	−7.521	−7.521
Sh_3	−4.048	−4.048	−5.233	−5.233	−4.184	−4.184	−5.641	−5.641

trajectory are organized as follows. If the game process does not terminate at node n_t, then Player i gets $\beta_i^c(x_{n_t}^*)$, but if it terminates at this node, Player i gets $\beta_i^\tau(x_{n_t}^*)$. Let us show the decomposition of the component of the Shapley value of Player 1 who is entitled to a payoff of 226.090 in the whole game (see Table 6.4). The imputation distribution procedure decomposes this amount over nodes, that is, it is equal to the expected sum of discounted payments (the components of the IDP represented in Fig. 6.2) received in the subgame emanating from this node (here the whole game). Indeed, we have

$$
226.090 = 122.601
$$
$$
+ 0.9 \cdot \{(0.3 - 0.05) \cdot 99.982 - 0.05 \cdot 3.735 + 0.6 \cdot 92.278 - 0.1 \cdot 3.735\}
$$
$$
+ 0.9^2 \cdot \{0.15 \cdot 88.1986 - 0.05 \cdot 4.9359 + 0.04 \cdot 86.833 - 0.01 \cdot 4.9359
$$
$$
+ 0.2 \cdot 57.6617 - 0.1 \cdot 6.0703 + 0.25 \cdot 57.2975 - 0.05 \cdot 6.0703\}
$$
$$
- 0.9^3 \cdot \{0.05 \cdot 6.7465 + 0.1 \cdot 6.7465 + 0.01 \cdot 8.7209 + 0.03 \cdot 8.7209
$$
$$
+ 0.1 \cdot 6.9734 + 0.1 \cdot 6.9734 + 0.15 \cdot 9.4015 + 0.1 \cdot 9.4015\}.
$$

Similar calculations can easily be done for Players 2 and 3. This decomposition of the Shapley value guarantees its node consistency. Consequently, players' payoffs in any subgame will be the corresponding components of the Shapley values calculated for these subgames.

6.4 Additional Readings

The node consistency of the Shapley value in a DGPET was first studied by Reddy et al. (2013) for a given duration, and the results were extended to a random terminal date by Parilina and Zaccour (2017). The DGPETs when players have asymmetric

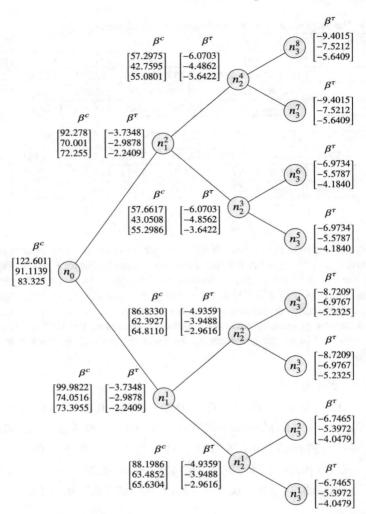

Fig. 6.2 Imputation distribution procedure

beliefs about transition probabilities though nodes are considered in Parilina and Zaccour (2022), where the authors propose two consistent distribution procedures of the payoffs based on the Nash bargaining solution. For a survey of the sustainability of cooperation in DGPETs, see Zaccour (2017).

In state-space dynamic games, an unknown terminal date may be due to events external to the game, for instance a *force majeure* such as a natural disaster or because the terminal date is endogenous, as in patent-race games or pursuit-evasion games (see, e.g., Haurie et al. 2012). Unforeseen events can obviously occur, and ideally, a contract should have provisions to deal with them, notwithstanding the initial intention to cooperate during the entire planning horizon.

The idea of random termination and the analysis of its influence on the results has been around awhile in dynamic games. For instance, Petrosyan and Murzov (1996) analyze a differential game model of random duration. Petrosyan and Shevko-plyas (2003) construct cooperative solutions for differential games with deterministic dynamics and a random duration, whereas Marín-Solano and Shevkoplyas (2011) consider a differential game with a random duration and nonconstant discounting. Yeung and Petrosyan (2011) study cooperation in discrete-time dynamic games with a random duration but without state dynamics. Gromova and Plekhanova (2018) discuss time-consistent solutions in a general class of dynamic games of random duration.

In the construction of a multistage project, Green and Taylor (2016) show how the uncertainty of the termination process influences Nash and cooperative equilibrium strategies and payoffs. In repeated games, it is an established result that not knowing the game's terminal date beforehand, that is, the number of stages to be played, can lead to sustainable cooperation (or collusion in a game à la Cournot). Normann and Wallace (2012) consider a repeated prisoner's dilemma game with four different termination rules, including the random case, and show how this affects the players' behavior.

There is also an extensive literature on dynamic games where the terminal time is endogenous, that is, influenced by the players' strategies. Typical classes of games that fall into this category are pursuit-evasion games (see, e.g., Cardaliaguet and Rainer 2018; Kumkov et al. 2017); patent races (see, e.g., Fudenberg et al. 1983; Harris and Vickers 1985); and optimal stopping games (see, e.g., Mazalov and Falko 2008; Ramsey and Szajowski 2005; Krasnosielska-Kobos 2016).

6.5 Exercises

Exercise 6.1 Generalize the results in Sect. 6.2.1 to $m > 2$ players.

Exercise 6.2 Generalize the results in Sect. 6.2.1 to the case where the terminal time is random.

Exercise 6.3 Let $M = \{1, 2, 3\}$ be the set of players involved in a pollution control game, and $\mathbb{T} = \{0, 1, 2, 3\}$ the set of periods. Let $u_{n_t} = (u_{n_t}^1, u_{n_t}^2, u_{n_t}^3)$ be the vector of players' emissions of some pollutant at node n_t in time period t, and denote by x_{n_t} the stock of pollution at this node. The evolution of this stock is governed by the following difference equation:

$$x_{n_t} = (1 - \delta_{n_t^-})x_{n_t^-} + \sum_{i \in M} u_{n_t^-}^i, \tag{6.36}$$

$$x_{n_0} = x_0, \tag{6.37}$$

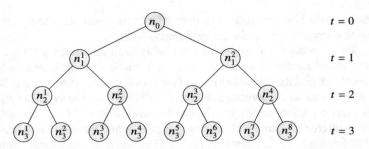

Fig. 6.3 Event tree graph for Exercise 6.3

with the initial stock x_0 at root node n_0 given, and δ_{n_t} $(0 < \delta_{n_t} < 1)$ being the stochastic rate of pollution absorption by nature at node n_t. Suppose that δ_{n_t} can take two possible values, namely, $\delta_{n_t} \in \{\underline{\delta}, \overline{\delta}\}$, with $\underline{\delta} < \overline{\delta}$. The event tree is depicted in Fig. 6.3. Let nodes $n_2^1, n_2^2, n_2^4, n_3^3, n_4^3, n_6^3, n_8^3$ correspond to the low level $\underline{\delta}$, and nodes n_1^1, n_1^2, $n_3^2, n_1^3, n_3^3, n_5^3, n_7^3$ correspond to the high level $\overline{\delta}$.

The damage cost is an increasing convex function in the pollution stock and given by $D_i(x_{n_t}) = \alpha_i x_{n_t}^2$, $i \in M$, where $\alpha_i > 0$. The cost of emissions is also given by a quadratic function, $C_i(u_{n_t}^i) = \frac{\beta_i}{2}\left(u_{n_t}^i - e\right)^2$, where e and β_i are strictly positive constants. At any node $n_t \in \mathbf{n}_t$, $t = 0, \ldots, T - 1$, Player i's total cost is then given by $\phi_i(n_t, x_{n_t}, u_{n_t}) = C_i(u_{n_t}^i) + D_i(x_{n_t})$, and by $\Phi_i(n_T, x_{n_T}) = D_i(x_{n_T})$ for any node $n_T \in \mathbf{n}_T$. There is a constraint on the player's control: $u_{n_t}^i \in [0, e]$, for all $i \in M$ and any node $n_t \in \mathbf{n}_t$, $t = 0, 1, 2$.

Suppose that each player aims at minimizing her expected total discounted cost over the event tree. The parameter values are as follows:

$$\text{Damage cost}: \ \alpha_1 = 0.3, \ \alpha_2 = 0.2, \ \alpha_3 = 0.5,$$
$$\text{Emissions cost}: \ \beta_1 = 0.9, \ \beta_2 = 0.8, \ \beta_3 = 0.5, \ e = 25,$$
$$\text{Nature absorption rate}: \ \underline{\delta} = 0.4, \ \overline{\delta} = 0.9,$$
$$\text{Discount rate}: \ \lambda_i = 0.95, \ \text{for } i = 1, 2, 3,$$
$$\text{Transition probabilities}: \ \pi_{n_1^1} = 0.6, \ \pi_{n_2^1} = 0.4, \ \pi_{n_1^2} = \pi_{n_2^2} = 0.3, \ \pi_{n_3^2} = 0.25,$$
$$\pi_{n_4^2} = 0.15, \ \pi_{n_1^3} = \pi_{n_2^3} = 0.15, \ \pi_{n_3^3} = 0.05, \ \pi_{n_4^3} = 0.25,$$
$$\pi_{n_5^3} = 0.05, \ \pi_{n_6^3} = 0.20, \ \pi_{n_7^3} = 0.10, \ \pi_{n_8^3} = 0.05.$$

1. Compute the cooperative controls for the whole game and the corresponding cooperative state trajectory.
2. Compute the γ characteristic function values in all subgames (including the whole game).
3. Compute the Shapley values in all subgames.
4. Determine the node-consistent IDP of the Shapley values using Theorem 6.1.
5. Find the Nash bargaining solution in the whole game and in any subgame.

6. Determine the node-consistent distribution procedure of the Nash bargaining solution using Proposition 6.2.

Exercise 6.4 Consider the three-player game in Exercise 6.3 but with a random duration. The transition probabilities over the event tree are given by

$$\pi_{n_1^1} = 0.6, \ \pi_{n_2^1} = 0.4, \ \pi_{n_1^2} = 0.3, \ \pi_{n_2^2} = \pi_{n_3^2} = 0.25, \ \pi_{n_4^2} = 0.1,$$
$$\pi_{n_1^3} = 0.05, \ \pi_{n_2^3} = 0.15, \ \pi_{n_3^3} = 0.05, \ \pi_{n_4^3} = 0.15,$$
$$\pi_{n_5^3} = 0.05, \ \pi_{n_6^3} = 0.15, \ \pi_{n_7^3} = 0.025, \ \pi_{n_8^3} = 0.025.$$

The probabilities that the game will terminate at a given node are as follows:

$$\tau_{n_1^1} = \tau_{n_2^1} = 0.05, \ \tau_{n_1^2} = 0.1, \ \tau_{n_2^2} = \tau_{n_3^2} = \tau_{n_4^2} = 0.05,$$
$$\tau_{n_1^3} = 0.05, \ \tau_{n_2^3} = 0.15, \ \tau_{n_3^3} = 0.05, \ \tau_{n_4^3} = 0.15,$$
$$\tau_{n_5^3} = 0.05, \ \tau_{n_6^3} = 0.15, \ \tau_{n_7^3} = \tau_{n_8^3} = 0.025.$$

The other parameter values remain the same as in Exercise 6.3.

The reward functions are given by $\phi_i(n_t, x_{n_t}, u_{n_t}) = C_i(u_{n_t}^i) + D_i(x_{n_t})$ for any node $n_t \in \mathbf{n}_t$, $t = 0, \ldots, T-1$, and by $\Phi_i(n_t, x_{n_t}) = D_i(x_{n_t})$ for any node $n_t \in \mathbf{n}_t$, $t = 1, \ldots, T$. As in Exercise 1, we have $D_i(x_{n_t}) = \alpha_i x_{n_t}^2$ and $C_i(u_{n_t}^i) = \frac{\beta_i}{2}\left(u_{n_t}^i - e\right)^2$.

1. Compute the cooperative controls for the whole game and the corresponding cooperative state trajectory.
2. Compute the γ characteristic function values in all subgames (including the whole game).
3. Compute the Shapley values in all subgames.
4. Determine the node-consistent IDP of the Shapley values using Eqs. (6.27) and (6.28) in Theorem 6.2.
5. Based on the IDP of the Shapley value, define the side payment to any player in any node of the event tree.

References

Cardaliaguet, P. and Rainer, C. (2018). Zero-sum differential games. In Başar, T. and Zaccour, G., editors, *Handbook of Dynamic Game Theory*, pages 373–430, Cham. Springer International Publishing.

Fudenberg, D., Gilbert, R., Stiglitz, J., and Tirole, J. (1983). Preemption, leapfrogging and competition in patent races. *European Economic Review*, 22(1):3–31. Market Competition, Conflict and Collusion.

Green, B. and Taylor, C. (2016). Breakthroughs, deadlines, and self-reported progress: Contracting for multistage projects. *American Economic Review*, 106(12):3660–3699.

Gromova, E. and Plekhanova, T. (2018). On the regularization of a cooperative solution in a multi-stage game with random time horizon. *Discrete Applied Mathematics*, 255.

Harris, C. and Vickers, J. (1985). Perfect equilibrium in a model of a race. *The Review of Economic Studies*, 52(2):193–209.

Haurie, A., Krawczyk, J. B., and Zaccour, G. (2012). *Games and Dynamic Games*. World Scientific, Singapore.

Krasnosielska-Kobos, A. (2016). Construction of Nash equilibrium based on multiple stopping problem in multi-person game. *Mathematical Methods of Operations Research*, 83(1):53–70.

Kumkov, S. S., Le Ménec, S., and Patsko, V. S. (2017). Zero-sum pursuit-evasion differential games with many objects: Survey of publications. *Dynamic Games and Applications*, 7(4):609–633.

Marín-Solano, J. and Shevkoplyas, E. V. (2011). Non-constant discounting and differential games with random time horizon. *Automatica*, 47(12):2626–2638.

Mazalov, V. and Falko, A. (2008). Nash equilibrium in two-sided mate choice problem. *International Game Theory Review*, 10(04):421–435.

Normann, H.-T. and Wallace, B. (2012). The impact of the termination rule on cooperation in a prisoner's dilemma experiment. *International Journal of Game Theory*, 41(3):707–718.

Parilina, E. M. and Zaccour, G. (2017). Node-consistent Shapley value for games played over event trees with random terminal time. *Journal of Optimization Theory and Applications*, 175(1):236–254.

Parilina, E. M. and Zaccour, G. (2022). Sustainable cooperation in dynamic games on event trees with players' asymmetric beliefs. *Journal of Optimization Theory and Applications*.

Petrosyan, L. and Murzov, N. V. (1996). Game theoretical problems of mechanics. *Lithuanian Mathematical Journal*, (6):423–432.

Petrosyan, L. and Shevkoplyas, E. (2003). Cooperative solutions for games with random duration. *Game Theory and Applications*, 9:125–139.

Ramsey, D. and Szajowski, K. (2005). Bilateral approach to the secretary problem. In Nowak, A. S. and Szajowski, K., editors, *Advances in Dynamic Games: Applications to Economics, Finance, Optimization, and Stochastic Control*, pages 271–284, Boston, MA. Birkhäuser Boston.

Reddy, P. V., Shevkoplyas, E., and Zaccour, G. (2013). Time-consistent Shapley value for games played over event trees. *Automatica*, 49(6):1521–1527.

Yeung, D. and Petrosyan, L. (2011). Subgame consistent cooperative solution of dynamic games with random horizon. *Journal of Optimization Theory and Applications*, 150:78–97.

Zaccour, G. (2017). Sustainability of cooperation in dynamic games played over event trees. In Melnik, R., Makarov, R., and Belair, J., editors, *Recent Progress and Modern Challenges in Applied Mathematics, Modeling and Computational Science*, pages 419–437, New York, NY. Springer New York.

Chapter 7
Node-Consistent Core in DGPETs

This chapter deals with the node consistency of the core in cooperative dynamic games played over event trees. Being a set-valued solution, the core does not (in general) uniquely define the profile of payoffs to the players. This is the first major difference from the single-valued solutions discussed in the previous chapter. The second difference is that the core may be empty. To overcome the latter problem, we consider the ε-core, which is always nonempty for some value ε. The node-consistency results provided in this chapter can easily be adapted to any set-valued cooperative solution.

7.1 Cooperative DGPETs

We recall the elements of the game and some of the notation needed previously given in the other sections.

Denote by $M = \{1, \dots, m\}$ the set of players. Player $i \in M$ chooses a control $u^i_{n_t} \in U^i_{n_t}$ at node $n_t \in \mathbf{n}_t$, $t \in \mathbb{T} \setminus \{T\}$. The state equations are given as

$$x_{n_t} = f^{n_t^-}\left(x_{n_t^-}, u_{n_t^-}\right), \quad x_{n_0} = x_0, \tag{7.1}$$

where $u_{n_t^-} \in U_{n_t^-}$, $n_t \in \mathbf{n}_t$, $t \in \mathbb{T} \setminus \{0\}$ and x_0 is the initial state at root node n_0.

The payoff functional of Player $i \in M$ is given by

$$J_i(\mathbf{x}, \mathbf{u}) = \sum_{t=0}^{T-1} \rho^t \sum_{n_t \in \mathbf{n}_t} \pi_{n_t} \phi_i(n_t, x_{n_t}, u_{n_t}) + \rho^T \sum_{n_T \in \mathbf{n}_T} \pi_{n_T} \Phi_i(n_T, x_{n_T}), \tag{7.2}$$

© The Author(s), under exclusive license to Springer Nature Switzerland AG 2022   191
E. Parilina et al., *Theory and Applications of Dynamic Games*, Theory and Decision
Library C 51, https://doi.org/10.1007/978-3-031-16455-2_7

where $\phi_i(n_t, x_{n_t}, u_{n_t})$ is the reward at node n_t; π_{n_t} is the probability of passing through that node; $\Phi_i(n_T, x_{n_T})$ is the payoff at terminal node n_T; π_{n_T} is the probability that the game ends at node n_T; and ρ is the discount factor $(0 < \rho < 1)$.

The players use the S-adapted open-loop strategies introduced in Chap. 5. The profile $\mathbf{u} = (\mathbf{u}^i : i \in M)$ is the S-adapted strategy vector of the m players. The game in normal form is defined by the payoffs $W_i(x_0, \mathbf{u}) = J_i(\mathbf{x}, \mathbf{u})$, $i \in M$, where $\mathbf{x}$ is obtained from $\mathbf{u}$ as the unique solution of the state equations (7.1) that emanate from the initial state x_0. The profile $\mathbf{u}^{nc} = (\mathbf{u}^{i,nc} : i \in M)$ is the Nash equilibrium in the game with S-adapted open-loop strategies.

If the players decide to cooperate in the entire game, then they choose the strategies that maximize their total discounted payoffs, that is,

$$\max_{\mathbf{u}^i, i \in M} \sum_{i \in M} W_i(x_0, \mathbf{u}).$$

Denote by $\mathbf{u}^*$ the resulting vector of cooperative controls, i.e.,

$$\mathbf{u}^* = \arg \max_{\mathbf{u}^i, i \in M} \sum_{i \in M} W_i(x_0, \mathbf{u}).$$

Denote by $\mathbf{x}^* = \{x_{n_\tau}^* : n_\tau \in \mathbf{n}_0^{++}\}$ the cooperative state trajectory generated by $\mathbf{u}^*$.

We need in the sequel to distinguish between cooperative and noncooperative strategies, and between outcomes realized in different subgames along the cooperative state trajectory $\mathbf{x}^*$. Denote by $\mathbf{u}^*[n_t, x_{n_t}^*]$ the profile of cooperative controls restricted to the subgame starting from node $n_t \in \mathbf{n}_t$ with state $x_{n_t}^*$, and by $\mathbf{u}^{nc}[n_t, x_{n_t}^*]$ the Nash equilibrium in the subgame starting from node n_t with cooperative state $x_{n_t}^*$.

Remark 7.1 As the objective is to design sustainable cooperative agreements, the analysis must be carried out along the cooperative state trajectory. That is, at any intermediate node $n_t \in \mathbf{n}_t$, the state is $x_{n_t}^*$, meaning that cooperation has prevailed along the path $[n_0, n_t]$.

As in the previous chapter, we suppose that the joint-optimization solution and the Nash equilibrium in the whole game and in any subgame are unique. The sufficient conditions for the existence and uniqueness of the joint-optimization solution and the Nash equilibrium are discussed in Chap. 1. Also, we recall that, in any subgame, including the whole game, the characteristic function (CF) is defined by

$$v\left(\cdot; x_{n_t}^*\right) : 2^M \to \mathbb{R}, \tag{7.3}$$

with $v\left(\varnothing; x_{n_t}^*\right) = 0$ for any $n_t \in \mathbf{n}_0^{++}$. Obviously, for the grand coalition M, it holds true that $v\left(M; x_{n_t}^*\right) = \sum_{i \in M} W_i(x_{n_t}^*, \mathbf{u}^*[n_t, x_{n_t}^*])$ for any node $n_t \in \mathbf{n}_0^{++}$.

The imputation set $Y(x_{n_t}^*)$ in the subgame starting from node n_t and the state $x_{n_t}^*$ is given by

$$Y(x_{n_t}^*) = \left\{ (y_1, \ldots, y_m) \mid y_i \geq v\left(\{i\}; x_{n_t}^*\right), \forall i \in M, \text{ and } \sum_{i \in M} y_i = v\left(M; x_{n_t}^*\right) \right\}. \tag{7.4}$$

Remark 7.2 The theoretical developments in this chapter apply to any approach used to compute the CF values, that is, the α, β, γ, δ, and potentially other approaches. Consequently, we do not opt for a specific one. However, in the example in Sect. 4, the values are computed assuming a γ-CF.

7.2 Node Consistency of the Core

Suppose that the players agree to cooperate for the whole duration of the game, and adopt the core as an allocation mechanism for the players' total cooperative outcome given by $\sum_{i \in M} W_i(x_0, \mathbf{u}^*)$. First, we recall the definition of an unstable imputation in the terminology of DGPETs.

Definition 7.1 An imputation $y(x_{n_t}^*) = (y_1(x_{n_t}^*), \ldots, y_m(x_{n_t}^*))$ is said to be unstable through a coalition $K \subset M$ if $v(K; x_{n_t}^*) > \sum_{i \in K} y_i(x_{n_t}^*)$. The imputation $y(x_{n_t}^*)$ is unstable if there is a coalition K such that $y(x_{n_t}^*)$ is unstable through K; otherwise, $y(x_{n_t}^*)$ is stable.

Second, we define the core in the subgame starting from any node n_t.

Definition 7.2 The set $C(x_{n_t}^*), n_t \in \mathbf{n}_0^{++} \setminus \{n_0\}$, of stable imputations $y(x_{n_t}^*)$ is called the core of the subgame starting from node n_t with state $x_{n_t}^*$.

The core of the subgame starting from node n_t with state $x_{n_t}^*$ is

$$C(x_{n_t}^*) = \Big\{ (y_1(x_{n_t}^*), \ldots, y_m(x_{n_t}^*)) \mid \sum_{i \in K} y_i(x_{n_t}^*) \geq v(K; x_{n_t}^*)$$

$$\forall K \subset M, \text{ and } \sum_{i \in M} y_i(x_{n_t}^*) = v(M; x_{n_t}^*) \Big\}. \tag{7.5}$$

To define the imputation set and core of the whole game, it suffices to set $n_t = n_0$ in the above definition and in (7.4) and (7.5).

Remark 7.3 The core may contain a large number of imputations or it may be empty.

1. We assume that the set $C(x_{n_t}^*)$ is nonempty for all $n_t \in \mathbf{n}_t$, $t \in \mathbb{T}$. Otherwise, the idea of designing node-consistent payments becomes irrelevant. The conditions of the nonemptiness of the core are discussed in Theorem 2.3.
2. The problem of selecting one specific imputation from a core that contains more than one is an open problem and far beyond the objectives of this book. For illustration purposes, in Sect. 7.4, we use the nucleolus[1] (see Schmeidler 1969) as a basis for constructing the IDP. The nucleolus is a single-valued solution concept and it belongs to the core if it is not empty (see Peleg and Sudhölter 2003).

---

[1] The nucleolus is related to the idea of excess in a coalition K, which is the difference between the sum of the payoffs of K's members and the characteristic function value of K. If we rank all the excesses from largest to smallest, and find the smallest possible vector of excesses over the set of imputations, we get the nucleolus (see Peleg and Sudhölter 2003).

Now, we consider the problem of the node consistency of the core. First, we define the decomposition over nodes of any agreed-upon imputation from the core of a DGPET. Second, we provide the conditions for the node consistency of that imputation and of the core in general.

Definition 7.3 We call $\left(\{\beta_i(x^*_{n_t})\}_{n_t \in \mathbf{n}_0^{++}} : i \in M\right)$ an imputation distribution procedure of the imputation $y(x_0) = (y_1(x_0), \ldots, y_m(x_0)) \in C(x_0)$ if, for all $i \in M$, the following condition holds:

$$y_i(x_0) = \sum_{t=0}^{T} \rho^t \sum_{n_t \in \mathbf{n}_t} \pi_{n_t} \beta_i\left(x^*_{n_t}\right), \tag{7.6}$$

where $\beta_i(x^*_{n_t})$ is the payment to Player i at node n_t with the cooperative state $x^*_{n_t}$.

Equation (7.6) means that the expected sum of the discounted payments to Player $i \in M$ is equal to her component in the imputation from the core that the players chose as an allocation mechanism for the entire game.

Remark 7.4 The quantities $v\left(K; x^*_{n_t}\right)$, $y(x^*_{n_t})$, $C(x^*_{n_t})$, and $\beta\left(x^*_{n_t}\right)$ depend on the node n_t. As it already appears in $x^*_{n_t}$, and to keep the notation simple, we refrain from adding n_t as an argument.

Remark 7.5 For our purposes, we only need to compute the core $C(x^*_{n_t})$ and the IDP $\beta\left(x^*_{n_t}\right)$ along the cooperative state $x^*_{n_t}$, which is obtained as a unique solution of the state dynamics equations after substitution for the cooperative strategies $\mathbf{u}^{i*}$. The core can also be defined for any other state trajectory.

Definition 7.4 The imputation $y(x_0) = (y_1(x_0), \ldots, y_m(x_0)) \in C(x_0)$ and the corresponding IDP

$$\left(\{\beta_i(x^*_{n_t})\}_{n_t \in \mathbf{n}_0^{++}} : i \in M\right)$$

are called node consistent in the DGPET if, for any $t = 0, \ldots, T$, there exists $y(x^*_{n_t}) = (y_1(x^*_{n_t}), \ldots, y_m(x^*_{n_t})) \in C(x^*_{n_t})$ satisfying the following condition:

$$\sum_{\tau=0}^{t-1} \sum_{n_\tau \in \mathbf{n}_\tau} \rho^\tau \pi_{n_\tau} \beta_i(x^*_{n_\tau}) + \rho^t \sum_{n_t \in \mathbf{n}_t} \pi_{n_t} y_i\left(x^*_{n_t}\right) = y_i(x_0), \tag{7.7}$$

for any player $i \in M$.

The interpretation of the above definition is as follows: Suppose that the players agreed on the imputation $y(x_0)$ and the IDP $\left(\{\beta_i(x^*_{n_t})\}_{n_t \in \mathbf{n}_0^{++}} : i \in M\right)$. Consider the subgame that starts at any intermediate node n_t, with a state value $x^*_{n_t}$. At this node,

Player $i \in M$ would have thus far collected $\sum_{\tau=0}^{t-1} \sum_{n_\tau \in \mathbf{n}_\tau} \rho^\tau \pi_{n_\tau} \beta_i(x_{n_\tau}^*)$. The node consistency of imputation $y(x_0) \in C(x_0)$ means that we can define an allocation $y_i\left(x_{n_t}^*\right)$ in the continuation game starting at n_t, which is in the core of the subgame, with the equality in (7.7) holding true.

Definition 7.5 The core $C(x_0)$ of a cooperative DGPET is a node-consistent allocation mechanism if any imputation $y(x_0) \in C(x_0)$ is node consistent.

Denote by $\pi_{n_t}^{n_{t+1}}$ the conditional probability that node $n_{t+1} \in \mathbf{n}_t^+$ is reached if node n_t has already been reached. The following theorem provides an explicit formula for the IDP $\left(\{\beta_i(x_{n_t}^*)\}_{n_t \in \mathbf{n}_0^{++}} : i \in M\right)$.

Theorem 7.1 *Suppose that the core $C(x_{n_t}^*)$ of the subgame starting from any node $n_t \in \mathbf{n}_t, t = 0, \ldots, T$ is nonempty. If Player $i \in M$ is paid*

$$\beta_i(x_{n_t}^*) = y_i(x_{n_t}^*) - \rho \sum_{n_{t+1} \in \mathbf{n}_t^+} \pi_{n_t}^{n_{t+1}} y_i\left(x_{n_{t+1}}^*\right), \quad \text{at any } n_t \in \mathbf{n}_0^{++} \setminus \mathbf{n}_T, \tag{7.8}$$

and

$$\beta_i(x_{n_T}^*) = y_i(x_{n_T}^*), \quad \text{at } n_T \in \mathbf{n}_T, \tag{7.9}$$

where $y(x_{n_t}^*) = (y_1(x_{n_t}^*), \ldots, y_m(x_{n_t}^*)) \in C(x_{n_t}^*)$ for any $n_t \in \mathbf{n}_t, t = 0, \ldots, T$, then the core $C(x_0)$ is node consistent.

Proof First, we prove that the components $\beta_i(x_{n_t}^*)$ determined by (7.8) and (7.9) form an IDP of an imputation $y(x_0) \in C(x_0)$. The satisfaction of (7.6) follows from the IDP construction by (7.8) and (7.9) and taking into account that

$$\pi_{n_t} \pi_{n_t}^{n_{t+1}} = \pi_{n_{t+1}}$$

for any $n_{t+1} \in \mathbf{n}_t^+, t = 0, \ldots, T-1$. Indeed, we have

$$\sum_{\theta=0}^{T} \sum_{n_\theta \in \mathbf{n}_\theta} \rho^\theta \pi_{n_\theta} \beta_i(x_{n_\theta}^*) = \sum_{\theta=0}^{T-1} \rho^\theta \sum_{n_\theta \in \mathbf{n}_\theta} \pi_{n_\theta} \left\{ y_i(x_{n_\theta}^*) - \rho \sum_{n_{\theta+1} \in \mathbf{n}_\theta^+} \pi_{n_\theta}^{n_{\theta+1}} y_i\left(x_{n_{\theta+1}}^*\right) \right\}$$

$$+ \rho^T \sum_{n_T \in \mathbf{n}_T} \pi_{n_T} y_i(x_{n_T}^*)$$

$$= \pi_{n_0} y_i(x_0) - \rho \sum_{n_1 \in \mathbf{n}_0^+} \pi_{n_0} \pi_{n_0}^{n_1} y_i\left(x_{n_1}^*\right)$$

$$+ \rho \sum_{n_1 \in \mathbf{n}_0^+} \pi_{n_1} y_i\left(x_{n_1}^*\right) - \ldots + \rho^T \sum_{n_T \in \mathbf{n}_{T-1}^+} \pi_{n_T} y_i(x_{n_T}^*)$$

$$- \rho^T \sum_{n_T \in \mathbf{n}_{T-1}^+} \pi_{n_{T-1}} \pi_{n_{T-1}}^{n_T} y_i(x_{n_T}^*)$$

$$= y_i(x_0).$$

If $C(x_0)$ and $C(x_{n_t}^*)$ are nonempty for any node n_t, then we can always find at least one imputation $y(x_{n_t}^*) \in C(x_{n_t}^*)$. Using $y(x_{n_t}^*)$, we construct the IDP $\left(\{\beta_i(x_{n_t}^*)\}_{n_t \in \mathbf{n}_0^{++}} : i \in M \right)$ with formulas (7.8) and (7.9) for any imputation from the core $C(x_0)$. The proof that this IDP is node consistent, i.e., satisfies equation (7.7), can be obtained by substituting $\beta_i(x_{n_\theta}^*)$ from (7.8) and (7.9) into (7.7) and taking into account that $\pi_{n_0} = 1$ and $\pi_{n_t}\pi_{n_t}^{n_{t+1}} = \pi_{n_{t+1}}$ for any $n_{t+1} \in \mathbf{n}_t^+$, $t = 0, \ldots, T-1$. Indeed, we have

$$\sum_{\theta=0}^{t-1} \sum_{n_\theta \in \mathbf{n}_\theta} \rho^\theta \pi_{n_\theta} \beta_i(x_{n_\theta}^*) + \rho^t \sum_{n_t \in \mathbf{n}_t} \pi_{n_t} y_i\left(x_{n_t}^*\right)$$

$$= \pi_{n_0} y_i(x_0) - \rho \sum_{n_1 \in \mathbf{n}_0^+} \pi_{n_0} \pi_{n_0}^{n_1} y_i\left(x_{n_1}^*\right) + \rho \sum_{n_1 \in \mathbf{n}_0^+} \pi_{n_1} y_i\left(x_{n_1}^*\right) - \cdots$$

$$- \rho^t \sum_{n_t \in \mathbf{n}_{t-1}^+} \pi_{n_{t-1}} \pi_{n_{t-1}}^{n_t} y_i\left(x_{n_t}^*\right) + \rho^t \sum_{n_t \in \mathbf{n}_t} \pi_{n_t} y_i\left(x_{n_t}^*\right)$$

$$= y_i(x_0),$$

and we easily get (7.7) using that $\pi_{n_0} = 1$ and $\pi_{n_t}\pi_{n_t}^{n_{t+1}} = \pi_{n_{t+1}}$ for any $n_{t+1} \in \mathbf{n}_t^+$, $t = 0, \ldots, T-1$. $\square$

Corollary 7.1 *The IDP defined by (7.8) and (7.9) satisfies the following equalities:*

$$\sum_{i \in M} \beta_i\left(x_{n_t}^*\right) = \sum_{i \in M} \phi_i(n_t, x_{n_t}^*, u_{n_t}^*), \text{ for all } n_t \in \mathbf{n}_0^{++} \setminus \mathbf{n}_T, \tag{7.10}$$

$$\sum_{i \in M} \beta_i\left(x_{n_T}^*\right) = \sum_{i \in M} \Phi_i(n_T, x_{n_T}^*), \text{ for any } n_T \in \mathbf{n}_T. \tag{7.11}$$

Proof We show that conditions (7.10) and (7.11) are satisfied if we use the IDP defined by (7.8) and (7.9). For terminal nodes, condition (7.11) is obviously satisfied. Consider condition (7.10) for any nonterminal node n_t:

$$\sum_{i \in M} \beta_i\left(x_{n_t}^*\right) = \sum_{i \in M} y_i(x_{n_t}^*) - \sum_{i \in M} \rho \sum_{n_{t+1} \in \mathbf{n}_t^+} \pi_{n_t}^{n_{t+1}} y_i(x_{n_{t+1}}^*)$$

$$= v(M; x_{n_t}^*) - \left\{ \sum_{i \in M} W_i(x_{n_t}^*, \mathbf{u}^*[n_t, x_{n_t}^*]) - \sum_{i \in M} \phi_i(n_t, x_{n_t}^*, u_{n_t}^*) \right\},$$

and given that

$$v(M; x_{n_t}^*) = \sum_{i \in M} W_i(x_{n_t}^*, \mathbf{u}^*[n_t, x_{n_t}^*]),$$

and based on the characteristic function construction, we get equality (7.10). $\square$

Equations in (7.10) and (7.11) state that the sum of payments to the players in any intermediate node, respectively, terminal node, is equal to the total realized payoff in this node using the cooperative controls $\mathbf{u}^{i*}$.

Implementing the imputation distribution procedure

$$\left(\{\beta_i(x^*_{n_t})\}_{n_t \in \mathbf{n}_0^{++}} : i \in M \right)$$

requires the definition of side payments or transfers between the players according to the following rule for any nonterminal node $n_t \in \mathbf{n}_t$, $t = 0, \ldots, T-1$:

$$\zeta_i(n_t) = \beta_i(x^*_{n_t}) - \phi_i(n_t, x^*_{n_t}, u^*_{n_t}), \tag{7.12}$$

where $\zeta_i(n_t)$ is the transfer payment that Player i makes at node n_t over the cooperative trajectory $x^*_{n_t}$, such that $\sum_{i \in M} \zeta_i(n_t) = 0$ for any node n_t. Similarly, for any terminal node $n_T \in \mathbf{n}_T$, we have

$$\zeta_i(n_T) = \beta_i(x^*_{n_T}) - \Phi_i(n_T, x^*_{n_T}). \tag{7.13}$$

Clearly, $\zeta_i(n_t)$ and $\zeta_i(n_T)$ can assume any sign depending on the sign of the difference in the right-hand sides of (7.12)–(7.13). We should notice that if the IDP is defined by (7.8) and (7.9), i.e., it is node consistent, then $\zeta_i(n_T) = 0$ for any terminal node $n_T \in \mathbf{n}_T$.

7.3 The ε-Core and the Least Core

As justification for the idea of the ε-core, consider a three-player cooperative game having the following CF values[2]:

$$v(\{1\}) = v(\{2\}) = v(\{3\}) = 0,$$
$$v(\{1, 2\}) = v(\{1, 3\}) = v(\{2, 3\}) = a, \quad v(\{1, 2, 3\}) = 1.$$

An imputation $(y_i : i \in K)$ is in the core if it satisfies the conditions

$$\sum_{i \in K} y_i \geq v(K) \text{ for all } K \subset M, \text{ and } \sum_{i \in M} y_i = 1.$$

Let $a = \frac{2}{3}$. It is easy to verify that the core is the allocation $\left(\frac{1}{3}, \frac{1}{3}, \frac{1}{3} \right)$, i.e., a singleton, which is the best outcome that one can envision. Suppose now that $a = \frac{2}{3} + \varepsilon$, where ε is a small positive number. The core is now empty, which is a huge qualitative difference from the case where $a = \frac{2}{3}$. In this example, any two-player coalition is

---

[2] This example was discussed in Example 2.5.

willing to deviate from the grand coalition even if the benefit is infinitesimal. If we
imagine that it is only worth deviating from the grand coalition, which allocates the
payoff using the core element, if the benefit is larger than a certain threshold, here ε,
then we would have defined the ε-core. Put differently, it is as though deviating from
the core costs ε, and no coalition can benefit from deviation if it has to pay this cost.

Definition 7.6 The set $C_\varepsilon(x_0)$ is called the ε-core of a DGPET if

$$C_\varepsilon(x_0) = \left\{ (y_1(x_0), \ldots, y_m(x_0)) | \sum_{i \in K} y_i(x_0) \geq v(K; x_0) - \varepsilon, \right.$$

$$\left. \forall K \subset M, \ K \neq \varnothing, M, \ \text{and} \ \sum_{i \in M} y_i(x_0) = v(M; x_0) \right\}, \quad (7.14)$$

where $\varepsilon \in \mathbb{R}$.

The core of a game is its ε-core when $\varepsilon = 0$. The ε-core is the set of efficient payoff
vectors that cannot be improved upon by any coalition if a coalition formation or
coalition deviation entails a cost of ε, if ε is positive, or a bonus of $-\varepsilon$, if ε is negative.
Naturally, we are interested in finding the smallest value of ε such that the ε-core is
nonempty.

Definition 7.7 The least core of a DGPET, denoted by $LC(x_0)$, is the intersection
of all nonempty ε-cores of the DGPET.

Let ε^* be the smallest ε for which $C_\varepsilon(x_0) \neq \varnothing$. Then, the least core $LC(x_0)$ is
$C_{\varepsilon^*}(x_0)$. Notice that ε^* can be positive, negative, or zero.

We mention some properties and offer some remarks about ε-cores and the least
core:

1. If $C(x_0) \neq \varnothing$, which means that $\varepsilon^* \leq 0$, then the least core occupies a central
   position within the core. If $C(x_0) = \varnothing$, that is, when $\varepsilon^* > 0$, then the least core
   may be regarded as revealing the "latent" position of the core (see Peleg and
   Sudhölter 2003).
2. In Shapley and Shubik (1966), the ε-core is called the strong ε-core, in order to
   distinguish it from another generalization of the core called the weak ε-core.
3. The excess or difference $v(K; x_0) - \sum_{i \in K} y_i(x_0)$ can be interpreted as a
   measure of coalition K's dissatisfaction with the imputation $y(x_0)$. If $y(x_0) \in$
   $C(x_0)$, this means that all coalitions are satisfied with $y(x_0)$. If $y(x_0) \in C_\varepsilon(x_0)$,
   then the measure of any coalition's dissatisfaction with imputation $y(x_0)$ is limited to ε. The value of ε determines the measure of dissatisfaction as $v(K; x_0) -$
   $\sum_{i \in K} y_i(x_0) \leq \varepsilon$.
4. It is clear that the set $C_\varepsilon(y_0)$ becomes larger as ε increases, and $C_\varepsilon(y_0)$ gets
   smaller as ε decreases. In other words, $C_{\varepsilon'}(y_0) \subseteq C_\varepsilon(y_0)$ if $\varepsilon' < \varepsilon$, and moreover,
   if $C_\varepsilon(y_0) \neq \varnothing$, then there is a strict inclusion $C_{\varepsilon'}(y_0) \subset C_\varepsilon(y_0)$. Therefore, we
   can shrink (or expand if necessary) the set $C_\varepsilon(y_0)$ by adjusting ε until we get one,
   and only one, imputation in it, if possible.

5. Let $\varepsilon \in \mathbb{R}$. Then, the ε-core $C_\varepsilon(x_0)$ is a convex compact polyhedron bounded by no more than $2m - 2$ hyperplanes of the form

$$\left\{(y_1(x_0), \ldots, y_m(x_0)) \middle| \sum_{i \in K} y_i(x_0) = v(K; x_0) - \varepsilon, \forall K \subset M, \ K \neq \varnothing, M \right.$$

$$\left. \text{and} \sum_{i \in M} y_i(x_0) = v(M; x_0) \right\}.$$

$$(7.15)$$

The computation of the ε-core is reduced to solving the following linear programming problem[3]:

$$\min \varepsilon \qquad\qquad (7.16)$$

$$\text{s.t.}$$

$$v(K; x_0) - \sum_{i \in K} y_i(x_0) \leq \varepsilon, \ \forall K \subset M, \ K \neq M.$$

We illustrate the determination of the ε-core and the least core with two examples. Although the core is nonempty in Example 7.1, we provide this example for the geometrical representation of the results.

Example 7.1 Let the cooperative game be given by the characteristic function

$$v(\{1\}) = 2, \ v(\{2\}) = 3, \ v(\{3\}) = 4,$$
$$v(\{1, 2\}) = 6, \ v(\{2, 3\}) = 7, \ v(\{1, 3\}) = 8,$$
$$v(\{1, 2, 3\}) = 15.$$

The nonempty core of the game is the set

$$C = \{(y_1, y_2, y_3) : y_1 \geq 2, \ y_2 \geq 3, \ y_3 \geq 4, \ y_1 + y_2 \geq 6,$$
$$y_2 + y_3 \geq 7, \ y_1 + y_3 \geq 8, \ y_1 + y_2 + y_3 = 15\}.$$

The ε-core of the game is the set

$$C_\varepsilon = \{(y_1, y_2, y_3) : y_1 \geq 2 - \varepsilon, \ y_2 \geq 3 - \varepsilon, \ y_3 \geq 4 - \varepsilon, \ y_1 + y_2 \geq 6 - \varepsilon,$$
$$y_2 + y_3 \geq 7 - \varepsilon, \ y_1 + y_3 \geq 8 - \varepsilon, \ y_1 + y_2 + y_3 = 15\}.$$

The core and ε-core, $\varepsilon = -1$ are represented in Fig. 7.1. To find the least core, we solve the following linear programming problem:

---

[3] For examples of calculations of the ε-core and the least core, see Barron (2013).

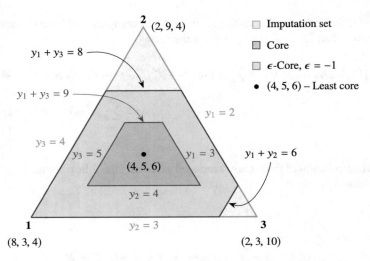

Fig. 7.1 Core, ε-core, and least core of the cooperative game from Example 7.1

$$\min \varepsilon$$

$$\text{s.t.}$$

$$y_1 \geq 2 - \varepsilon, \quad y_2 \geq 3 - \varepsilon, \quad y_3 \geq 4 - \varepsilon,$$

$$y_1 + y_2 \geq 6 - \varepsilon, \quad y_2 + y_3 \geq 7 - \varepsilon, \quad y_1 + y_3 \geq 8 - \varepsilon,$$

$$y_1 + y_2 + y_3 = 15.$$

The unique solution is given by $y = (4, 5, 6)$ and $\varepsilon^* = -2$.

Example 7.2 Consider the cooperative game from Example 7.1 and only change $v(\{1, 2\}) = 6$ to $v(\{1, 2\}) = 12$.

The core of the game is empty. The ε-core is the set

$$C_\varepsilon = \{(y_1, y_2, y_3) : y_1 \geq 2 - \varepsilon, \ y_2 \geq 3 - \varepsilon, \ y_3 \geq 4 - \varepsilon, \ y_1 + y_2 \geq 12 - \varepsilon,$$
$$y_2 + y_3 \geq 7 - \varepsilon, \ y_1 + y_3 \geq 8 - \varepsilon, \ y_1 + y_2 + y_3 = 15\},$$

and it is nonempty only if $\varepsilon > 0$. The ε-core when $\varepsilon = 1$ is shown in Fig. 7.2. We notice that the ε-core is not contained in the imputation set. For this reason, we refer to the elements of the ε-core as payoff vectors because (in general) they may not satisfy the individual rationality conditions. Solving the linear programming problem

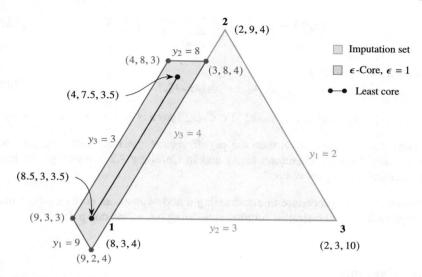

Fig. 7.2 The imputation set, ε-core, and least core of the cooperative game from Example 7.2

$$\min \varepsilon$$

s.t.

$$y_1 \geq 2 - \varepsilon, \quad y_2 \geq 3 - \varepsilon, \quad y_3 \geq 4 - \varepsilon,$$

$$y_1 + y_2 \geq 12 - \varepsilon, \quad y_2 + y_3 \geq 7 - \varepsilon, \quad y_1 + y_3 \geq 8 - \varepsilon,$$

$$y_1 + y_2 + y_3 = 15,$$

we find the least core, that is, the line segment $y = \{(y_1, 11.5 - y_1, 3.5), 4 \leq y_1 \leq 8.5\}$ with $\varepsilon^* = 0.5$ (see Fig. 7.2). In this case, the least core contains more than one imputation. Note that Player 3 gets 3.50 in the least core, which is less than what she could obtain by acting alone, that is, $v(\{1\}) = 4$.

It seems reasonable to consider the ε-core as a cooperative solution if the core is empty in one or more subgames. In such cases, the players must agree on the value of ε for which the ε-core is nonempty in all subgames. Consequently, the node-consistent ε-core may be constructed from the elements of the least core. This means that, when we define the distribution procedure for any payoff vector from the ε-core by (7.8)–(7.9), we can adopt the least core as $y(\cdot)$ if it is a singleton or choose one particular element if it has more than one.

Corollary 7.2 *Let $\varepsilon \in \mathbb{R}$ be such that the ε-core $C_\varepsilon(x_0)$ of the game and the ε-core $C_\varepsilon(x_{n_t}^*)$ of the subgame starting from any nonterminal node $n_t \in \mathbf{n}_t, t = 1, \ldots, T - 1$ are nonempty. Then, the ε-core $C_\varepsilon(x_0)$ is node consistent if the players are paid with the corresponding distribution procedure for each payoff vector $y(x_0) \in C_\varepsilon(x_0)$ satisfying the following conditions for any $n_t \in \mathbf{n}_0^{++} \setminus \mathbf{n}_T$:*

$$\beta_i(x_{n_t}^*) = y_i(x_{n_t}^*) - \rho \sum_{n_{t+1} \in \mathbf{n}_t^+} \pi_{n_t}^{n_{t+1}} y_i\left(x_{n_{t+1}}^*\right), \tag{7.17}$$

and for $n_T \in \mathbf{n}_T$:

$$\beta_i(x_{n_T}^*) = \phi_i(n_T, x_{n_T}^*), \tag{7.18}$$

where $y(x_{n_t}^*) = (y_1(x_{n_t}^*), \ldots, y_m(x_{n_t}^*)) \in C_\varepsilon(x_{n_t}^*)$ *for any* $n_t \in \mathbf{n}_t, t = 0, \ldots, T - 1$.

Note that if ε is positive, then the payoff vectors from the ε-core may be not individually rational. Therefore, above and in Corollary 7.2, we change the term "imputation" to "payoff vector."

Remark 7.6 The procedure of constructing a node-consistent core or ε-core may be applied for any cooperative solution containing more than one point.

7.4   Example

We illustrate the concepts seen in this chapter with an example of transboundary pollution control.[4]

Denote by $M = \{1, \ldots, m\}$ the set of players representing countries and by $\mathbb{T} = \{0, \ldots, T\}$ the set of periods. Production activities in each country generate revenues and, as a by-product, pollutant emissions, e.g., CO_2. Denote by $q_{n_t}^i$ Player i's production of goods and services at node $n_t \in \mathbf{n}_t$, $t \in \mathbb{T} \setminus \{T\}$, and by $u_{n_t}^i$ the resulting emissions, with $u_{n_t}^i = h_i\left(q_{n_t}^i\right)$, where $h_i(\cdot)$ is an increasing function satisfying $h_i(0) = 0$. Assuming a monotone increasing relationship between production and revenues, we can express the revenues, denoted by R_i, directly as a function of emissions. To keep it simple, we adopt the following specification:

$$R_i\left(u_{n_t}^i\right) = \alpha u_{n_t}^i - \frac{1}{2}\left(u_{n_t}^i\right)^2, \tag{7.19}$$

where α is a positive parameter. We constrain the revenues to be nonnegative for any $n_t \in \mathbf{n}_0^{++} \setminus \mathbf{n}_T$ and any player i, which will be reflected in the Karush-Kuhn-Tucker (KKT) conditions stated later on.

Denote by $u_{n_t} = (u_{n_t}^1, \ldots, u_{n_t}^m)$ the vector of countries' emissions at node $n_t \in \mathbf{n}_t$, $t \in \mathbb{T} \setminus \{T\}$, and by x_{n_t} the stock of pollution at this node. The evolution of this stock is governed by the following difference equation:

$$x_{n_t} = (1 - \delta_{n_t^-})x_{n_t^-} + \sum_{i \in M} u_{n_t^-}^i, \ n_t \in \mathbf{n}_0^{++} \setminus \{n_0\}, \ x_{n_0} = x_0, \tag{7.20}$$

---

[4] For background on this class of games, see the surveys in Jørgensen et al. (2010), De Zeeuw (2018).

where the initial state x_0 is given for the root node n_0, and δ_{n_t} $(0 < \delta_{n_t} < 1)$ is a stochastic rate of pollution absorption by Mother Nature at node n_t. We suppose that δ_{n_t} can take two possible values, namely, $\delta_{n_t} \in \{\underline{\delta}, \overline{\delta}\}$, with $\underline{\delta} < \overline{\delta}$.

Each country suffers an environmental damage cost due to pollution accumulation. We assume that this cost is an increasing convex function in this stock and we retain the quadratic form $D_i(x_{n_t}) = c_i x_{n_t}^2$, $i \in M$, where c_i is a strictly positive parameter. Denote by $\rho \in (0, 1)$ the discount factor. Assume that Player $i \in M$ maximizes the following objective functional:

$$J_i(\mathbf{x}, \mathbf{u}) = \sum_{t=0}^{T-1} \rho^t \sum_{n_t \in \mathbf{n}_t} \pi_{n_t} \left(R_i(u_{n_t}^i) - D_i(x_{n_t}) \right) + \rho^T \sum_{n_T \in \mathbf{n}_T} \pi_{n_T} \Phi_i(n_T, x_{n_T}),$$

subject to (7.20), and $u_{n_t}^i \geq 0$ for all $i \in M$ and any $n_t \in \mathbf{n}_t, t = 0, \ldots, T - 1$. Suppose that the payoff function of Player i at a terminal node n_T is

$$\Phi_i(n_T, x_{n_T}) = -c_i x_{n_T}.$$

Substituting $\Phi_i(n_T, x_{n_T})$, $R_i\left(u_{n_t}^i\right)$, and $D_i(x_{n_t})$ by their values, we get

$$J_i(\mathbf{x}, \mathbf{u}) = \sum_{t=0}^{T-1} \rho^t \sum_{n_t \in \mathbf{n}_t} \pi_{n_t} \left(\alpha u_{n_t}^i - \frac{1}{2}(u_{n_t}^i)^2 - c_i x_{n_t}^2 \right) - \rho^T \sum_{n_T \in \mathbf{n}_T} \pi_{n_T} c_i x_{n_T}.$$

We use the following parameter values in the numerical illustration:

$$M = \{1, 2, 3\}, \ \mathbb{T} = \{0, 1, 2, 3\}, \ x_0 = 0, \ \rho = 0.9, \ \alpha = 35,$$
$$c_1 = 0.15, \ c_2 = 0.10, \ c_3 = 0.05, \ \underline{\delta} = 0.10, \ \overline{\delta} = 0.45.$$

The event tree is depicted in Fig. 7.3, where nodes n_2^1, n_2^3, n_2^4 correspond to the high level of pollution reduction $\overline{\delta} = 0.45$, and nodes n_0, n_1^1, n_2^2, n_2^3 correspond to the low level of pollution reduction $\underline{\delta} = 0.10$.

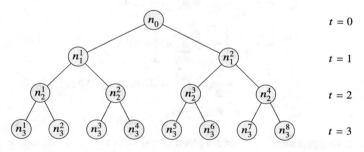

Fig. 7.3 Event tree for $T = 3$

The transition probabilities are as follows:

$$\pi_{n_1^1} = 0.3, \ \pi_{n_2^1} = 0.7,$$

$$\pi_{n_1^2} = 0.2, \ \pi_{n_2^2} = 0.1, \ \pi_{n_3^2} = 0.3, \ \pi_{n_4^2} = 0.4,$$

$$\pi_{n_1^3} = 0.05, \ \pi_{n_2^3} = 0.15, \ \pi_{n_3^3} = 0.01, \ \pi_{n_4^3} = 0.09,$$

$$\pi_{n_5^3} = 0.2, \ \pi_{n_6^3} = 0.1, \ \pi_{n_7^3} = 0.3, \ \pi_{n_8^3} = 0.1.$$

To design a node-consistent core, we first compute the cooperative controls in all subgames, including the whole game. To do this, we formulate the joint-optimization problem and write the necessary conditions using Pontryagin's maximum principle and the KKT conditions (see Chap. 5). When the players cooperate, they form the grand coalition $M = \{1, 2, 3\}$. To simplify the notation, we refer to this coalition by 123. The pseudo-Lagrangian (5.11) is given by

$$\bar{\mathcal{L}}_{123}(n_t, x_{n_t}, u_{n_t}, \lambda_{\mathbf{n}_t^+}^{123}, \theta_{n_t}^{123}) = \sum_{i \in M} \left[\alpha u_{n_t}^i - \frac{1}{2}(u_{n_t}^i)^2 \right] - x_{n_t}^2 \sum_{i \in M} c_i$$

$$+ \rho \sum_{\nu \in \mathbf{n}_t^+} \pi_{n_t}^\nu \lambda_\nu^{123} \left[(1 - \delta_{n_t}) x_{n_t} + \sum_{i \in M} u_{n_t}^i \right]$$

$$- \sum_{i \in M} \theta_{n_t}^{i,123} \left[\alpha u_{n_t}^i - \frac{1}{2}(u_{n_t}^i)^2 \right], \ n_t \in \mathbf{n}_t, \ t = 0, 1, 2,$$

where $\lambda_{\mathbf{n}_t^+}^{123} = \left(\lambda_\nu^{123} : \nu \in \mathbf{n}_t^+ \right)$ are the costate variables indexed over the set of nodes following node n_t, and $\pi_{n_t}^\nu = \pi_\nu / \pi_{n_t}$, and $\theta_{n_t}^{123} = \left(\theta_{n_t}^{i,123} : i \in M \right)$ are functions of the nodes.

The necessary conditions for the maximization problem are

$$\frac{\partial \bar{\mathcal{L}}_{123}}{\partial u_{n_t}^i} = 0, \ i \in M, \ n_t \in \mathbf{n}_0^{++} \backslash \mathbf{n}_T,$$

$$\lambda_{n_t}^{123} = \frac{\partial \bar{\mathcal{L}}_{123}}{\partial x_{n_t}}, \ n_t \in \mathbf{n}_0^{++} \backslash \{n_0 \cup \mathbf{n}_T\},$$

$$\lambda_{n_T}^{123} = \frac{\partial}{\partial x_{n_T}} \left[-x_{n_T} \sum_{i \in M} c_i \right], \ n_T \in \mathbf{n}_T,$$

$$\theta_{n_t}^{i,123} \left[\alpha u_{n_t}^i - \frac{1}{2}(u_{n_t}^i)^2 \right] = 0, \ n_t \in \mathbf{n}_0^{++} \backslash \mathbf{n}_T,$$

$$\theta_{n_t}^{i,123} \geq 0, \ n_t \in \mathbf{n}_0^{++} \backslash \mathbf{n}_T,$$

such that state equations (7.20) and a given initial state $x_{n_0} = x_0$. Writing in full the above equations, the system becomes

$$\alpha_i - u^i_{n_t} + \rho \sum_{\nu \in \mathbf{n}^+_t} \pi^\nu_{n_t} \lambda^{123}_\nu - \theta^{i,123}_{n_t}(\alpha_i - u^i_{n_t}) = 0, \ i \in M, \ n_t \in \mathbf{n}_0, \mathbf{n}_1, \mathbf{n}_2,$$

$$\lambda^{123}_{n_t} = -2x_{n_t} \left(\sum_{i \in M} c_i \right) + \rho(1 - \delta_{n_t}) \sum_{\nu \in \mathbf{n}^+_t} \pi^\nu_{n_t} \lambda^{123}_\nu, \ n_t \in \mathbf{n}_1, \mathbf{n}_2,$$

$$\lambda^{123}_{n_T} = - \left(\sum_{i \in M} c_i \right), \ n_T \in \mathbf{n}_3,$$

$$\theta^{i,123}_{n_t} \left[\alpha u^i_{n_t} - \frac{1}{2} \left(u^i_{n_t} \right)^2 \right] = 0, \ n_t \in \mathbf{n}_0, \mathbf{n}_1, \mathbf{n}_2,$$

$$\theta^{i,123}_{n_t} \geq 0, \ n_t \in \mathbf{n}_0, \mathbf{n}_1, \mathbf{n}_2,$$

$$x_{n_t} = (1 - \delta_{n^-_t})x_{n^-_t} + \sum_{i \in M} u^i_{n^-_t}, \ n_t \in \mathbf{n}_1, \mathbf{n}_2, \mathbf{n}_3,$$

$$x_{n_0} = x_0.$$

Solving this system, we find the cooperative control (emissions) values $\mathbf{u}^* = \mathbf{u}^{i*}$, $i = 1, 2, 3$, where $\mathbf{u}^{i*} = (u^{i*}_{n_t} : n_t \in \mathbf{n}_t, t = 0, 1, 2)$, and the corresponding state (pollution stock) values $\mathbf{x}^* = (x^*_{n_t} : n_t \in \mathbf{n}_t, t = 0, 1, 2, 3)$, and $x^*_{n_0} = x_0$. Using the parameter values given above, we obtain the following cooperative state trajectory:

$x^*_{n_0}$	$x^*_{n^1_1}$	$x^*_{n^2_1}$	$x^*_{n^1_2}$	$x^*_{n^2_2}$	$x^*_{n^3_2}$	$x^*_{n^4_2}$
0	23.1833	23.1833	47.8221	47.8221	44.7483	44.7483

$x^*_{n^1_3}$	$x^*_{n^2_3}$	$x^*_{n^3_3}$	$x^*_{n^4_3}$	$x^*_{n^5_3}$	$x^*_{n^6_3}$	$x^*_{n^7_3}$	$x^*_{n^8_3}$
147.230	147.230	130.492	130.492	144.463	144.463	128.802	128.802

Note that nodes having the same parent have the same state value due to the one-period lag in the state dynamics.

The value of the characteristic function $v(123; x_0)$ is equal to the sum of players' payoffs, that is, $\sum_{i \in M} J_i(\mathbf{x}^*, \mathbf{u}^*)$. The values $v(123; x_0)$ for the game starting from the root node n_0 and the subgame starting from any node n_t from cooperative state $x^*_{n_t}$ are given in Table 7.1 (last row in the table).

We use the γ-approach to define the other coalitions' CF values. Therefore, Player i's value $v(\{i\}; x_{n_t}), i \in M$ in the subgame starting from $n_t \in \mathbf{n}_t, t = 0, 1, 2$, is her payoff in the S-adapted Nash equilibrium in that subgame. To obtain this equilibrium, we write down the KKT conditions and solve the corresponding system of equations.

Table 7.1 Characteristic functions for the DGPET and subgames

Time	$t = 0$	$t = 1$		$t = 2$			
Node	n_0	n_1^1	n_1^2	n_2^1	n_2^2	n_2^3	n_2^4
$v(1)$	−305.351	−81.274	62.680	249.499	251.759	292.554	294.668
$v(2)$	493.969	388.146	477.315	370.501	372.008	399.205	400.614
$v(3)$	1149.040	803.154	844.338	491.502	492.255	505.853	506.558
$v(12)$	414.849	452.190	667.150	620.013	623.779	691.772	695.296
$v(13)$	1023.880	836.805	1007.580	741.011	744.024	798.418	801.237
$v(23)$	1640.010	1223.210	1349.570	862.008	864.268	905.063	907.178
$v(123)$	2347.730	1662.900	1868.100	1111.550	1116.070	1197.660	1201.890

Time	$t = 3$							
Node	n_3^1	n_3^2	n_3^3	n_3^4	n_3^5	n_3^6	n_3^7	n_3^8
$v(1)$	−22.085	−22.085	−19.574	−19.574	−21.670	−21.670	−19.320	−19.320
$v(2)$	−14.723	−14.723	−13.049	−13.049	−14.446	−14.446	−12.880	−12.880
$v(3)$	−7.362	−7.362	−6.525	−6.525	−7.223	−7.223	−6.440	−6.440
$v(12)$	−36.808	−36.808	−32.623	−32.623	−36.116	−36.116	−32.200	−32.200
$v(13)$	−29.446	−29.446	−26.098	−26.098	−28.893	−28.893	−25.760	−25.760
$v(23)$	−22.085	−22.085	−19.574	−19.574	−21.670	−21.670	−19.320	−19.320
$v(123)$	−44.169	−44.169	−39.148	−39.148	−43.339	−43.339	−38.641	−38.641

The pseudo-Lagrangian corresponding to Player $i = 1, 2, 3$ takes the following form:

$$\bar{\mathcal{L}}_i(n_t, x_{n_t}, u_{n_t}, \lambda_{\mathbf{n}_t^+}^i, \theta_{n_t}^{i,nc}) = \alpha u_{n_t}^i - \frac{1}{2}(u_{n_t}^i)^2 - c_i x_{n_t}^2$$

$$+ \rho \sum_{\nu \in \mathbf{n}_t^+} \pi_{n_t}^\nu \lambda_\nu^i \left[(1 - \delta_{n_t}) x_{n_t} + \sum_{i \in M} u_{n_t}^i \right]$$

$$- \theta_{n_t}^{i,nc} \left[\alpha u_{n_t}^i - \frac{1}{2}(u_{n_t}^i)^2 \right], \ n_t \in \mathbf{n}_t, \ t = 0, 1, 2,$$

where $\lambda_{\mathbf{n}_t^+}^i = (\lambda_\nu^i : \nu \in \mathbf{n}_t^+)$ are Player i's costate variables indexed over the set of nodes that follow node n_t.

The necessary conditions for the maximization problem are

$$\frac{\partial \bar{\mathcal{L}}_i}{\partial u^i_{n_t}} = 0, \ i \in M, \ n_t \in \mathbf{n}^{++}_0 \backslash \mathbf{n}_T,$$

$$\lambda^i_{n_t} = \frac{\partial \bar{\mathcal{L}}_i}{\partial x_{n_t}}, \ i \in M, \ n_t \in \mathbf{n}^{++}_0 \backslash \{n_0 \cup \mathbf{n}_T\},$$

$$\lambda^i_{n_T} = \frac{\partial}{\partial x_{n_T}} \left(-c_i x_{n_T} \right), \ i \in M, \ n_T \in \mathbf{n}_T,$$

$$\theta^{i,nc}_{n_t} \left[\alpha u^i_{n_t} - \frac{1}{2} \left(u^i_{n_t} \right)^2 \right] = 0, \ \theta^{i,nc}_{n_t} \geq 0, \ i \in M, \ n_t \in \mathbf{n}^{++}_0 \backslash \mathbf{n}_T,$$

along with the state equations in (7.20), given the initial state $x_{n_0} = x_0$.

Substituting the expression of the pseudo-Lagrangian into the last system, we obtain

$$\alpha - u^i_{n_t} + \rho \sum_{\nu \in \mathbf{n}^+_t} \pi^\nu_{n_t} \lambda^i_\nu - \theta^{i,nc}_{n_t} \left[\alpha - u^i_{n_t} \right] = 0, \ i \in M, \ n_t \in \mathbf{n}_0, \mathbf{n}_1, \mathbf{n}_2,$$

$$\lambda^i_{n_t} = -2 c_i x_{n_t} + \rho (1 - \delta_{n_t}) \sum_{\nu \in \mathbf{n}^+_t} \pi^\nu_{n_t} \lambda^i_\nu, \ i \in M, \ n_t \in \mathbf{n}_1, \mathbf{n}_2,$$

$$\lambda^i_{n_T} = -c_i, \ i \in M, \ n_T \in \mathbf{n}_3,$$

$$\theta^{i,nc}_{n_t} \left[\alpha u^i_{n_t} - \frac{1}{2} \left(u^i_{n_t} \right)^2 \right] = 0, \ i \in M, \ n_t \in \mathbf{n}_0, \mathbf{n}_1, \mathbf{n}_2,$$

$$\theta^{i,nc}_{n_t} \geq 0, \ i \in M, \ n_t \in \mathbf{n}_0, \mathbf{n}_1, \mathbf{n}_2,$$

$$x_{n_t} = (1 - \delta_{n^-_t}) x_{n^-_t} + \sum_{i \in M} u^i_{n^-_t}, \ n_t \in \mathbf{n}_1, \mathbf{n}_2, \mathbf{n}_3,$$

$$x_{n_0} = x_0.$$

Solving this system, we find the Nash equilibrium $\mathbf{u}^{nc} = \mathbf{u}^{i,nc}$, $i = 1, 2, 3$, where $\mathbf{u}^{i,nc} = (u^{i,nc}_{n_t} : n_t \in \mathbf{n}_t, t = 0, 1, 2)$ and $\mathbf{x}^{nc} = (x^{nc}_{n_t} : n_t \in \mathbf{n}_t, t = 0, 1, 2, 3)$ with $x^{nc}_{n_0} = x_0$.

The CF value $v(\{i\}; x_0)$ corresponds to Player i's Nash equilibrium payoff in the entire game, that is, $J_i(\mathbf{x}^{nc}, \mathbf{u}^{nc})$. Similarly, we compute the value $v(\{i\}; x^*_{n_t})$ for the subgame starting from node n_t with cooperative state $x^*_{n_t}$. The results are reported in Table 7.1 (rows $v(1), v(2), v(3)$).

What remains to be computed are the CF values for intermediate coalitions, that is, coalitions of two players ij. Following the γ-CF approach, to obtain $v(ij; x_0)$, we need to find the Nash equilibrium in the two-player DGPET, that is, the coalition ij acting as one player and the left-out player k as the other. We show the steps for $ij = 12$. The pseudo-Lagrangian of coalition 12 is given by

$$\bar{\mathcal{L}}_{12}(n_t, x_{n_t}, u_{n_t}, \lambda^{12}_{\mathbf{n}_t^+}, \theta^{12}_{n_t}) = \sum_{i \in \{1,2\}} \left[\alpha u^i_{n_t} - \frac{1}{2} (u^i_{n_t})^2 \right] - x^2_{n_t} \sum_{i \in \{1,2\}} c_i$$

$$+ \rho \sum_{\nu \in \mathbf{n}_t^+} \pi^\nu_{n_t} \lambda^{12}_\nu \left[(1 - \delta_{n_t}) x_{n_t} + \sum_{i \in M} u^i_{n_t} \right]$$

$$- \sum_{i \in \{1,2\}} \theta^{i,12}_{n_t} \left[\alpha u^i_{n_t} - \frac{1}{2} (u^i_{n_t})^2 \right], \; n_t \in \mathbf{n}_t, \; t = 0, 1, 2,$$

and the pseudo-Lagrangian of Player 3 by

$$\bar{\mathcal{L}}_3(n_t, x_{n_t}, u_{n_t}, \lambda^3_{\mathbf{n}_t^+}, \theta^3_{n_t}) = \alpha u^3_{n_t} - \frac{1}{2} (u^3_{n_t})^2 - c_3 x^2_{n_t}$$

$$+ \rho \sum_{\nu \in \mathbf{n}_t^+} \pi^\nu_{n_t} \lambda^3_\nu \left[(1 - \delta_{n_t}) x_{n_t} + \sum_{i \in M} u^i_{n_t} \right]$$

$$- \theta^3_{n_t} \left[\alpha u^3_{n_t} - \frac{1}{2} (u^3_{n_t})^2 \right], \; n_t \in \mathbf{n}_t, \; t = 0, 1, 2,$$

where $\lambda^{12}_{\mathbf{n}_t^+} = (\lambda^{12}_\nu : \nu \in \mathbf{n}_t^+)$, $\lambda^3_{\mathbf{n}_t^+} = (\lambda^3_\nu : \nu \in \mathbf{n}_t^+)$ are the costate variables corresponding to coalition 12 and Player 3, respectively, and $\theta^{12}_{n_t} = (\theta^{i,12}_{n_t} : i \in \{1, 2\})$. Putting together the necessary conditions of the two optimization problems, we have

$$\frac{\partial \bar{\mathcal{L}}_{12}}{\partial u^i_{n_t}} = 0, \; i \in \{1, 2\}, \quad \frac{\partial \bar{\mathcal{L}}_3}{\partial u^3_{n_t}} = 0, \; n_t \in \mathbf{n}_0^{++} \backslash \mathbf{n}_T,$$

$$\lambda^{12}_{n_t} = \frac{\partial \bar{\mathcal{L}}_{12}}{\partial x_{n_t}}, \; n_t \in \mathbf{n}_0^{++} \backslash (n_0 \cup \mathbf{n}_T),$$

$$\lambda^3_{n_t} = \frac{\partial \bar{\mathcal{L}}_3}{\partial x_{n_t}}, \; n_t \in \mathbf{n}_0^{++} \backslash (n_0 \cup \mathbf{n}_T),$$

$$\lambda^{12}_{n_T} = \frac{\partial}{\partial x_{n_T}} \left(-(c_1 + c_2) x_{n_T} \right), \; n_T \in \mathbf{n}_T,$$

$$\lambda^3_{n_T} = \frac{\partial}{\partial x_{n_T}} \left(-c_3 x_{n_T} \right), \; n_T \in \mathbf{n}_T,$$

$$\theta^{i,12}_{n_t} \left[\alpha u^i_{n_t} - \frac{1}{2} (u^i_{n_t})^2 \right] = 0, \; \theta^{i,12}_{n_t} \geq 0, \; i = 1, 2, \; n_t \in \mathbf{n}_0, \mathbf{n}_1, \mathbf{n}_2,$$

$$\theta^3_{n_t} \left[\alpha u^3_{n_t} - \frac{1}{2} (u^3_{n_t})^2 \right] = 0, \; \theta^3_{n_t} \geq 0, \; n_t \in \mathbf{n}_0, \mathbf{n}_1, \mathbf{n}_2,$$

along with the state equations in (7.20), and a given initial state $x_{n_0} = x_0$. Substituting the expressions of pseudo-Lagrangians $\bar{\mathcal{L}}_{12}$ and $\bar{\mathcal{L}}_3$ into the last system, we obtain

$$\alpha - u_{n_t}^i + \rho \sum_{\nu \in \mathbf{n}_t^+} \pi_{n_t}^\nu \lambda_\nu^{12} - \theta_{n_t}^{i,nc} \left[\alpha - u_{n_t}^i\right] = 0, \ i \in \{1, 2\}, \ n_t \in \mathbf{n}_0, \mathbf{n}_1, \mathbf{n}_2,$$

$$\alpha - u_{n_t}^3 + \rho \sum_{\nu \in \mathbf{n}_t^+} \pi_{n_t}^\nu \lambda_\nu^3 - \theta_{n_t}^{3,nc} \left[\alpha - u_{n_t}^3\right] = 0, \ n_t \in \mathbf{n}_0, \mathbf{n}_1, \mathbf{n}_2,$$

$$\lambda_{n_t}^{12} = -2(c_1 + c_2)x_{n_t} + \rho(1 - \delta_{n_t}) \sum_{\nu \in \mathbf{n}_t^+} \pi_{n_t}^\nu \lambda_\nu^{12}, \ n_t \in \mathbf{n}_1, \mathbf{n}_2,$$

$$\lambda_{n_t}^3 = -2c_3 x_{n_t} + \rho(1 - \delta_{n_t}) \sum_{\nu \in \mathbf{n}_t^+} \pi_{n_t}^\nu \lambda_\nu^3, \ n_t \in \mathbf{n}_1, \mathbf{n}_2,$$

$$\lambda_{n_T}^{12} = -(c_1 + c_2), \ n_T \in \mathbf{n}_3,$$

$$\lambda_{n_T}^3 = -c_3, \ n_T \in \mathbf{n}_3,$$

$$\theta_{n_t}^{i,12} \left[\alpha u_{n_t}^i - \frac{1}{2}\left(u_{n_t}^i\right)^2\right] = 0, \ \theta_{n_t}^{i,12} \geq 0, \ i = 1, 2, \ n_t \in \mathbf{n}_0, \mathbf{n}_1, \mathbf{n}_2,$$

$$\theta_{n_t}^3 \left[\alpha u_{n_t}^3 - \frac{1}{2}\left(u_{n_t}^3\right)^2\right] = 0, \ \theta_{n_t}^3 \geq 0, \ n_t \in \mathbf{n}_0, \mathbf{n}_1, \mathbf{n}_2,$$

$$x_{n_t} = (1 - \delta_{n_t^-})x_{n_t^-} + \sum_{i \in M} u_{n_t^-}^i, \ n_t \in \mathbf{n}_1, \mathbf{n}_2, \mathbf{n}_3, \ x_{n_0} = x_0.$$

Solving this system, we find the Nash equilibrium, and the value of characteristic function $v(12; x_0)$ is equal to the payoff of Player (coalition) 12 in the Nash equilibrium, that is, $J_1(\cdot) + J_2(\cdot)$. The values $v(12; x_{n_t}^*)$ for the subgame starting at node n_t with cooperative state $x_{n_t}^*$ can be calculated in the same way and are given in Table 7.1 (row $v(12)$ in the table). The same steps are repeated to find the values $v(13)$ and $v(23)$ for the game and any subgame. These values are also reported in Table 7.1. It is easy to verify that the γ characteristic function is superadditive at each node, i.e., $v(S \cup T) \geq v(S) + v(T)$ for any disjoint coalitions $S, T \subset M$. Using the CF values in Table 7.1, we can define the nonempty cores of the whole game and of subgames, which are determined using (7.5). The results are given in Table 7.2. The cores of the subgames starting at terminal nodes consist of single points because the players do not make decisions at these nodes. In this last stage, the characteristic functions are additive.

To define the distribution procedure for any imputation from the core, we need to select a single imputation from the corresponding core for any subgame. We suppose that the players adopt the nucleolus to select this imputation. (We recall that the nucleolus is in the core when it is not empty.) The nucleoli of the game and all subgames are given in Table 7.3. In Table 7.4, we give the Shapley values of all subgames and note that, in each subgame, the Shapley value is in the core. Interestingly, it is not clear which solution, the nucleolus or the Shapley value, would be chosen by the players. Indeed, payoff-wise, the ranking is not the same for all players, with Player 1 preferring the Shapley value, while Players 2 and 3 would vote for the nucleolus.

Table 7.2 Cores of the DGPET and subgames

Node n_0
$\{y = (y_1, y_2, y_3) : y_1 \geq -305.351, y_2 \geq 493.969, y_3 \geq 1149.04,$
$y_1 + y_2 \geq 414.849, y_1 + y_3 \geq 1023.88, y_2 + y_3 \geq 1640.01,$
$y_1 + y_2 + y_3 = 2347.73\}$
Node n_1^1
$\{y = (y_1, y_2, y_3) : y_1 \geq -81.274, y_2 \geq 388.146, y_3 \geq 803.154,$
$y_1 + y_2 \geq 452.19, y_1 + y_3 \geq 836.805, y_2 + y_3 \geq 1223.21,$
$y_1 + y_2 + y_3 = 1662.9\}$
Node n_1^2
$\{y = (y_1, y_2, y_3) : y_1 \geq 62.6804, y_2 \geq 477.315, y_3 \geq 844.338,$
$y_1 + y_2 \geq 667.15, y_1 + y_3 \geq 1007.58, y_2 + y_3 \geq 1349.57,$
$y_1 + y_2 + y_3 = 1868.1\}$
Node n_2^1
$\{y = (y_1, y_2, y_3) : y_1 \geq 249.499, y_2 \geq 370.501, y_3 \geq 491.502,$
$y_1 + y_2 \geq 620.013, y_1 + y_3 \geq 741.011, y_2 + y_3 \geq 862.008,$
$y_1 + y_2 + y_3 = 1111.55\}$
Node n_2^2
$\{y = (y_1, y_2, y_3) : y_1 \geq 251.759, y_2 \geq 372.008, y_3 \geq 492.255,$
$y_1 + y_2 \geq 623.779, y_1 + y_3 \geq 744.024, y_2 + y_3 \geq 864.268,$
$y_1 + y_2 + y_3 = 1116.07\}$
Node n_2^3
$\{y = (y_1, y_2, y_3) : y_1 \geq 292.554, y_2 \geq 399.205, y_3 \geq 505.853,$
$y_1 + y_2 \geq 691.772, y_1 + y_3 \geq 798.418, y_2 + y_3 \geq 905.063,$
$y_1 + y_2 + y_3 = 1197.66\}$
Node n_2^4
$\{y = (y_1, y_2, y_3) : y_1 \geq 294.668, y_2 \geq 400.614, y_3 \geq 506.558,$
$y_1 + y_2 \geq 695.296, y_1 + y_3 \geq 801.237, y_2 + y_3 \geq 907.178,$
$y_1 + y_2 + y_3 = 1201.89\}$
Nodes n_3^1, n_3^2
$y = (y_1, y_2, y_3) = (-22.0845, -14.723, -7.36149)$
Nodes n_3^3, n_3^4
$y = (y_1, y_2, y_3) = (-19.5738, -13.0492, -6.52461)$
Nodes n_3^5, n_3^6
$y = (y_1, y_2, y_3) = (-21.6695, -14.4463, -7.22317)$
Nodes n_3^7, n_3^8
$y = (y_1, y_2, y_3) = (-19.3202, -12.8802, -6.44008)$

Table 7.3 Nucleoli of the DGPET and its subgames

$t=0$		$t=1$	
n_0		n_1^1	n_1^2
$\begin{bmatrix} 31.337 \\ 830.657 \\ 1485.730 \end{bmatrix}$		$\begin{bmatrix} 103.017 \\ 572.437 \\ 987.445 \end{bmatrix}$	$\begin{bmatrix} 223.937 \\ 638.572 \\ 1005.590 \end{bmatrix}$

$t=2$			
n_2^1	n_2^2	n_2^3	n_2^4
$\begin{bmatrix} 249.516 \\ 370.518 \\ 491.519 \end{bmatrix}$	$\begin{bmatrix} 251.775 \\ 372.025 \\ 492.272 \end{bmatrix}$	$\begin{bmatrix} 292.571 \\ 399.222 \\ 505.870 \end{bmatrix}$	$\begin{bmatrix} 294.685 \\ 400.631 \\ 506.575 \end{bmatrix}$

$t=3$			
n_3^1, n_3^2	n_3^3, n_3^4	n_3^5, n_3^6	n_3^7, n_3^8
$\begin{bmatrix} -22.085 \\ -14.723 \\ -7.361 \end{bmatrix}$	$\begin{bmatrix} -19.574 \\ -13.049 \\ -6.525 \end{bmatrix}$	$\begin{bmatrix} -21.670 \\ -14.446 \\ -7.223 \end{bmatrix}$	$\begin{bmatrix} -19.320 \\ -12.880 \\ -6.440 \end{bmatrix}$

Table 7.4 The Shapley values of the DGPET and its subgames

$t=0$		$t=1$	
n_0		n_1^1	n_1^2
$\begin{bmatrix} 100.073 \\ 807.799 \\ 1439.850 \end{bmatrix}$		$\begin{bmatrix} 135.755 \\ 563.666 \\ 963.478 \end{bmatrix}$	$\begin{bmatrix} 252.584 \\ 630.897 \\ 984.623 \end{bmatrix}$

$t=2$			
n_2^1	n_2^2	n_2^3	n_2^4
$\begin{bmatrix} 249.518 \\ 370.518 \\ 491.517 \end{bmatrix}$	$\begin{bmatrix} 251.778 \\ 372.024 \\ 492.270 \end{bmatrix}$	$\begin{bmatrix} 292.573 \\ 399.221 \\ 505.868 \end{bmatrix}$	$\begin{bmatrix} 294.687 \\ 400.631 \\ 506.573 \end{bmatrix}$

$t=3$			
n_3^1, n_3^2	n_3^3, n_3^4	n_3^5, n_3^6	n_3^7, n_3^8
$\begin{bmatrix} -22.085 \\ -14.723 \\ -7.361 \end{bmatrix}$	$\begin{bmatrix} -19.574 \\ -13.049 \\ -6.525 \end{bmatrix}$	$\begin{bmatrix} -21.670 \\ -14.446 \\ -7.223 \end{bmatrix}$	$\begin{bmatrix} -19.320 \\ -12.880 \\ -6.440 \end{bmatrix}$

Finally, assuming that the nucleolus is adopted (simply for the sake of illustration), we determine the node-consistent IDP using Eqs. (7.8) and (7.9) in Theorem 7.1. The results, that is, the $\beta_i(x_{n_t}^*)$, $i \in M$, are shown in Fig. 7.4. To interpret the results, let us consider the payoff of Player 1. According to the nucleolus, Player 1 should get 31.3372 in the whole game (see Table 7.3). Our IDP decomposes this amount over nodes as follows:

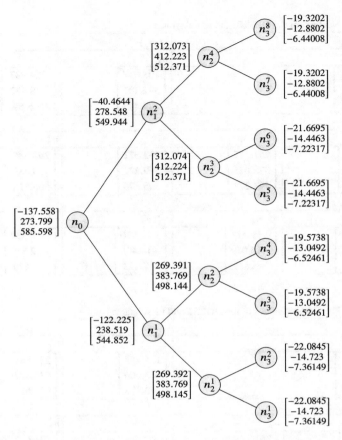

Fig. 7.4 Distribution procedure for the nucleolus (31.3372, 830.657, 1485.73) from the core of the DGPET

$$31.3372 = -137.558$$
$$+0.3 \cdot 0.9 \cdot (-122.225) + 0.7 \cdot 0.9 \cdot (-40.4641)$$
$$+0.2 \cdot (0.9)^2 \cdot 269.392 + 0.1 \cdot (0.9)^2 \cdot 269.391$$
$$+0.3 \cdot (0.9)^2 \cdot 312.074 + 0.4 \cdot (0.9)^2 \cdot 312.073$$
$$+0.05 \cdot (0.9)^3 \cdot (-22.0845) + 0.15 \cdot (0.9)^3 \cdot (-22.0845)$$
$$+0.01 \cdot (0.9)^3 \cdot (-19.5738) + 0.09 \cdot (0.9)^3 \cdot (-19.5738)$$
$$+0.2 \cdot (0.9)^3 \cdot (-21.6695) + 0.1 \cdot (0.9)^3 \cdot (-21.6695)$$
$$+0.3 \cdot (0.9)^3 \cdot (-19.3202) + 0.1 \cdot (0.9)^3 \cdot (-19.3202).$$

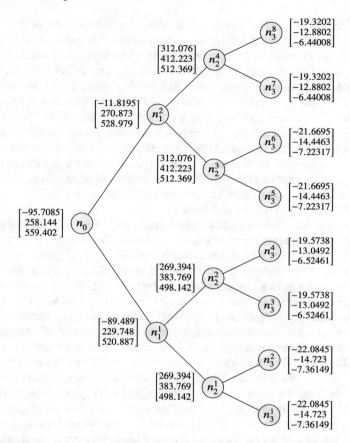

Fig. 7.5 Distribution procedure for the Shapley value (100.073, 807.799, 1439.85) in the DGPET

Similar computations give the decomposition over nodes of the total payoffs of Players 2 and 3. The distribution procedure for the Shapley value (100.073, 807.799, 1439.85) belonging to the core of the DGPET is represented in Fig. 7.5.

7.5 Additional Readings

There is a literature dealing with the time consistency of the core in dynamic cooperative games. It, however, does not necessarily use the same terminology as the one adopted here.

The node-consistent core in DGPETs is considered in Parilina and Zaccour (2015). A few papers assume a stochastic environment. Predtetchinski (2007) (see also

Predtetchinski et al. 2002; Herings et al. 2006) consider a class of discrete-time stochastic dynamic games where one nontransferable utility game (NTU game) from a finite set can be realized in each time period. The author introduces the strong sequential core for stationary cooperative games, i.e., defines the set of utilities that are robust to deviation by any coalition, and provides conditions for nonemptiness of this core. As the game considered is of the NTU variety, there is no issue of determining transfers to maintain time consistency. Xu and Veinott Jr. (2013) analyze the time consistency of the core in a setting where the players' profits depend on a (same) random variable. As in a DGPET, the realization of the random variable does not depend on the players' actions, and all players face the same randomness. The retained game is without externalities, that is, noncoalition members do not influence the coalition's payoff. This implies that the CF values are obtained by solving optimization (not equilibrium) problems.

In a deterministic environment, Germain et al. (2003) deal with the sustainability of the core in a dynamic game of pollution control. The transfer payment, which was initially proposed in Chander and Tulkens (1997) for static games, is defined by the difference between the noncooperative and cooperative costs of a player, plus a share of the surplus induced by extending the cooperation in the current period. These transfers are balanced, i.e., their sum in each period is equal to zero. Germain et al. (2003) establish the necessary and sufficient conditions for an imputation to belong to the γ-core of the cooperative static game at period t. In Filar and Petrosyan (2000), the characteristic function changes over time according to some dynamics, and the values of the CF for the whole game are determined by the sum of the corresponding values of the stage characteristic functions. In Lehrer and Scarsini (2013), the payoff of a coalition depends on the history of allocations, and any coalition is allowed to deviate at any time. Interestingly, a deviation induces a structural change, i.e., the deviating coalition becomes the grand coalition in a new dynamic game, which leads to the introduction of a new solution concept (the intertemporal core) that insures stability against such deviations.

Petrosyan (1992) proposes the concept of the strong subgame consistency of the core (see also Petrosyan 1993 for the multicriteria case). Suppose that the cores of the dynamic game and all subgames are nonempty. When the players cooperate, they agree on the cooperative strategy profile and expect to receive the components of the imputation belonging to the core of the game. At an intermediate node, the players can add the payoffs they have obtained so far to any imputation from the core of the remaining subgame. If this sum belongs to the core of the game, then the core is strongly subgame consistent. The difference from the concept of subgame consistency lies in the fact that, here, the players can choose any imputation from the core of the subgame. Further, the summation of the previously obtained payoffs with any imputation from the core of the subgame will finally give a set of vectors, not only one vector as in the definition of subgame consistency. Consequently, the principle of strong subgame consistency requires stronger sufficient conditions than does the subgame consistency of the core. Such conditions are given in Parilina and Petrosyan (2018, 2020) for stochastic games, and in Sedakov (2018) for finite multistage games.

7.6 Exercises

Exercise 7.1 Consider a three-player Cournot competition game, with capacity constraints, played over four periods, that is, $\mathbb{T} = \{0, 1, 2, 3\}$. Let $q_{n_t}^i$ be the quantity produced by Player i at node $n_t \in \mathbf{n}_t$, $t \in \mathbb{T} \setminus \{T\}$, and $q_{n_t} = \sum_{i \in M} q_{n_t}^i$ be the total quantity at that node. The demand is stochastic and the event tree is described in Fig. 7.6. The stochastic inverse demand is given by the following linear function:

$$P(q_{n_t}) = A_{n_t} - B_{n_t} q_{n_t}, \ n_t \in \mathbf{n}_t, \ t \in \mathbb{T}, \tag{7.21}$$

where A_{n_t} and B_{n_t} are positive parameters.

Let $K_{n_t}^i$ be the available production capacity, and $I_{n_t}^i$ the investment in this capacity made by Player i at node n_t, $t \in \mathbb{T}$. A vector of production capacities $(K_{n_t}^i : i \in M, n_t \in \mathbf{n}_t, t \in \mathbb{T})$ represents the state variables. Assuming a one-period lag before an investment becomes productive, the evolution of the production capacity of Player i is then described by the following state equation:

$$K_{n_t}^i = (1 - \delta) K_{n_t^-}^i + I_{n_t^-}^i, \ K_{n_0}^i = k_0^i, \tag{7.22}$$

where k_0^i denotes the initial capacity of Player i, and $\delta, 0 \le \delta < 1$ is the depreciation rate of the capacity. The quantity produced is subject to the available capacity, i.e.,

$$q_{n_t}^i \le K_{n_t}^i, \ n_t \in \mathbf{n}_t, \ t \in \mathbb{T}. \tag{7.23}$$

The strategy of Player i at node $n_t \in \mathbf{n}_t$, $t = 0, 1, 2$ is a vector $(q_{n_t}^i, I_{n_t}^i)$ satisfying (7.23). The production and investment costs are given by the following quadratic functions:

$$C_i(q_{n_t}^i) = \frac{c_i}{2} (q_{n_t}^i)^2, \ c_i > 0, \tag{7.24}$$

$$D_i(I_{n_t}^i) = \frac{d_i}{2} (I_{n_t}^i)^2, \ d_i > 0. \tag{7.25}$$

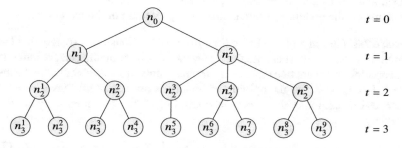

Fig. 7.6 Event tree representing the uncertain demand for Exercise 7.1

Denote by $S_i\left(K_{n_T}^i\right)$ the salvage value of Player i's production capacity at terminal nodes $n_T \in \mathbf{n}_T$ in time $T = 3$, given by

$$S_i\left(K_{n_T}^i\right) = \frac{v_i}{2}\left(K_{n_T}^i\right)^2, \quad v_i > 0.$$

Function $S_i(\cdot)$ is Player i's payoff function defined at the terminal nodes. Her payoff function at nonterminal node n_t is

$$H_i(n_t) = q_{n_t}^i P(q_{n_t}) - C_i(q_{n_t}^i) - D_i(I_{n_t}^i),$$

where $P(q_{n_t})$, $C_i(q_{n_t}^i)$, and $D_i(I_{n_t}^i)$ are given in (7.21), (7.24), and (7.25), respectively. Each player aims at maximizing the expected sum of discounted payoffs. The discount factor is denoted by ρ. The parameters are given by the following values:

Node	n_0	n_1^1	n_1^2	n_2^1	n_2^2	n_2^3	n_2^4	n_2^5
A_{n_t}	100	120	110	90	100	130	100	105
B_{n_t}	3.0	3.1	2.9	2.8	3.3	3.1	3.5	2.9

$c_1 = 0.5$, $c_2 = 0.3$, $c_3 = 0.25$, $d_1 = 10$, $d_2 = 9$, $d_3 = 6$, $\delta = 0.2$, $\rho = 0.9$,

$k_0^1 = 15$, $k_0^2 = 18$, $k_0^3 = 16$, $v_1 = 0.5$, $v_2 = 0.35$, $v_3 = 0.25$,

$\pi_{n_1^1} = 0.40$, $\pi_{n_1^2} = 0.60$, $\pi_{n_2^1} = 0.15$, $\pi_{n_2^2} = 0.25$, $\pi_{n_2^3} = 0.30$,

$\pi_{n_2^4} = 0.20$, $\pi_{n_2^5} = 0.10$, $\pi_{n_3^1} = 0.10$, $\pi_{n_3^2} = 0.05$, $\pi_{n_3^3} = 0.15$,

$\pi_{n_3^4} = 0.10$, $\pi_{n_3^5} = 0.30$, $\pi_{n_3^6} = 0.05$, $\pi_{n_3^7} = 0.15$, $\pi_{n_3^8} = 0.05$,

$\pi_{n_3^9} = 0.05$.

1. Find the cooperative state trajectory in the dynamic game played over an event tree.
2. Calculate the values of the γ characteristic function for any subgame.
3. Determine the core of the game and any subgame on the cooperative state trajectory.
4. Define the least core of the entire game and any subgame.
5. Determine the node-consistent distribution procedure of the least core.

Exercise 7.2 (*Based on* Mazalov and Parilina 2019). Consider a society that has an opinion on some event. There are three players, which are media centers influencing social opinion. An event tree represents the uncertainty of society's belief in the media centers. Let $x_{n_t} \in \mathbb{R}^1$ be the opinion of society at node $n_t \in \mathbf{n}_t$. Player $i = 1, 2, 3$ influences the opinion of the society by control $u_{n_t}^i \in \mathbb{R}^1$. The opinion dynamics are defined by

$$x_{n_t} = Ax_{n_t^-} + \sum_{i=1}^{3} \alpha_{n_t^-}^i u_{n_t^-}^i, \tag{7.26}$$

where A is a positive constant, $\alpha^i_{n_t} \in [0, 1]$ for any player i and any node n_t, and $\sum_{i=1}^{3} \alpha^i_{n_t} = 1$. Constant $\alpha^i_{n_t}$ measures society's belief in center i's opinion. The beliefs are assumed to be stochastic. The initial opinion is $x_{n_0} = x_0$.

Player i's costs at nonterminal node n_t are defined by a linear-quadratic function, i.e.,

$$K_i(n_t, x_{n_t}, u_{n_t}) = (x_{n_t} - \bar{x}^i)^2 + \frac{\gamma_i}{2}(u^i_{n_t})^2,$$

and, at the terminal node, defined by function

$$S_i(n_T, x_{n_T}) = (x_{n_T} - \bar{x}^i)^2,$$

where $\bar{x}^i$ is a target opinion of Player i, and γ_i is a positive constant for any i.

1. Consider the cooperative version of the game, supposing that, under cooperation, players agree on the average opinion

$$\hat{x} = \frac{\sum_{i=1}^{3} \bar{x}^i}{3},$$

and then the costs of Player i in the cooperative version of the game are defined by

$$K_i(n_t, x_{n_t}, u_{n_t}) = (x_{n_t} - \hat{x})^2 + \frac{\gamma_i}{2}(u^i_{n_t})^2,$$

and, at the terminal node, defined by a function

$$S_i(n_T, x_{n_T}) = (x_{n_T} - \hat{x})^2.$$

Find the cooperative control profile and cooperative state trajectory.
2. Determine the values of the γ characteristic function for any subgame played over the event tree in Fig. 7.7. The parameter values are given by

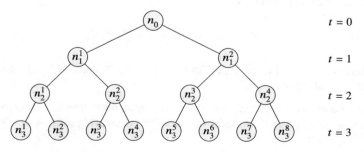

Fig. 7.7 Event tree representing the uncertain beliefs

$$a = 0.5, \ x_0 = 0.3, \ \delta = 0.9,$$
$$\gamma_1 = 0.6, \ \gamma_2 = 0.3, \ \gamma_3 = 0.09,$$
$$\bar{x}^1 = 0.7, \ \bar{x}^2 = 0.9, \ \bar{x}^3 = 1.0,$$
$$\pi_{n_1^1} = 0.20, \ \pi_{n_1^2} = 0.80, \ \pi_{n_2^1} = 0.15, \ \pi_{n_2^2} = 0.05, \ \pi_{n_2^3} = 0.30,$$
$$\pi_{n_2^4} = 0.50, \ \pi_{n_3^1} = 0.10, \ \pi_{n_3^2} = 0.05, \ \pi_{n_3^3} = 0.03, \ \pi_{n_3^4} = 0.02,$$
$$\pi_{n_3^5} = 0.20, \ \pi_{n_3^6} = 0.10, \ \pi_{n_3^7} = 0.30, \ \pi_{n_3^8} = 0.20.$$

Node	n_0	n_1^1	n_1^2	n_2^1	n_2^2	n_2^3	n_2^4
$\alpha_{n_t}^1$	0.33	0.50	0.25	0.65	0.50	0.15	0.25
$\alpha_{n_t}^2$	0.33	0.25	0.50	0.20	0.25	0.25	0.20
$\alpha_{n_t}^3$	0.34	0.25	0.25	0.15	0.25	0.60	0.55

3. Define the core and the Shapley value of the game and any subgame on the cooperative state trajectory. Interpret the results.
4. If the cores of the game and any subgame on the cooperative state trajectory are nonempty, determine the node-consistent distribution procedure for an element from the core of the game.

Exercise 7.3 Consider a market with $m \geq 2$ firms that compete in quantity over a finite horizon. A firm $i \in M$ selects its quantity $q_{n_t}^i \in \mathbb{R}_+$ to be produced at node n_t. The market price is sticky.[5] At node n_t, the price $p_{n_t} \in \mathbb{R}_+$ is defined by the following state equation:

$$p_{n_t} = s_{n_t^-} p_{n_t^-} + (1 - s_{n_t^-}) \left(a - b \sum_{i=1}^n q_{n_t}^i \right), \qquad (7.27)$$

where $s_{n_t^-} \in [0, 1]$, and a and b are positive constants. The first term in the right-hand side of (7.27) represents the inertia in price, while the second term reflects the price adjustment. The initial market price is given by $p_{n_0} = p_0$. The profit of firm i at a nonterminal node n_t is given by

$$K_i(n_t, p_{n_t}, q_{n_t}) = p_{n_t} q_{n_t}^i - \frac{c_i}{2} (q_{n_t}^i)^2,$$

where parameter $c_i > 0$ represents firm i's unit cost. The profit at a terminal node n_T is zero for all players.

1. Define a cooperative dynamic game played over an event tree. Consider the cooperative version of the game and find the cooperative control profile and cooperative state trajectory.

---

[5] The model in this exercise is a stochastic and discrete-time version of the sticky-price continuous-time model in Fershtman and Kamien (1987). For a deterministic and discrete-time version, see Parilina and Sedakov (2020).

2. Determine the values of the γ characteristic function for any subgame played over the event tree in Fig. 7.6, with the probabilities π_{n_t} given for the event tree in Example 7.1, and the following parameter values:

$$m = 3, \ a = 100, \ b = 0.9,$$
$$p_0 = 10, \ c_1 = 1.8, \ c_2 = 4.8, \ c_3 = 3.4.$$

Node	n_0	n_1^1	n_1^2	n_2^1	n_2^2	n_2^3	n_2^4	n_2^5
s_{n_t}	0.30	0.20	0.50	0.30	0.15	0.24	0.78	0.65

3. Define the core of the game and any subgame on the cooperative state trajectory.
4. If the cores of the game and any subgame on the cooperative state trajectory are nonempty, determine the node-consistent distribution procedure for an element belonging to the core of the game.

Exercise 7.4 Consider a modified version of the game described in Example 7.2. Assume that the society is divided into two subsocieties. The first subsociety is influenced by the first and second players, and the second subsociety by the second and third players. Also, as in Example 7.2, an event tree represents the uncertainty of society's beliefs in the media centers (players). Let $x_{n_t}^i \in \mathbb{R}^1, i = 1, 2$, be the opinion of subsociety i at node $n_t \in \mathbf{n}_t$. Player 1 influences the opinion of the first subsociety by control $u_{n_t}^1 \in \mathbb{R}^1$, and Player 3 influences the opinion of the second subsociety by control $u_{n_t}^3 \in \mathbb{R}^1$. Player 2 influences the opinion of both subsocieties 1 and 2 by controls $u_{n_t}^2 = (u_{n_t}^{21}, u_{n_t}^{22}) \in \mathbb{R}^2$, respectively. The opinion dynamics are defined by

$$x_{n_t}^1 = \psi x_{n_t^-}^1 + (1 - \psi)x_{n_t^-}^2 + \alpha_{n_t^-}^1 u_{n_t^-}^1 + \alpha_{n_t^-}^2 u_{n_t^-}^{21},$$
$$x_{n_t}^2 = \psi x_{n_t^-}^1 + (1 - \psi)x_{n_t^-}^2 + \zeta_{n_t^-}^2 u_{n_t^-}^{22} + \zeta_{n_t^-}^3 u_{n_t^-}^3,$$

where $\psi \in (0, 1)$ is a share of subsociety 1 in the whole society, $\alpha_{n_t}^i \in [0, 1]$ for any player $i = 1, 2$ and any node n_t, and $\sum_{i=1}^2 \alpha_{n_t}^i = 1$; $\zeta_{n_t}^i$ for any player $i = 2, 3$ and any node n_t and $\sum_{i=2}^3 \zeta_{n_t}^i = 1$. Constant $\alpha_{n_t}^i$ measures subsociety 1's belief in media center i's opinion, $i = 1, 2$. Constant $\zeta_{n_t}^i$ measures subsociety 2's belief in media center i's opinion, $i = 2, 3$. The beliefs are assumed to be stochastic and defined on an event tree. The initial opinion is $x_{n_0} = (x_{n_0}^1, x_{n_0}^2) = (x_0^1, x_0^2)$.

Player j's costs at nonterminal node n_t are defined by the linear-quadratic function

$$K_j(n_t, x_{n_t}, u_{n_t}) = \psi(x_{n_t}^1 - \bar{x}^j)^2 + (1 - \psi)(x_{n_t}^2 - \bar{x}^j)^2 + \frac{\gamma_j}{2}(u_{n_t}^j)^2,$$

for Players 1 and 3, and for Player 2 by

$$K_2(n_t, x_{n_t}, u_{n_t}) = \psi(x_{n_t}^1 - \bar{x}^2)^2 + (1 - \psi)(x_{n_t}^2 - \bar{x}^2)^2 + \frac{\gamma_2}{2}((u_{n_t}^{21})^2 + (u_{n_t}^{22})^2).$$

At a terminal node, the costs of any player $j = 1, 2, 3$ are given by

$$S_j(n_T, x_{n_T}) = \psi(x_{n_t}^1 - \bar{x}^j)^2 + (1 - \psi)(x_{n_t}^2 - \bar{x}^j)^2.$$

1. Consider the cooperative version of the game, supposing that, under cooperation, players agree on the average opinion

$$\hat{x} = \frac{\sum_{j=1}^{3} \bar{x}^j}{3},$$

   and then use $\hat{x}$ in the cost functions of all players instead of $\bar{x}^j$ when they cooperate.
2. Find the cooperative control profile and the cooperative state trajectory.
3. Determine the values of the γ characteristic function for any subgame played over the event tree in Figure. 7.7, with the following parameter values:

$$\psi = 0.2, \quad x_0 = (0.3, 0.2), \quad \delta = 0.95,$$
$$\gamma_1 = 0.05, \quad \gamma_2 = 0.10, \quad \gamma_3 = 0.08,$$
$$\bar{x}^1 = 0.7, \quad \bar{x}^2 = 0.9, \quad \bar{x}^3 = 0.6,$$

Node	n_0	n_1^1	n_1^2	n_2^1	n_2^2	n_2^3	n_2^4
$\alpha_{n_t}^1$	0.30	0.50	0.25	0.80	0.75	0.15	0.80
$\alpha_{n_t}^2$	0.70	0.50	0.75	0.20	0.25	0.85	0.20
$\zeta_{n_t}^2$	0.40	0.20	0.30	0.35	0.25	0.60	0.50
$\zeta_{n_t}^3$	0.60	0.80	0.70	0.65	0.75	0.40	0.50

   and the same probabilities π_{n_t} as in Example 7.2.
4. Define the core and ε-core of the game for $\varepsilon = 4.5$ and any subgame on the cooperative state trajectory.
5. If the cores of the game and any subgame on the cooperative state trajectory are nonempty, determine the node-consistent distribution procedure for the elements of the least core.

References

Barron, E. N. (2013). *Game Theory: An Introduction.* Wiley Series in Operations Research and Management Science. Wiley.

Chander, P. and Tulkens, H. (1997). The core of an economy with multilateral environmental externalities. *International Journal of Game Theory*, 26(3):379–401.

De Zeeuw, A. (2018). *Dynamic Games of International Pollution Control: A Selective Review*, pages 703–728. Springer International Publishing, Cham.

Fershtman, C. and Kamien, M. I. (1987). Dynamic duopolistic competition with sticky prices. *Econometrica*, 55(5):1151–1164.

Filar, J. A. and Petrosyan, L. (2000). Dynamic cooperative games. *International Game Theory Review*, 02(01):47–65.

Germain, M., Toint, P., Tulkens, H., and de Zeeuw, A. (2003). Transfers to sustain dynamic core-theoretic cooperation in international stock pollutant control. *Journal of Economic Dynamics and Control*, 28(1):79–99.

Herings, P. J.-J., Predtetchinski, A., and Perea, A. (2006). The weak sequential core for two-period economies. *International Journal of Game Theory*, 34(1):55–65.

Jørgensen, S., Martín-Herrán, G., and Zaccour, G. (2010). Dynamic games in the economics and management of pollution. *Environmental Modeling & Assessment*, 15:433–467.

Lehrer, E. and Scarsini, M. (2013). On the Core of Dynamic Cooperative Games. *Dynamic Games and Applications*, 3(3):359–373.

Mazalov, V. and Parilina, E. M. (2019). Game of competition for opinion with two centers of influence. In Khachay, M., Kochetov, Y., and Pardalos, P., editors, *Mathematical Optimization Theory and Operations Research*, pages 673–684, Cham. Springer International Publishing.

Parilina, E. M. and Petrosyan, L. (2018). Strongly subgame-consistent core in stochastic games. *Automation and Remote Control*, 79(8):1515–1527.

Parilina, E. M. and Petrosyan, L. (2020). On a simplified method of defining characteristic function in stochastic games. *Mathematics*, 8(7).

Parilina, E. M. and Sedakov, A. (2020). Stable Coalition Structures in Dynamic Competitive Environment. In Pineau, P.-O., Sigué, S., and Taboubi, S., editors, *Games in Management Science: Essays in Honor of Georges Zaccour*, pages 381–396, Cham. Springer International Publishing.

Parilina, E. M. and Zaccour, G. (2015). Node-consistent core for games played over event trees. *Automatica*, 53:304–311.

Peleg, B. and Sudhölter, P. (2003). *Introduction to the Theory of Cooperative Games*. Theory and Decision Library. Kluwer Academic Publishers.

Petrosyan, L. (1992). Construction of strongly dynamically stable solutions in cooperative differential games. *Vestnik Leningradskogo universiteta. Ser. 1: Matematika, mekhanika i astronomiya*, pages 33–38.

Petrosyan, L. (1993). High-dynamically stable principles of optimality in multicriterion problems of optimal control. *Izvestiya Akademii Nauk. Teoriya i Sistemy Upravleniya*, (1):169–174.

Predtetchinski, A. (2007). The strong sequential core for stationary cooperative games. *Games and Economic Behavior*, 61(1):50–66.

Predtetchinski, A., Herings, P. J.-J., and Peters, H. (2002). The strong sequential core for two-period economies. *Journal of Mathematical Economics*, 38(4):465–482. Special Issue on the Brown-Maastricht conferences.

Schmeidler, D. (1969). The nucleolus of a characteristic function game. *SIAM Journal on Applied Mathematics*, 17(6):1163–1170.

Sedakov, A. A. (2018). On the strong time consistency of the core. *Automation and Remote Control*, 79(4):757–767.

Shapley, L. S. and Shubik, M. (1966). Quasi-cores in a monetary economy with nonconvex preferences. *Econometrica*, 34(4):805–827.

Xu, N. and Veinott Jr., A. (2013). Sequential stochastic core of a cooperative stochastic programming game. *Operations Research Letters*, 41:430–435.

Chapter 8
Cooperative Equilibria in DGPETs

In this chapter, we extend to the class of dynamic games played over event trees (DGPETs) the concepts of subgame perfect ε-equilibrium and incentive equilibrium discussed in a deterministic framework in Chap. 4. In Sect. 8.1, we recall the elements of noncooperative and cooperative DGPETs and introduce some needed additional notation. In Sect. 8.2, we describe the concept of approximated or ε-equilibria and provide conditions for the existence of a subgame perfect ε-equilibrium in DGPETs. Section 8.3 deals with incentive equilibria in linear-quadratic DGPETs.

8.1 Preliminaries

We give a brief description of DGPETs and introduce some additional notation that will be needed.

Let $\mathbb{T} = \{0, 1, \ldots, T\}$ be the set of periods and $M = \{1, 2, \ldots, m\}$ the set of players. Player $i \in M$ chooses a vector of control (or decision) variables $u_{n_t}^i \in U_{n_t}^i \subseteq \mathbb{R}^{\ell_i}$ at node $n_t \in \mathbf{n}_t, t \in \mathbb{T}\backslash\{T\}$, where $U_{n_t}^i$ is the set of admissible control values for Player i at this node. Denote by $U_{n_t} = \prod_{i \in M} U_{n_t}^i$ the joint decision set at node n_t, and by $\mathbf{U}^i = \prod_{\nu \in \mathbf{n}_0^{++}\backslash\mathbf{n}_T} U_\nu^i$ the product of the sets of decision variables of Player i.

The vector of state variables at node $n_t \in \mathbf{n}_t, t \in \mathbb{T}$ is $x_{n_t} \in X \subset \mathbb{R}^q$, where X is the set of admissible states and q an integer number. The evolution over time of the state is given by

$$x_{n_t} = f^{n_t^-}\left(x_{n_t^-}, u_{n_t^-}\right), \quad x_{n_0} = x_0, \tag{8.1}$$

where $u_{n_t^-} \in U_{n_t^-}, n_t \in \mathbf{n}_t, t \in \mathbb{T} \setminus \{0\}$ and x_0 is the initial state at root node n_0.

The payoff functional of Player $i \in M$ is

$$J_i(\mathbf{x}, \mathbf{u}) = \sum_{t=0}^{T-1} \rho^t \sum_{n_t \in \mathbf{n}_t} \pi_{n_t} \phi_i(n_t, x_{n_t}, u_{n_t}) + \rho^T \sum_{n_T \in \mathbf{n}_T} \pi_{n_T} \Phi_i(n_T, x_{n_T}), \tag{8.2}$$

E. Parilina et al., *Theory and Applications of Dynamic Games*, Theory and Decision Library C 51, https://doi.org/10.1007/978-3-031-16455-2_8

where $\rho \in (0, 1)$ is the discount factor, and $\mathbf{x}$ and $\mathbf{u}$ are given by

$$\mathbf{x} = \{x_\nu : \nu \in \mathbf{n}_0^{++}\}, \tag{8.3}$$

$$\mathbf{u} = \{u_\nu^i \in U_\nu^i : \nu \in \mathbf{n}_0^{++} \setminus \mathbf{n}_T, \, i \in M\}, \tag{8.4}$$

$\phi_i(n_t, x_{n_t}, u_{n_t})$ is the reward to Player i at node $n_t \in \mathbf{n}_t, t \in \mathbb{T} \setminus \{T\}$, and $\Phi_i(n_T, x_{n_T})$ is the reward to Player i at terminal node $n_T \in \mathbf{n}_T$.

We assume an S-adapted open-loop information structure in which each player designs a rule that selects values of the decision variables that depend on the node of the event tree n_t and initial state x_0, that is, $u_{n_t}^i = u_{n_t}^i(x_0) \in U_{n_t}^i$. As the initial state is a given constant, we do not distinguish between the control $u_{n_t}^i$ and the decision rule $u_{n_t}^i(x_0)$.

Denote by

$\mathbf{u}^i = \{u_\nu^i : \nu \in \mathbf{n}_0^{++} \setminus \mathbf{n}_T\}$: an admissible S-adapted strategy or the collection of decisions of Player $i \in M$ for all nodes;

$\mathbf{u} = \{\mathbf{u}^i, \, i \in M\}$: the admissible S-adapted strategy profile;

$W_i(x_0, \mathbf{u}) = J_i(\mathbf{x}, \mathbf{u})$: the payoff of Player $i \in M$, where $\mathbf{x}$ is obtained from $\mathbf{u}$ as the unique solution of the state equations emanating from the initial state x_0;

$\mathbf{u}^{nc}$: a Nash equilibrium in the game in S-adapted strategies;

$(\mathbf{u}^i, \mathbf{u}^{-i,nc})$: an S-adapted strategy profile when all players $j \neq i, j \in M$, use their Nash equilibrium policies;

$\mathbf{u}^*$: the vector of cooperative strategies such that

$$\mathbf{u}^* = \arg \max_{\mathbf{u}^i : i \in M} \sum_{i \in M} W_i(x_0, \mathbf{u}); \tag{8.5}$$

$\mathbf{x}^* = \{x_\nu^* : \nu \in \mathbf{n}_0^{++}\}$: the cooperative state trajectory generated by the cooperative controls $\mathbf{u}^*(x_0)$;

$\mathbf{n}_t^{++}$: the subtree of the event tree starting from node n_t;

$\mathbf{u}^i[n_t, x_{n_t}^*] = \{u_\nu^i : \nu \in \mathbf{n}_t^{++} \setminus \mathbf{n}_T\}$: an admissible S-adapted strategy of Player i in the subgame starting from node n_t, with initial state $x_{n_t}^*$;

$\mathbf{u}[n_t, x_{n_t}^*] = (\mathbf{u}^i[n_t, x_{n_t}^*] : i \in M)$: an S-adapted strategy profile in the subgame starting from node n_t, with initial state $x_{n_t}^*$;

$W_i(x_{n_t}^*, \mathbf{u}[n_t, x_{n_t}^*])$: the payoff to Player i in the subgame starting from node n_t with cooperative state $x_{n_t}^*$, which is given by

$$W_i\left(x_{n_t}^*, \mathbf{u}[n_t, x_{n_t}^*]\right) = \sum_{\theta=t}^{T-1} \rho^{\theta-t} \sum_{n_\theta \in \mathbf{n}_\theta \cap \mathbf{n}_t^{++}} \pi_{n_t}^{n_\theta} \phi_i(n_\theta, x_{n_\theta}, u_{n_\theta}) \tag{8.6}$$

$$+ \rho^{T-t} \sum_{n_T \in \mathbf{n}_T \cap \mathbf{n}_t^{++}} \pi_{n_t}^{n_T} \Phi_i(n_T, x_{n_T});$$

$\mathbf{u}^{nc}[n_t, x_{n_t}^*] = (\mathbf{u}^{i,nc}[n_t, x_{n_t}^*] : i \in M)$: one S-adapted Nash equilibrium in the subgame starting from node n_t with cooperative state $x_{n_t}^*$;

$W_i\left(x_{n_t}^*, \mathbf{u}^{nc}[n_t, x_{n_t}^*]\right)$: the payoff to Player i in the Nash equilibrium in the subgame starting from node n_t with cooperative state $x_{n_t}^*$;

$W_i\left(x_{n_t}^*, \mathbf{u}^*[n_t, x_{n_t}^*]\right)$: the cooperative payoff of Player i in the subgame starting from node n_t with cooperative state $x_{n_t}^*$.

8.2 Approximated Cooperative Equilibria in DGPETs

Being collectively optimal and self-supported, a cooperative equilibrium is the best of two worlds. However, constructing a subgame perfect cooperative equilibrium in a finite-horizon dynamic game runs up against the fact that each player has an incentive not to cooperate in the last stage, and by a backward-induction argument, to also defect in all stages.[1] As a second-best option, a subgame perfect ε-equilibrium can be constructed in any finite-horizon dynamic game where ε is a number, to be determined endogenously, that is larger than the gain that any player can achieve by deviating from cooperation. In this section, we propose the use of trigger strategies to reach such an equilibrium in DGPETs. Our approach is in line with what has been done in repeated games, modulo the fact that here we additionally have a vector of state variables evolving over time.

We describe the DGPET as a game in extensive form. This means that each player knows not only the current node $n_t \in \mathbf{n}_t, t = 0, \ldots, T$ and what she played along the (unique) path leading from the initial node n_0 to n_t^-, but also what the other players did in all previous periods. Let this path be $(n_0, n_1, \ldots, n_{t-1}, n_t)$ and denoted by $n_0 \rightsquigarrow n_t$. The collection of nodes and corresponding strategy profiles realized on the path $n_0 \rightsquigarrow n_t$, except node n_t, is called the *history of node* n_t and is denoted by $\mathcal{H}_{n_t}^u = \left(u_{n_0}, u_{n_1}, \ldots, u_{n_{t-1}}\right)$.

Definition 8.1 A strategy $\sigma^i = \{\sigma_{n_t}^i : n_t \in \mathbf{n}_0^{++} \setminus \mathbf{n}_T\}$ of Player $i \in M$ in the game played over an event tree is a mapping that associates to each node n_t an action $u_{n_t}^i \in U_{n_t}^i$ with each history $\mathcal{H}_{n_t}^u$, that is,

$$\sigma_{n_t}^i : \mathcal{H}_{n_t}^u \longrightarrow U_{n_t}^i.$$

A strategy tells a player what to do at each node n_t of the event tree. Denote by Σ^i the set of strategies of Player i, by $\sigma = (\sigma^1, \ldots, \sigma^m)$ a strategy profile, and by $\Sigma = \Sigma^1 \times \cdots \times \Sigma^m$ the set of possible strategy profiles. For a given strategy profile, we can compute the expected payoff in all subgames, including the whole game, for

---

[1] If the reader is not familiar with this established result, a good exercise is to consider a repeated version of the prisoner's dilemma example and show that, regardless of how (finitely) many times the game is repeated, the only subgame perfect equilibrium is to defect at each stage.

any given initial state. To avoid adding new notations, we denote the payoff to Player i in the subgame starting at node n_t with state x_{n_t} as a function of the strategy profile, by

$$W_i(x_{n_t}, \sigma) = W_i(x_{n_t}, \mathbf{u}[n_t, x_{n_t}]),$$

where $\mathbf{u}[n_t, x_{n_t}]$ is a trajectory of controls in the subgame starting at node n_t and determined by profile σ.

Definition 8.2 A strategy profile $\hat{\sigma}$ is an ε-equilibrium if, for each player $i \in M$ and each strategy $\sigma^i \in \Sigma^i$, the following inequality holds:

$$W_i(x_0, \hat{\sigma}) \geqslant W_i(x_0, (\sigma^i, \hat{\sigma}^{-i})) - \varepsilon. \tag{8.7}$$

The ε-equilibrium guarantees that the individual deviation will not increase the payoff of any player in the game by more than ε. If ε equals zero, then we have a Nash equilibrium. Definition 8.2 considers only a deviation in the initial node. Other, stronger definition of the ε-equilibrium can be formulated when deviations can occur in any subgame, and not only along the cooperative state trajectory.

Definition 8.3 A strategy profile $\hat{\sigma}$ is a subgame perfect ε-equilibrium if, for each player $i \in M$, each node $n_t \in \mathbf{n}_t$, each strategy $\sigma^i \in \Sigma^i$, and each history $\mathcal{H}_{n_t}^u$, the following inequality holds:

$$W_i(x_{n_t}, \hat{\sigma}|\mathcal{H}_{n_t}^u) \geqslant W_i(x_{n_t}, (\sigma^i, \hat{\sigma}^{-i})|\mathcal{H}_{n_t}^u) - \varepsilon,$$

where $W_i(x_{n_t}, \hat{\sigma}|\mathcal{H}_{n_t}^u)$ is Player i's payoff in the subgame starting at node n_t with state x_{n_t}, given by history $\mathcal{H}_{n_t}^u$, when the players use strategy profile $\hat{\sigma}$.

To construct an approximated equilibrium in the class of strategies introduced in Definition 8.1, we make two assumptions. First, the players agree in a preplay arrangement on implementing the cooperative trajectory $\mathbf{u}^*$ from (8.5). Second, if Player i deviates from cooperation at node n_t^- by choosing a control $u_{n_t^-}^i \neq u_{n_t^-}^{i*}$, then cooperation breaks down and all players switch to their Nash equilibrium strategies in the subgame starting at node n_t in state $x_{n_t} = f^{n_t^-}\left(x_{n_t^-}^*, (u_{n_t^-}^i, u_{n_t^-}^{-i*})\right)$, and Player i obtains the reward $W_i(x_{n_t}, \mathbf{u}^{nc}[n_t, x_{n_t}])$, $i \in M$.

Denote by $\hat{\sigma} = (\hat{\sigma}^i : i \in M)$ a strategy profile that prescribes that Player i implement the cooperative control $u_{n_t}^{i*}$ at node n_t if in the history of this node no deviation from the cooperative trajectory has been observed, and to choose $u_{n_t}^{i,nc}$ otherwise. Denote by $\hat{\mathbf{u}}[n_t, x_{n_t}] = \{\hat{u}_{n_\theta}^i : n_\theta \in \mathbf{n}_t^{++} \setminus \mathbf{n}_T, i \in M\}$ the collection of controls corresponding to strategy profile $\hat{\sigma}$ such that the trigger mode of the strategy is implemented in the subgame starting at node n_t in state x_{n_t}.

A player's trigger strategy consists of two behavior types or two modes:

The nominal mode. If the history of node n_t has been

$$\mathcal{H}_{n_t}^{u*} = \left(u_{n_0}^*, u_{n_1}^*, \dots, u_{n_{t-1}}^*\right), \tag{8.8}$$

i.e., all players used their cooperative controls on the path $n_0 \rightsquigarrow n_{t-1}$, then any player $i \in M$ implements the cooperative control $u_{n_t}^{i*}$ at node n_t.

The trigger mode. If the history of node n_t is such that there exists a node ν on the path $n_0 \rightsquigarrow n_{t-1}$ where $u_\nu \neq u_\nu^*$, then Player i implements her Nash equilibrium strategy in the subgame starting from the successor of ν and the corresponding state. Here, the history of node n_t is such that there exists a node ν and at least one deviating player $j \in M$, that is, the history $\mathcal{H}_\nu^u$ of node ν is part of $\mathcal{H}_{n_t}^{u*}$, and u_ν is not a part of $\mathcal{H}_{n_t}^{u*}$, but if we replace the control u_ν^j of Player j in node ν by the cooperative control u_ν^{j*}, then the pair (u_ν^{j*}, u_ν^{-j}) will be u_ν^*, which corresponds to the history $\mathcal{H}_{n_t}^{u*}$.

Formally, the trigger strategy of Player i is defined as follows:

$$\hat{\sigma}^i(\mathcal{H}_{n_t}^u) = \begin{cases} u_{n_t}^{i*}, & \text{if } \mathcal{H}_{n_t}^u = \mathcal{H}_{n_t}^{u*}, \\ \hat{u}_{n_t}^i, & \text{if there exists a node } \nu \text{ on path } n_0 \rightsquigarrow n_t \\ & \text{such that } u_\nu \neq u_\nu^*, \end{cases} \tag{8.9}$$

where $\hat{u}_{n_t}^i$ is Player i's control at node n_t, which implements the punishing strategy in the subgame starting in the unique node belonging to the set $\nu^+ \cap (n_0 \rightsquigarrow n_t)$. The control $\hat{u}_{n_t}^i$ coincides with $u_{n_t}^{i,nc}$ calculated as part of the Nash equilibrium for the subgame starting at node $n = \nu^+ \cap (n_0 \rightsquigarrow n_t)$ in state x_n.

Remark 8.1 To avoid further complicating the notation, we omitted the state argument in the punishing control and the trigger strategy, but we stress that they depend on the state value.

Let node n be a direct successor of node ν in which Player i deviates. The collection of controls (u_ν^i, u_ν^{-i*}) is then realized, and the state value in node n can be computed using the state dynamics $x_n = f^\nu(x_\nu^*, (u_\nu^i, u_\nu^{-i*}))$.

To prove the existence of a subgame perfect ε-equilibrium, we need some additional notations. In the subgame starting from node $n_t \in \mathbf{n}_t$ in state x_{n_t}, the collection of controls punishing Player i's individual deviation is given by

$$\hat{\mathbf{u}}^{(i)}[n_t, x_{n_t}] = (\hat{\mathbf{u}}^{j(i)}[n_t, x_{n_t}] : j \in M),$$

where $\hat{\mathbf{u}}^{j(i)}[n_t, x_{n_t}] = \left\{ \hat{u}_{n_\theta}^{j(i)} : n_\theta \in \mathbf{n}_t^{++} \setminus \mathbf{n}_T \right\}$, which leads to the following Nash equilibrium state trajectory:

$$\hat{\mathbf{x}}^{(i)}(n_t) = \{ \hat{x}_{n_\theta}^{(i)} : n_\theta \in \mathbf{n}_t^{++} \}.$$

To construct the trigger strategies, we need to find m punishing strategy profiles for each subgame on the cooperative state trajectory.

Theorem 8.1 *For any $\varepsilon \geqslant \tilde{\varepsilon}$ in the game played over an event tree, there exists a subgame perfect ε-equilibrium with players' payoffs $(W_1(x_0, \mathbf{u}^*), \ldots, W_m(x_0, \mathbf{u}^*))$, and*

$$\tilde{\varepsilon} = \max_{\substack{i \in M}} \max_{\substack{n_t \in \mathbf{n}_t \\ t=0,\dots,T-1}} \varepsilon_{n_t}^i, \tag{8.10}$$

where

$$
\begin{aligned}
\varepsilon_{n_t}^i = \max_{u_{n_t}^i \in U_{n_t}^i} \Big\{ & \phi_i(n_t, x_{n_t}^*, (u_{n_t}^i, u_{n_t}^{-i*})) - \phi_i(n_t, x_{n_t}^*, u_{n_t}^*) \\
& + \sum_{\theta=t+1}^{T-1} \rho^{\theta-t} \sum_{n_\theta \in \mathbf{n}_\theta \cap \mathbf{n}_t^{++}} \pi_{n_t}^{n_\theta} \left(\phi_i(n_\theta, \hat{x}_{n_\theta}^{(i)}, \hat{u}_{n_\theta}^{(i)}) - \phi_i(n_\theta, x_{n_\theta}^*, u_{n_\theta}^*) \right) \\
& + \rho^{T-t} \sum_{n_T \in \mathbf{n}_T \cap \mathbf{n}_t^{++}} \pi_{n_t}^{n_T} \left(\Phi_i(n_T, \hat{x}_{n_T}^{(i)}) - \Phi_i(n_T, x_{n_T}^*) \right) \Big\},
\end{aligned}
\tag{8.11}
$$

and $\hat{u}_{n_\theta}^{(i)}$ is a control profile in node n_θ corresponding to a strategy profile $\hat{\sigma}$ determined by (8.9) and when the trigger mode of the strategy begins in the subgame starting at the node belonging to the set $\mathbf{n}_t^+$ and in state $f^{n_t}\left(x_{n_t}, (u_{n_t}^i, u_{n_t}^{-i*})\right)$. Therefore, the differences $\left(\phi_i(n_\theta, \hat{x}_{n_\theta}^{(i)}, \hat{u}_{n_\theta}^{(i)}) - \phi_i(n_\theta, x_{n_\theta}^*, u_{n_\theta}^*) \right)$ and $\left(\Phi_i(n_T, \hat{x}_{n_T}^{(i)}) - \Phi_i(n_T, x_{n_T}^*) \right)$ also depend on the control $u_{n_t}^i$. The state $\hat{x}_{n_\theta}^{(i)}$, $n_\theta \in \mathbf{n}_\theta \cap \mathbf{n}_t^{++}$ is a state trajectory corresponding to $\hat{\mathbf{u}}^{(i)}[n_t, x_{n_t}^*]$.

Proof Consider the trigger strategy $\hat{\sigma} = (\hat{\sigma}^i : i \in M)$ defined in (8.9), and the subgame starting from any node $n_t \in \mathbf{n}_t, t = 0, \dots, T - 1$. Consider possible histories of the node n_t and compute the benefit of Player i deviating at node n_t. Her payoff in this subgame in cooperation will be given by

$$
\begin{aligned}
W_i(x_{n_t}^*, \mathbf{u}^*[n_t, x_{n_t}^*]) = & \phi_i(n_t, x_{n_t}^*, u_{n_t}^*) + \sum_{\theta=t+1}^{T-1} \rho^{\theta-t} \sum_{n_\theta \in \mathbf{n}_\theta \cap \mathbf{n}_t^{++}} \pi_{n_t}^{n_\theta} \phi_i(n_\theta, x_{n_\theta}^*, u_{n_\theta}^*) \\
& + \rho^{T-t} \sum_{n_T \in \mathbf{n}_T \cap \mathbf{n}_t^{++}} \pi_{n_t}^{n_T} \Phi_i(n_T, x_{n_T}^*).
\end{aligned}
\tag{8.12}
$$

First, let the history of node n_t be $\mathcal{H}_{n_t}^{u*}$. Suppose Player i deviates in node n_t from the cooperative trajectory. In this case, she may secure the following payoff in the subgame starting at node n_t, given the information that the strategy profile $\hat{\sigma} = (\hat{\sigma}^j(\cdot) : j \in M)$ determined by (8.9) will materialize:

$$
\begin{aligned}
\max_{u_{n_t}^i \in U_{n_t}^i} \Big\{ & \phi_i(n_t, x_{n_t}^*, (u_{n_t}^i, u_{n_t}^{-i*})) + \sum_{\theta=t+1}^{T-1} \rho^{\theta-t} \sum_{n_\theta \in \mathbf{n}_\theta \cap \mathbf{n}_t^{++}} \pi_{n_t}^{n_\theta} \phi_i(n_\theta, \hat{x}_{n_\theta}^{(i)}, \hat{u}_{n_\theta}^{(i)}) \\
& + \rho^{T-t} \sum_{n_T \in \mathbf{n}_T \cap \mathbf{n}_t^{++}} \pi_{n_t}^{n_T} \Phi_i(n_T, \hat{x}_{n_T}^{(i)}) \Big\},
\end{aligned}
\tag{8.13}
$$

where a punishing Nash strategy starts to be implemented in nodes from $\mathbf{n}_t^+$. Then, we can compute the benefit Player i derives from deviating at node n_t as the difference between (8.13) and (8.12), namely,

$$
\begin{aligned}
\varepsilon_{n_t}^i = \max_{u_{n_t}^i \in U_{n_t}^i} & \left\{ \phi_i(n_t, x_{n_t}^*, (u_{n_t}^i, u_{n_t}^{-i*})) - \phi_i(n_t, x_{n_t}^*, u_{n_t}^*) \right. \\
& + \sum_{\theta=t+1}^{T-1} \rho^{\theta-t} \sum_{n_\theta \in \mathbf{n}_\theta \cap \mathbf{n}_t^{++}} \pi_{n_t}^{n_\theta} \left(\phi_i(n_\theta, \hat{x}_{n_\theta}^{(i)}, \hat{u}_{n_\theta}^{(i)}) - \phi_i(n_\theta, x_{n_\theta}^*, u_{n_\theta}^*) \right) \\
& \left. + \rho^{T-t} \sum_{n_T \in \mathbf{n}_T \cap \mathbf{n}_t^{++}} \pi_{n_t}^{n_T} \left(\Phi_i(n_T, \hat{x}_{n_T}^{(i)}) - \Phi_i(n_T, x_{n_T}^*) \right) \right\}.
\end{aligned}
\tag{8.14}
$$

Second, suppose that the history of node n_t does not coincide with the cooperative history $\mathcal{H}_{n_t}^{u*}$. This means that all players have switched from the nominal mode to the trigger one. Player i will have no benefit from deviating in node n_t because the players implement their Nash equilibrium strategies regardless of which player (or group of players) has deviated in the previous nodes.

Computing the maximum benefit from deviation for any subgame and any player using (8.14), we determine

$$
\tilde{\varepsilon} = \max_{i \in M} \max_{\substack{n_t \in \mathbf{n}_t \\ t=0,\dots,T-1}} \varepsilon_{n_t}^i.
$$

And for any $\varepsilon \geq \tilde{\varepsilon}$ the strategy profile determined by (8.9) is a subgame perfect ε-equilibrium by construction. $\qquad\square$

8.2.1 Cost of Cooperation in DGPETs

In a subgame perfect ε-equilibrium, $\tilde{\varepsilon}$ represents the maximal benefit that a player could obtain in the whole game by deviating from cooperation individually. The value of $\tilde{\varepsilon}$ depends on the variables' measurement units, as well as on other game data such as the number of players and the duration of the game. For this reason, there is no obvious way to derive an upper bound for $\tilde{\varepsilon}$. In the repeated-game literature, the stage gain from deviation is compared to the stage-game payoff, that is, the comparison is done in relative terms. As these numbers (gain from deviation and stage payoff) are the same in all stages, this comparison is intuitive and meaningful. In state-space games like DGPETs, the payoffs can vary considerably across nodes and time periods, and consequently, the choice of any particular node payoff as a benchmark becomes an issue in itself. For instance, cooperation may require very heavy investments in the early stages, and the rewards come later.

The above elements suggest selecting a *relative* measure rather than an absolute one to interpret $\tilde{\varepsilon}$. Further, as $\tilde{\varepsilon}$ is computed for the whole game, we need to compare it to a total payoff that the players can also obtain in the whole game. For these reasons, we adopt the *cost of cooperation*, which is given by

$$CoC = \frac{\tilde{\varepsilon}}{\sum\limits_{j \in M} W_j(x_0, \mathbf{u}^*)},$$

where the denominator is the total cooperative payoff.

Remark 8.2 As discussed in Sect. 2.4.1, one can compute the price of anarchy (PoA) to assess the cost of a lack of coordination between the players. We determine the PoA in the next example.

8.2.2   An Illustrative Example

We illustrate the theoretical development with an example of pollution control.

Denote by $M = \{1, \ldots, m\}$ the set of players representing countries, and by $\mathbb{T} = \{0, \ldots, T\}$ the set of periods. Production activities in each country generate revenues and, as a by-product, pollutant emissions, e.g., CO_2. Denote by $u_{n_t}^i$ the emissions of Player i. Assuming a monotone increasing relationship between production and revenues, we can express the revenues, denoted by f_i, directly as a function of emissions. To keep it simple, we adopt the following specification:

$$f_i\left(u_{n_t}^i\right) = \alpha_i u_{n_t}^i - \frac{1}{2}\beta_i\left(u_{n_t}^i\right)^2, \tag{8.15}$$

where α_i, β_i are positive parameters. We constrain the revenues to be nonnegative for any $n_t \in \mathbf{n}_0^{++} \setminus \mathbf{n}_T$ and any player i.

Denote by $u_{n_t} = (u_{n_t}^1, \ldots, u_{n_t}^m)$ the vector of countries' emissions at node $n_t \in \mathbf{n}_t$, $t \in \mathbb{T} \setminus \{T\}$, and by x_{n_t} the stock of pollution at this node. The evolution of this stock is governed by the following difference equation:

$$x_{n_t} = (1 - \delta_{n_t^-})x_{n_t^-} + \sum_{i \in M} u_{n_t^-}^i, \quad n_t \in \mathbf{n}_0^{++}\setminus\{n_0\}, \tag{8.16}$$

where the initial state x_0 is given for the root node n_0, and δ_{n_t} ($0 < \delta_{n_t} < 1$) is a stochastic rate of pollution absorption by Mother Nature at node n_t. We suppose that δ_{n_t} can take two possible values, that is, $\delta_{n_t} \in \{\underline{\delta}, \overline{\delta}\}$, with $\underline{\delta} < \overline{\delta}$.

Each country suffers an environmental damage cost due to pollution accumulation. We assume that this cost is an increasing convex function in this stock and retain the quadratic form $D_i(x_{n_t}) = \frac{1}{2}d_i x_{n_t}^2$, $i \in M$, where d_i is a strictly positive parameter.

Denote by $\rho \in (0, 1)$ the discount factor. Assume that Player $i \in M$ maximizes the following objective functional:

$$J_i(\mathbf{x}, \mathbf{u}) = \sum_{t=0}^{T-1} \rho^t \sum_{n_t \in \mathbf{n}_t} \pi_{n_t} \left(f_i(u_{n_t}^i) - D_i(x_{n_t}) \right) + \rho^T \sum_{n_T \in \mathbf{n}_T} \pi_{n_T} \Phi_i(n_T, x_{n_T}),$$

subject to (8.16), and $u_{n_t}^i \geq 0$ for all $i \in M$ and any $n_t \in \mathbf{n}_t, t = 0, \ldots, T-1$. Suppose that the payoff function of Player i at a terminal node n_T is given by

$$\Phi_i(n_T, x_{n_T}) = -\gamma_i x_{n_T}.$$

Substituting for $\Phi_i(n_T, x_{n_T})$, $f_i(u_{n_t}^i)$, and $D_i(x_{n_t})$ by their values, we get

$$J_i(\mathbf{x}, \mathbf{u}) = \sum_{t=0}^{T-1} \rho^t \sum_{n_t \in \mathbf{n}_t} \pi_{n_t} \left(\alpha_i u_{n_t}^i - \frac{1}{2}\beta_i (u_{n_t}^i)^2 - \frac{1}{2}d_i x_{n_t}^2 \right) - \rho^T \sum_{n_T \in \mathbf{n}_T} \pi_{n_T} \gamma_i x_{n_T}.$$

We use the following parameter values in the numerical illustration:

$$M = \{1, \ldots, 5\}, \ \mathbb{T} = \{0, 1, 2, 3\}, \ x_0 = 0, \ \rho = 0.9,$$
$$\alpha_1 = 50, \ \alpha_2 = 49, \ \alpha_3 = 48, \ \alpha_4 = 47, \ \alpha_5 = 46,$$
$$\gamma_1 = 1.0, \ \gamma_2 = 0.9, \ \gamma_3 = 0.8, \ \gamma_4 = 0.7, \ \gamma_5 = 0.6,$$
$$d_1 = 0.05, \ d_2 = 0.04, \ d_3 = 0.03, \ d_4 = 0.02, \ d_5 = 0.01,$$
$$\beta_i = 1, \ i \in M,$$
$$\underline{\delta} = 0.1, \ \overline{\delta} = 0.3.$$

The event tree is depicted in Fig. 8.1, where the root node n_0 and all left-hand successor nodes have the low rate $\underline{\delta} = 0.1$ of pollution absorption by Nature, and all right-hand successor nodes have the high level $\overline{\delta} = 0.3$ of pollution absorption. It is a binary tree, i.e., each node in periods $t = 0, \ldots, 2$ has two successors. The conditional probability of realization of the upward (downward) successor of any node is $\frac{1}{2}$ ($\frac{1}{2}$). So, for instance, we have probabilities $\pi_{n_1^1} = \frac{1}{2}$ and $\pi_{n_1^2} = \frac{1}{2}$ in period 1, and probabilities $\pi_{n_2^1} = \pi_{n_2^2} = \pi_{n_2^3} = \pi_{n_2^4} = \frac{1}{4}$ for $t = 2$.

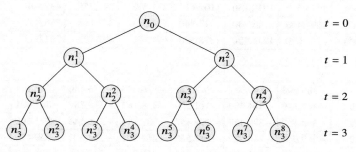

Fig. 8.1 Event tree for $T = 3$

Using (8.5), we compute the cooperative controls or emissions for each possible subgame and for the whole game. The cooperative state trajectory is given by

$x^*_{n_0}$	$x^*_{n^1_1}$	$x^*_{n^2_1}$	$x^*_{n^1_2}$	$x^*_{n^2_2}$	$x^*_{n^3_2}$	$x^*_{n^4_2}$	$x^*_{n^1_3}$
0.000	86.246	86.246	181.887	181.887	171.589	171.589	385.698

$x^*_{n^2_3}$	$x^*_{n^3_3}$	$x^*_{n^4_3}$	$x^*_{n^5_3}$	$x^*_{n^6_3}$	$x^*_{n^7_3}$	$x^*_{n^8_3}$
385.698	349.321	349.321	376.430	376.430	342.112	342.112

We use (8.6) to compute the players' payoffs in each of the subgames and report them in Table 8.1.

For each player $i \in M$ and any node $n_t \in \mathbf{n}_t$, $t = 0, \ldots, T-1$, we need to solve the optimization problem defined in (8.13). Once we obtain these payoffs, we compute the differences with the cooperative payoffs and give them in Table 8.2. Based on these differences, we determine the values $\varepsilon^i_{n_t}$ for $i \in M$ and $n_t \in \mathbf{n}_t$, $t = 0, \ldots, T-1$ (see Table 8.3). Note that in Table 8.2, the benefits from deviations are calculated when the players do not keep the cooperative controls. Otherwise, the values in Tables 8.2 and 8.3 coincide.

From Table 8.3, we see that, at root node n_0, only Players 4 and 5 benefit from deviating from cooperation. At time $t = 1$, all players can gain by deviating, and the same is true for period 2, where all players gain by deviating, as expected. We note that deviation in period $t = 2$ does not give players a large benefit because of the form of the terminal payoff $\Phi_i(n_T, x_{n_T})$. The largest benefit from deviation is realized for Player 5 at node n_0. Finally, we mention that, in this example, $\tilde{\varepsilon}$ is equal to

Table 8.1 Players' payoffs in cooperation

Time period	$t = 0$	$t = 1$		$t = 2$			
Node	n_0	n^1_1	n^2_1	n^1_2	n^2_2	n^3_2	n^4_2
$W_1(x^*, \mathbf{u}^*)$	1529.020	772.688	898.037	69.318	102.058	168.662	199.548
$W_2(x^*, \mathbf{u}^*)$	1581.450	894.471	1002.770	219.946	249.411	300.255	328.052
$W_3(x^*, \mathbf{u}^*)$	1636.590	1018.150	1109.410	371.573	397.765	432.848	457.556
$W_4(x^*, \mathbf{u}^*)$	1694.450	1143.740	1217.940	524.201	547.118	566.440	588.061
$W_5(x^*, \mathbf{u}^*)$	1755.010	1271.220	1328.380	677.828	697.472	701.033	719.565

$t = 3$							
n^1_3	n^2_3	n^3_3	n^4_3	n^5_3	n^6_3	n^7_3	n^8_3
−385.698	−385.698	−349.321	−349.321	−376.430	−376.430	−342.112	−342.112
−347.129	−347.129	−314.389	−314.389	−338.787	−338.787	−307.901	−307.901
−308.559	−308.559	−279.457	−279.457	−301.144	−301.144	−273.690	−273.690
−269.989	−269.989	−244.525	−244.525	−263.501	−263.501	−239.479	−239.479
−231.419	−231.419	−209.593	−209.593	−225.858	−225.858	−205.267	−205.267

Table 8.2 Maximum benefits from deviating from the cooperative control for any subgame

Time period	$t = 0$	$t = 1$		$t = 2$			
Node	n_0	n_1^1	n_1^2	n_2^1	n_2^2	n_2^3	n_2^4
Player 1	−504.721	154.280	138.449	3.645	3.645	3.645	3.645
Player 2	−254.934	188.391	169.196	3.892	3.892	3.892	3.892
Player 3	−4.169	225.891	202.974	4.147	4.147	4.147	4.147
Player 4	247.604	266.869	239.864	4.410	4.410	4.410	4.410
Player 5	500.410	311.420	279.949	4.682	4.682	4.682	4.682

Table 8.3 Values of $\varepsilon_{n_t}^i$ for any subgame and any player

Time period	$t = 0$	$t = 1$		$t = 2$			
Node	n_0	n_1^1	n_1^2	n_2^1	n_2^2	n_2^3	n_2^4
Player 1	0	154.280	138.449	3.645	3.645	3.645	3.645
Player 2	0	188.391	169.196	3.892	3.892	3.892	3.892
Player 3	0	225.891	202.974	4.147	4.147	4.147	4.147
Player 4	247.604	266.869	239.864	4.410	4.410	4.410	4.410
Player 5	500.410	311.420	279.949	4.682	4.682	4.682	4.682

Table 8.4 The maximal benefit from deviation in time period t calculated for 8 periods

t	0	1	2	3	4	5	6	7
Player 1	−9028.58	−7664.88	−5973.94	−4074.90	−2142.56	−547.32	202.41	3.65
Player 2	−5922.16	−5028.35	−3915.83	−2644.36	−1326.97	−224.58	246.78	3.89
Player 3	−3052.46	−2565.16	−1967.56	−1266.57	−508.53	120.56	295.64	4.15
Player 4	−424.81	−257.94	−90.80	107.43	326.38	469.61	349.09	4.41
Player 5	1955.05	1868.57	1671.33	1421.91	1145.03	822.68	407.27	4.68

500.410. The cost of cooperation in this game is $500.410/(1529.020 + 1581.450 + 1636.590 + 1694.450 + 1755.220)$, which is approximately 6.1%.

The game at hand belongs to the class of environmental games with negative externalities. Given the general result that cooperation is hard to achieve in such a setting (see, e.g., the survey in Jørgensen et al. 2010), the results in the above example with a short time horizon come as no surprise. To get an idea of the impact of having a longer time horizon, we now let the set of periods be $\mathbb{T} = \{0, 1, \ldots, 8\}$, with everything else being equal. The event tree is depicted in Fig. 8.2.

As before, we compute the cooperative payoffs as well as the benefits of deviating from the cooperative trajectory in all subgames. To save on space, in Tables 8.4 and 8.5, we only provide the results concerning the benefits of deviating from the agreement and the values of $\varepsilon_{n_t}^j$. (More precisely, we only give the maximum values

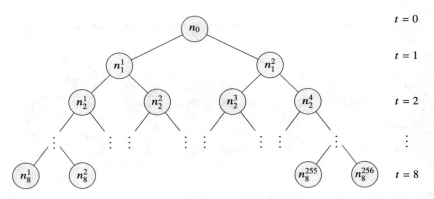

Fig. 8.2 Event tree for $T = 8$

Table 8.5 The maximal ε^j in time period t, denoted by $\varepsilon^{j*} := \max\limits_{n_t \in \mathbf{n}_t} \varepsilon^j_{n_t}$, calculated for an 8-period horizon

t	0	1	2	3	4	5	6	7
ε^{*1}	0	0	0	0	0	0	202.410	3.645
ε^{2*}	0	0	0	0	0	0	246.781	3.892
ε^{3*}	0	0	0	0	0	120.558	295.635	4.147
ε^{4*}	0	0	0	107.426	326.377	469.611	349.092	4.410
ε^{5*}	1955.050	1868.570	1671.330	1421.910	1145.030	822.679	407.274	4.682

for each time period.) These tables show that the first time a player can benefit from deviating from cooperation is Player 5 in period 0. Interestingly, Players 1 and 2 would deviate in period 6, Player 3 in period 5, and Player 4 in period 3. The lower the damage cost, the higher the incentive to deviate earlier. Players are ranked in decreasing order of damage costs (γ and d), i.e., the largest costs are for Player 1, and the smallest are for Player 5. Player 5 is the first who has an incentive to deviate, then Player 4, then Player 3, and finally Players 1 and 2. The value of $\tilde{\varepsilon}$ is equal to 1955.050, and the cost of cooperation is 26%. As under a cooperative regime, the total accumulated pollution is lower than under noncooperation. Thus, the results obtained here are encouraging, not only from an economic point of view, but also from an environmental one.

Figure 8.3 represents the relation between $\tilde{\varepsilon}$ and discount factor ρ for the game with 8 time periods. The larger the discount factor, the larger is $\tilde{\varepsilon}$.

Table 8.6 gives information on when each player has the first positive benefit from deviation. Player 5, with the lowest cost coefficients, will deviate in the root node for any discount factor. Player 1 will not deviate before period 6 if $\rho \geqslant 0.4$. The results clearly show that the higher the discount rate, the lower the incentive to deviate from cooperation. Also, the higher the discount rate, the later the players deviate from

Fig. 8.3 Value of $\bar{\varepsilon}$ as a function of discount factor ρ

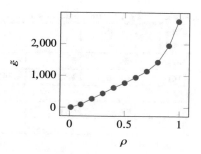

Table 8.6 The first time period when the player has an incentive to deviate

ρ	0.01	0.1	0.2	0.3	0.4	0.5	0.6	0.7	0.8	0.9	0.99
Player 1	0	1	4	5	6	6	6	6	6	6	6
Player 2	0	0	2	4	5	5	5	6	6	6	6
Player 3	0	0	0	1	2	4	5	5	5	5	5
Player 4	0	0	0	0	0	0	1	1	2	3	4
Player 5	0	0	0	0	0	0	0	0	0	0	0

Fig. 8.4 The cost of cooperation (in percents) as a function of discount factor ρ

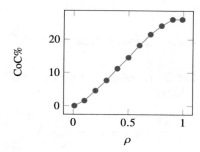

cooperation. As expected, they will all eventually defect from cooperation in the last period.

Figure 8.4 represents the cost of cooperation as a function of discount factor ρ. We note that this value does not exceed 27% for any ρ.

We calculate the price of anarchy in the game as

$$PoA = \frac{\max_{\mathbf{u}} \sum_{i \in M} J_i(x_0, \mathbf{u})}{\sum_{i \in M} J_i(x_0, \mathbf{u}^{nc})},$$

where $J_i(x_0, \mathbf{u}^{nc})$ is Player i's payoff in the Nash equilibrium Finally, Table 8.7 provides the price of anarchy in any subgame.

Table 8.7 PoA for the game with $T = 3$

Time period	$t = 0$	$t = 1$		$t = 2$			
Node	n_0	n_1^1	n_1^2	n_2^1	n_2^2	n_2^3	n_2^4
PoA	1.878	1.451	1.346	1.011	1.011	1.010	1.009

$t = 3$							
n_3^1	n_3^2	n_3^3	n_3^4	n_3^5	n_3^6	n_3^7	n_3^8
1	1	1	1	1	1	1	1

8.3 Incentive Equilibrium Strategies in DGPETs[2]

In this section, we extend to the class of DGPETs the concepts of incentive strategies and incentive equilibria introduced in Sect. 4.4 in a deterministic setup. To simplify the exposition and ease the computational burden, we focus on two-player linear-quadratic dynamic games. Our objective is to show how a cooperative solution can be sustained by implementing incentive strategies.

8.3.1 Linear-Quadratic DGPET

Let the set of players be $M = \{1, 2\}$. The optimization problem of Player i is as follows:

$$\max_{\mathbf{u}^i} W_i(x_0, \mathbf{u}) = \max_{\mathbf{u}^i \in \mathbf{U}^i} \sum_{t=0}^{T-1} \sum_{n_t \in \mathbf{n}_t} \pi_{n_t} \left(\frac{1}{2} x'_{n_t} Q^i_{n_t} x_{n_t} + p^{i\prime}_{n_t} x_{n_t} \right. \tag{8.17}$$

$$\left. + \frac{1}{2} \sum_{j=1}^2 u^{j\prime}_{n_t} R^{ij}_{n_t} u^j_{n_t} \right) + \sum_{n_T \in \mathbf{n}_T} \pi_{n_T} \left(\frac{1}{2} x'_{n_T} Q^i_{n_T} x_{n_T} + p^{i\prime}_{n_T} x_{n_T} \right),$$

subject to

$$x_{n_t} = A_{n_t^-} x_{n_t^-} + \sum_{j=1}^2 B^j_{n_t^-} u^j_{n_t^-}, \quad x_{n_0} = x_0, \tag{8.18}$$

---

[2] This section draws partly from Kanani Kuchesfehani and Zaccour (2016).

where $Q^i_{n_t} \in \mathbb{R}^{q \times q}$, $R^{ij}_{n_t} \in \mathbb{R}^{\ell_j \times \ell_j}$, $p^i_{n_t} \in \mathbb{R}^q$, $A_{n_t} \in \mathbb{R}^{q \times q}$ and $B^j_{n_t} \in \mathbb{R}^{q \times \ell_j}$ for all $n_t \in \mathbf{n}_t, t \in \mathbb{T}$.

Assumption 8.1 The matrices $Q^i_{n_t}$ and $Q^i_{n_T}$ are symmetric and negative semidefinite and $R^{ii}_{n_t}$ is negative definite. Additionally, the matrices $R^{ij}_{n_t}$ for $i \neq j$ are such that $R^{ii}_{n_t} + R^{ji}_{n_t}$ are negative definite as well.

By Assumption 8.1, the objective function (8.17) will be strictly concave in the control variables. Note that if one considers the dynamic game with nonstochastic transitions, i.e., the set $\mathbf{n}_t$ consists of one element for all t, then this optimization problem reduces to the one considered in a multistage game.

8.3.2 Cooperative Solution

Now we suppose that the two players agree to cooperate and maximize their joint payoff, that is, $\max_{\mathbf{u}} \sum_{i=1}^{2} W_i(x_0, \mathbf{u})$ subject to (8.18). The maximum principle can be applied to (8.17), taking into account state dynamics (8.18), to find the solution of the optimization problem.

The Lagrangian associated with the joint optimization problem is given by

$$
\mathcal{L}^c = \sum_{t=0}^{T-1} \sum_{n_t \in \mathbf{n}_t} \pi_{n_t} \left(\frac{1}{2} x'_{n_t} (Q^1_{n_t} + Q^2_{n_t}) x_{n_t} + (p^1_{n_t} + p^2_{n_t})' x_{n_t} \right. \tag{8.19}
$$
$$
\left. + \frac{1}{2} \begin{bmatrix} u^1_{n_t} \\ u^2_{n_t} \end{bmatrix}' \begin{bmatrix} R^{11}_{n_t} + R^{21}_{n_t} & 0 \\ 0 & R^{12}_{n_t} + R^{22}_{n_t} \end{bmatrix} \begin{bmatrix} u^1_{n_t} \\ u^2_{n_t} \end{bmatrix} \right)
$$
$$
+ \sum_{n_T \in \mathbf{n}_T} \pi_{n_T} \left(\frac{1}{2} x'_{n_T} (Q^1_{n_T} + Q^2_{n_T}) x_{n_T} + (p^1_{n_T} + p^2_{n_T})' x_{n_T} \right) + \lambda^c_{n_0}{}' (x_0 - x_{n_0})
$$
$$
+ \sum_{t=1}^{T} \sum_{n_t \in \mathbf{n}_t} \pi_{n_t} \lambda^c_{n_t}{}' \left(A_{n_t^-} x_{n_t^-} + \sum_{j=1}^{2} B^j_{n_t^-} u^j_{n_t^-} - x_{n_t} \right).
$$

For any process z_{n_t}, $n_t \in \mathbf{n}_0^{++}$, we denote by $z_{\pi, \mathbf{n}_t^+} := \sum_{\nu \in \mathbf{n}_t^+} \pi^\nu_{n_t} z_\nu$ the conditional sum of the successors of node $n_t \in \mathbf{n}_0^{++} \backslash \mathbf{n}_T$.

The optimality conditions associated with the above dynamic optimization problem are given by

$$\frac{\partial \mathcal{L}^c}{\partial u_{n_t}^i} = \pi_{n_t} u_{n_t}^{i\prime} \sum_{j=1}^{2} R_{n_t}^{ji} + \sum_{\nu \in \mathbf{n}_t^+} \pi_\nu \, \lambda_\nu^c B_{n_t}^i = 0,$$

$$\Rightarrow u_{n_t}^{i*} = -\left(\sum_{j=1}^{2} R_{n_t}^{ji} \right)^{-1} B_{n_t}^i \lambda_{\pi,\mathbf{n}_t^+}^c. \tag{8.20}$$

$$\frac{\partial \mathcal{L}^c}{\partial x_{n_t}} = \pi_{n_t} \sum_{j=1}^{2} (Q_{n_t}^j x_{n_t} + p_{n_t}^j) + A_{n_t}^\prime \sum_{\nu \in \mathbf{n}_t^+} \pi_\nu \, \lambda_\nu^c - \pi_{n_t} \lambda_{n_t}^c = 0,$$

$$\Rightarrow \lambda_{n_t}^c = A_{n_t}^\prime \lambda_{\pi,\mathbf{n}_t^+}^c + \sum_{j=1}^{2} (Q_{n_t}^j x_{n_t} + p_{n_t}^j). \tag{8.21}$$

$$\frac{\partial \mathcal{L}^c}{\partial x_{n_T}} = \pi_{n_T} \left(\sum_{j=1}^{2} (Q_{n_T}^j x_{n_T} + p_{n_T}^j) - \lambda_{n_T}^c \right) = 0,$$

$$\Rightarrow \lambda_{n_T}^c = \sum_{j=1}^{2} (Q_{n_T}^j x_{n_T} + p_{n_T}^j). \tag{8.22}$$

Let us define k_{n_t} and α_{n_t} recursively as follows:

$$k_{n_t} = \sum_{j=1}^{2} Q_{n_t}^j + A_{n_t}^\prime k_{\pi,\mathbf{n}_t^+} \left(I + \sum_{j=1}^{2} S_{n_t}^j k_{\pi,\mathbf{n}_t^+} \right)^{-1} A_{n_t}, \tag{8.23}$$

$$\alpha_{n_t} = \sum_{j=1}^{2} p_{n_t}^j + \left(I - A_{n_t}^\prime k_{\pi,\mathbf{n}_t^+} \left(I + \sum_{j=1}^{2} S_{n_t}^j k_{\pi,\mathbf{n}_t^+} \right)^{-1} \sum_{j=1}^{2} S_{n_t}^j \right) \alpha_{\pi,\mathbf{n}_t^+}, \tag{8.24}$$

where $S_{n_t}^i = B_{n_t}^i \left(\sum_{j=1}^{2} R_{n_t}^{ij} \right)^{-1} B_{n_t}^{i\prime}$ for $n_t \in \mathbf{n}_0^{++} \backslash \mathbf{n}_T$, and $k_{n_T} = \sum_{j=1}^{2} Q_{n_T}^j$ and $\alpha_{n_T} = \sum_{j=1}^{2} p_{n_T}^j$ for $n_T \in \mathbf{n}_T$.

Assumption 8.2 All matrices $\left(I + \sum_{j=1}^{2} S_{n_t}^j k_{\pi,\mathbf{n}_t^+} \right)$ for any node $n_t \in \mathbf{n}_0^{++} \backslash \mathbf{n}_T$ are invertible.

The satisfaction of Assumption 8.2 clearly depends on the parameter values. The following proposition gives the form of the cooperative controls and the corresponding state trajectory in two-player linear-quadratic DGPETs.

Proposition 8.1 *Under Assumptions 8.1 and 8.2, the cooperative (joint optimization) solution is given by*

$$u_{n_t}^{i*} = -\left(\sum_{j=1}^{2} R_{n_t}^{ji} \right)^{-1} B_{n_t}^i \left(k_{\pi,\mathbf{n}_t^+} x_\nu^* + \alpha_{\pi,\mathbf{n}_t^+} \right),$$

where x^* is the associated state trajectory determined by

$$x_\nu^* = (I + \sum_{j=1}^{2} S_{n_t}^j k_{\pi,\mathbf{n}_t^+})^{-1} A_{n_t} x_{n_t}^* - (I + \sum_{j=1}^{2} S_{n_t}^j k_{\pi,\mathbf{n}_t^+})^{-1} \sum_{j=1}^{2} S_{n_t}^j \alpha_{\pi,\mathbf{n}_t^+}. \quad (8.25)$$

Proof Due to the strict concavity of the objective function, we have the unique relation given by (8.20) for all n_t.

$$x_\nu^* = A_{n_t} x_{n_t}^* - \sum_{j=1}^{2} S_{n_t}^j \lambda_{\pi,\mathbf{n}_t^+}^c, \quad n_t \in \mathbf{n}_t, \ t = 1, \ldots, T, \ x_{n_0}^* = x_0.$$

Let us suppose that the costate variables are linear in the state (see Engwerda 2005), that is,

$$\lambda_{n_t}^c = k_{n_t} x_{n_t}^* + \alpha_{n_t}, \ n_t \in \mathbf{n}_t; \forall t,$$

which leads to

$$x_\nu^* = A_{n_t} x_{n_t}^* - \sum_{j=1}^{2} S_{n_t}^j (k_{\pi,\mathbf{n}_t^+} x_\nu + \alpha_{\pi,\mathbf{n}_t^+}).$$

The right-hand side of the above equation contains the expected value of the terms evaluated at the successor nodes $\nu \in \mathbf{n}_t^+$. We know that $x_{\nu_1}^* = x_{\nu_2}^*$; $\forall \nu_1, \nu_2 \in \mathbf{n}_t^+$, then $\sum_{\nu \in \mathbf{n}_t^+} \pi_\nu x_\nu^* = x_\nu^*$. Since the matrix $\left(I + \sum_{j=1}^{2} S_{n_t}^j k_{\pi,\mathbf{n}_t^+} \right)$ is assumed to be invertible, we have

$$x_\nu^* = \left(I + \sum_{j=1}^{2} S_{n_t}^j k_{\pi,\mathbf{n}_t^+} \right)^{-1} \left(A_{n_t} x_{n_t}^* - \sum_{j=1}^{2} S_{n_t}^j \alpha_{\pi,\mathbf{n}_t^+} \right). \quad (8.26)$$

From the optimality condition (8.21) and using the above we have

$$k_{n_t} x_{n_t}^* + \alpha_{n_t} = A_{n_t}' k_{\pi,\mathbf{n}_t^+} x_\nu^* + \alpha_{\pi,\mathbf{n}_t^+} + \sum_{j=1}^{2} Q_{n_t}^j x_{n_t} + \sum_{j=1}^{2} p_{n_t}^j$$

$$= \left(\sum_{j=1}^{2} Q_{n_t}^j + A_{n_t}' k_{\pi,\mathbf{n}_t^+} (I + \sum_{j=1}^{2} S_{n_t}^j k_{\pi,\mathbf{n}_t^+})^{-1} A_{n_t} \right) x_{n_t}^*$$

$$+ \sum_{j=1}^{2} p_{n_t}^j + \left(I - A_{n_t}' k_{\pi,\mathbf{n}_t^+} (I + \sum_{j=1}^{2} S_{n_t}^j k_{\pi,\mathbf{n}_t^+})^{-1} \sum_{j=1}^{2} S_{n_t}^j \right) \alpha_{\pi,\mathbf{n}_t^+}.$$

Collecting the coefficients of $x_{n_t}^*$, the relations (8.23) and (8.24) follow. The remaining statements follow from using the terminal conditions. $\qquad\square$

It should be clear by now that Assumptions 8.1 and 8.2 were made to guarantee the existence of a solution to the joint optimization problem. Denote by $\mathbf{u}^{i*} = \left(u_{n_t}^{i*} : n_t \in \mathbf{n}_0^{++} \setminus \mathbf{n}_T\right)$ the cooperative strategy of Player i in the game.

8.3.3 S-adapted Incentive Equilibria

In this section, we design incentive equilibrium strategies to support the cooperative (or coordinated) solution $u_{n_t}^* = (u_{n_t}^{1*}, u_{n_t}^{2*}) \in U_{n_t}^1 \times U_{n_t}^2$ for any node n_t. To do so, we suppose that each player adopts a strategy that is a function of the other player's strategy. Denote the set of such strategies of Player $i = 1, 2$ by

$$\Psi_i = \left\{\psi_i : \mathbf{U}^j \to \mathbf{U}^i\right\}, \quad i \neq j.$$

We define the incentive equilibrium in this class of strategies for a two-player DGPET.

Definition 8.4 A strategy profile (ψ_1, ψ_2) with $\psi_i \in \Psi_i$, $i = 1, 2$ is an incentive equilibrium at $(\mathbf{u}^{1*}, \mathbf{u}^{2*})$ if

$$W_1(x_0, (\mathbf{u}^{1*}, \mathbf{u}^{2*})) \geq W_1(x_0, (\mathbf{u}^1, \psi_2(\mathbf{u}^1))), \ \forall \mathbf{u}^1 \in \mathbf{U}^1,$$
$$W_2(x_0, (\mathbf{u}^{1*}, \mathbf{u}^{2*})) \geq W_2(x_0, (\psi_1(\mathbf{u}^2), \mathbf{u}^2)), \ \forall \mathbf{u}^2 \in \mathbf{U}^2,$$
$$\psi_1(u_{n_t}^{2*}) = u_{n_t}^{1*}, \ n_t \in \mathbf{n}_0^{++} \setminus \mathbf{n}_T,$$
$$\psi_2(u_{n_t}^{1*}) = u_{n_t}^{2*}, \ n_t \in \mathbf{n}_0^{++} \setminus \mathbf{n}_T.$$

The above definition states that if a player implements her part of the agreement, then the best response of the other player is to do the same. In this sense, each player's incentive strategy represents a threat to implement a different control than the optimal one if the other player deviates from her optimal strategy. To determine these incentive strategies, we need to solve two optimal control problems, in each of which one player assumes that the other player is using her incentive strategy. The optimization problem of Player i is given by (8.17)–(8.18) with the additional constraint stating that Player j is using her incentive strategy, that is,

$$u_{n_t}^j = \psi_j(u_{n_t}^i), \ i, j = 1, 2, i \neq j.$$

To show how it works, let us introduce the following corresponding Lagrangian to Player i's optimization problem, $i, j = 1, 2, i \neq j$:

$$\mathcal{L}_i(\lambda^i, \mathbf{x}, \mathbf{u}^i) = \frac{1}{2}\left(x'_{n_0} Q^i_{n_0} x_{n_0} + 2(p^i_{n_0})' x_{n_0} + (u^i_{n_0})' R^{ii}_{n_0} u^i_{n_0} + \psi'_j(u^i_{n_0}) R^{ij}_{n_0} \psi_j(u^i_{n_0})\right)$$

$$+ \sum_{t=1}^{T-1} \sum_{n_t \in \mathbf{n}_t} \frac{\pi_{n_t}}{2} \left(x'_{n_t} Q^i_{n_t} x_{n_t} + 2p^{i\prime}_{n_t} x_{n_t} + (u^i_{n_t})' R^{ii}_{n_t} u^i_{n_t} + \psi'_j(u^i_{n_t}) R^{ij}_{n_t} \psi_j(u^i_{n_t})\right)$$

$$+ \sum_{n_T \in \mathbf{n}_T} \frac{\pi_{n_T}}{2} \left(x'_{n_T} Q^i_{n_T} x_{n_T} + 2p^{i\prime}_{n_T} x_{n_T}\right) + \lambda^{i\,\prime}_{n_0}(x_0 - x_{n_0})$$

$$+ \sum_{t=1}^{T} \sum_{n_t \in \mathbf{n}_t} \pi_{n_t} \lambda^{i\,\prime}_{n_t} \left(A_{n_t^-} x_{n_t^-} + B^i_{n_t^-} u^i_{n_t^-} + B^j_{n_t^-} \psi_j(u^i_{n_t^-}) - x_{n_t}\right),$$

where $\lambda^i_{n_t}$ represents the Lagrange multiplier associated with the constraint (8.18). The first-order optimality conditions are

$$\frac{\partial \mathcal{L}_i}{\partial u^i_{n_t}} = \pi_{n_t} \left(R^{ii}_{n_t} u^i_{n_t} + \left(\frac{\partial \psi_j}{\partial u^i_{n_t}}\right)' R^{ij}_{n_t} \psi_j(u^i_{n_t})\right) + \left(B^i_{n_t} + B^j_{n_t} \frac{\partial \psi_j}{\partial u^i_{n_t}}\right)' \sum_{\nu \in \mathbf{n}_t^+} \pi_\nu \lambda^i_\nu = 0,$$

$$\Rightarrow u^i_{n_t} = -\left(R^{ii}_{n_t}\right)^{-1} \left(B^i_{n_t} + B^j_{n_t} \frac{\partial \psi_j}{\partial u^i_{n_t}}\right)' \lambda^i_{\pi, \mathbf{n}_t^+} - (R^{ii}_{n_t})^{-1} \left(\frac{\partial \psi_j}{\partial u^i_{n_t}}\right)' R^{ij}_{n_t} \psi_j(u^i_{n_t}).$$

$$(8.27)$$

$$\frac{\partial \mathcal{L}_i}{\partial x_{n_t}} = \pi_{n_t}(Q^i_{n_t} x_{n_t} + p^i_{n_t}) + A'_{n_t} \sum_{\nu \in \mathbf{n}_t^+} \pi_\nu \lambda^i_\nu - \pi_{n_t} \lambda^i_{n_t} = 0,$$

$$\Rightarrow \lambda^i_{n_t} = A'_{n_t} \lambda^i_{\pi, \mathbf{n}_t^+} + Q^i_{n_t} x_{n_t} + p^i_{n_t}. \tag{8.28}$$

$$\frac{\partial \mathcal{L}_i}{\partial x_{n_T}} = \pi_{n_T}(Q^i_{n_T} x_{n_T} + p^i_{n_T} - \lambda^i_{n_T}) = 0,$$

$$\Rightarrow \lambda_{n_T} = Q^i_{n_T} x_{n_T} + p^i_{n_T}. \tag{8.29}$$

The proposition below states the conditions to be satisfied by incentive strategies.

Proposition 8.2 *To be an incentive equilibrium at* $\mathbf{u}^*$, *a pair of strategies* $(\psi_1, \psi_2) \in \Psi_1 \times \Psi_2$ *must satisfy the following conditions:*

$$\left(\sum_{j=1}^{2} R^{ji}_{n_t}\right)^{-1} B^i_{n_t} \left(k_{\pi, \mathbf{n}_t^+} x^*_\nu + \alpha_{\pi, \mathbf{n}_t^+}\right)$$

$$= (R^{ii}_{n_t})^{-1} \left(B^i_{n_t} + B^j_{n_t} \frac{\partial \psi_j}{\partial u^i_{n_t}}\right)' \lambda^i_{\pi, \mathbf{n}_t^+} - (R^{ii}_{n_t})^{-1} \left(\frac{\partial \psi_j}{\partial u^i_{n_t}}\right)' R^{ij}_{n_t} \psi_j(u^i_{n_t}),$$

for $i, j = 1, 2$, $i \neq j$ *with all functions evaluated at* $(\mathbf{u}_1^*, \mathbf{u}_2^*)$. *The variables* $k^i_{n_t}$ *and* $\alpha^i_{n_t}$ *are recursively defined by*

$$k_{n_t}^i = Q_{n_t}^i + A_{n_t}' k_{\pi,\mathbf{n}_t^+}^i \left(I + \sum_{i=1}^{2}(S_{n_t}^i + l_{n_t}^i)k_{\pi,\mathbf{n}_t^+}^i \right)^{-1} A_{n_t}, \tag{8.30}$$

$$\alpha_{n_t}^i = p_{n_t}^i + A_{n_t}' \alpha_{\pi,\mathbf{n}_t^+}^i$$
$$- A_{n_t}' k_{\pi,\mathbf{n}_t^+}^i \left(I + \sum_{i=1}^{2}(S_{n_t}^i + l_{n_t}^i)k_{\pi,\mathbf{n}_t^+}^i \right)^{-1} \sum_{i=1}^{2}\left((S_{n_t}^i + l_{n_t}^i)\alpha_{\pi,\mathbf{n}_t^+}^i + h_{n_t}^i \right), \tag{8.31}$$

where

$$S_{n_t}^i = B_{n_t}^i (R_{n_t}^{ii})^{-1}(B_{n_t}^i)', \tag{8.32}$$

$$l_{n_t}^i = B_{n_t}^i (R_{n_t}^{ii})^{-1}\left(B_{n_t}^j \frac{\partial \psi_j}{\partial u_{n_t}^i} \right)', \tag{8.33}$$

$$h_{n_t}^i = -B_{n_t}^i (R_{n_t}^{ii})^{-1}\left(\frac{\partial \psi_j}{\partial u_{n_t}^i} \right)' R_{n_t}^{ij}\psi_j(u_{n_t}^i), \ i,j = 1,2, \ i \neq j. \tag{8.34}$$

Proof From the optimality condition (8.27) we get

$$x_\nu^* = A_{n_t} x_{n_t}^*$$
$$- \sum_{i=1}^{2} B_{n_t}^i \left((R_{n_t}^{ii})^{-1}\left(B_{n_t}^i + B_{n_t}^j \frac{\partial \psi_j}{\partial u_{n_t}^i} \right)' \lambda_{\pi,\mathbf{n}_t^+}^i - (R_{n_t}^{ii})^{-1}\left(\frac{\partial \psi_j}{\partial u_{n_t}^i} \right)' R_{n_t}^{ij}\psi_j(u_{n_t}^i) \right).$$

Using the definitions (8.32)–(8.34) we get

$$x_\nu^* = A_{n_t} x_{n_t}^* - \sum_{i=1}^{2}\left((S_{n_t}^i + l_{n_t}^i)\lambda_{\pi,\mathbf{n}_t^+}^i + h_{n_t}^i \right).$$

Now, define $\lambda_{n_t}^i = k_{n_t}^i x_{n_t}^* + \alpha_{n_t}^i, \ n_t \in \mathbf{n}_t$ which leads to

$$x_\nu^* = A_{n_t} x_{n_t}^* - \sum_{i=1}^{2}(S_{n_t}^i + l_{n_t}^i)k_{\pi,\mathbf{n}_t^+}^i x_\nu^* - \sum_{i=1}^{2}\left((S_{n_t}^i + l_{n_t}^i)\alpha_{\pi,\mathbf{n}_t^+}^i + h_{n_t}^i \right).$$

Assuming that the matrix $I + \sum_{i=1}^{2}(S_{n_t}^i + l_{n_t}^i)k_{\pi,\mathbf{n}_t^+}^i$ is invertible, we have

$$x_\nu^* = \left(I + \sum_{i=1}^{2}(S_{n_t}^i + l_{n_t}^i)k_{\pi,\mathbf{n}_t^+}^i \right)^{-1}\left(A_{n_t} x_{n_t}^* - \sum_{i=1}^{2}(S_{n_t}^i + l_{n_t}^i)\alpha_{\pi,\mathbf{n}_t^+}^i + h_{n_t}^i \right).$$
$$\tag{8.35}$$

Substituting the above in the (8.28) we obtain

$$k^i_{n_t} x^*_{n_t} + \alpha^i_{n_t}$$

$$= Q^i_{n_t} x^*_{n_t} + A'_{n_t} k^i_{\pi, \mathbf{n}^+_t} \left(I + \sum_{i=1}^{2} (S^i_{n_t} + l^i_{n_t}) k^i_{\pi, \mathbf{n}^+_t} \right)^{-1} A_{n_t} x^*_{n_t} + p^i_{n_t}$$

$$+ A'_{n_t} \left(\alpha^i_{\pi, \mathbf{n}^+_t} - k^i_{\pi, \mathbf{n}^+_t} \left(I + \sum_{i=1}^{2} (S^i_{n_t} + l^i_{n_t}) k^i_{\pi, \mathbf{n}^+_t} \right)^{-1} \sum_{i=1}^{2} \left((S^i_{n_t} + l^i_{n_t}) \alpha^i_{\pi, \mathbf{n}^+_t} + h^i_{n_t} \right) \right).$$

Collecting the coefficients of $x^*_{n_t}$ leads to the relations in (8.30). The remaining statements directly follow by using the terminal conditions. $\square$

One important concern with incentive strategies is their credibility. These strategies are said to be credible if it is in each player's best interest to implement her incentive strategy upon detecting a deviation by the other player from the agreed-upon solution. Otherwise, the threat is not believable, and a player can freely cheat on the agreement without any consequences. A formal definition of credibility follows.

Definition 8.5 The incentive equilibrium strategy $(\psi_i \in \Psi_i, i = 1, 2)$ is credible at $\mathbf{u}^* \in \mathbf{U}^1 \times \mathbf{U}^2$ if the following inequalities are satisfied:

$$W_1(x_0, (\psi_1(\mathbf{u}^2), \mathbf{u}^2)) \geq W_1(x_0, (\mathbf{u}^{1*}, \mathbf{u}^2)), \quad \forall \mathbf{u}^2 \in \mathbf{U}^2,$$
$$W_2(x_0, (\mathbf{u}^1, \psi_2(\mathbf{u}^1))) \geq W_2(x_0, (\mathbf{u}^1, \mathbf{u}^{2*})), \quad \forall \mathbf{u}^1 \in \mathbf{U}^1.$$

Note that the above definition characterizes the credibility of the equilibrium strategies for any possible deviation in the set $\mathbf{U}^1 \times \mathbf{U}^2$.

In order to push the computations and analysis farther, we need to assume a particular functional form for the incentive strategies. As we are focusing on linear-quadratic games, it is intuitive to suppose that these strategies have the following simple linear form:

$$\psi_i(u^j_{n_t}) = u^{i*}_{n_t} + p^i_{n_t} \left(u^{j*}_{n_t} - u^j_{n_t} \right), \quad i, j = 1, 2, i \neq j, \tag{8.36}$$

where $p^i_{n_t}$ is the penalty coefficient to be determined optimally. Note that if Player j does implement her cooperative control, then the difference $u^{j*}_{n_t} - u^j_{n_t}$ would be equal to zero, and Player i implements her part of the agreement, that is, $u^{i*}_{n_t}$.

Remark 8.3 Although the adoption of linear strategies is standard in linear-quadratic dynamic games (LQDGs), this does not exclude the possibility of using nonlinear strategies. In some instances, such nonlinear strategies can be attractive, as they lead to (i) higher payoffs for at least one player (see, e.g., Başar 1974 in the context of the design of incentive games); (ii) a Pareto solution under some conditions (see Dockner and Van Long 1993 for an example in environmental economics), or credible incentive strategies in linear-state deterministic differential games (see Martín-Herrán and Zaccour 2005).

Solving the necessary optimality conditions given in (8.27) yields the values of the incentive control and costate variables, that is, $u_{n_t}^{i,I}$ and $\lambda_{n_t}^{i,I}$, on which we impose the equality $u_{n_t}^{i,I} = u_{n_t}^{i*}$. This implies that $u_{n_t}^{i*}$ must satisfy its associated condition given in (8.27), and moreover, we have

$$\psi_i(u_{n_t}^{j*}) = u_{n_t}^{i*}, \text{ for } i = 1, 2,$$

which simplifies the arguments in condition (8.27). Additionally, $\mathbf{u}^*$ satisfies the condition in (8.20) that characterizes the cooperative solution. Using Eqs. (8.20) and the simplified (8.27), one may establish the necessary conditions that must be satisfied by the incentive equilibrium strategies.

8.3.4   An Example

To illustrate in the most parsimonious way the steps involved in the determination of an incentive equilibrium and the verification of its credibility, we consider a simplified version of the example in Sect. 8.2.2.

Denote by $M = \{1, 2\}$ the set of players representing countries, and by $\mathbb{T} = \{0, 1, 2\}$ the set of periods. The event tree is given in Fig. 8.5. Suppose that the conditional probability of realization of any successor of any node is $\frac{1}{2}$. Consequently, we have probabilities $\pi_{n_1^1} = \pi_{n_1^2} = \frac{1}{2}$ in $t = 1$, and probabilities $\pi_{n_2^1} = \pi_{n_2^2} = \pi_{n_2^3} = \pi_{n_2^4} = \frac{1}{4}$ for $t = 2$.

Denote by $u_{n_t}^i$ the emissions of Player i, and assume that the revenues can be expressed by the following function:

Fig. 8.5 Event tree

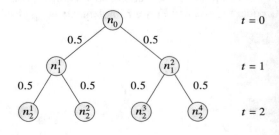

$$f_i\left(u^i_{n_t}\right) = \alpha_i u^i_{n_t} - \frac{1}{2}\beta_i\left(u^i_{n_t}\right)^2, \tag{8.37}$$

where α_i, β_i are positive parameters. We constrain the revenues to be nonnegative for any $n_t \in \mathbf{n}_0^{++} \setminus \mathbf{n}_T$ and any player i.

Denote by $u_{n_t} = (u^1_{n_t}, u^2_{n_t})$ the vector of countries' emissions at node $n_t \in \mathbf{n}_t$, $t \in \mathbb{T} \setminus \{T\}$, and by x_{n_t} the stock of pollution at this node. The evolution of this stock is governed by the following difference equation:

$$x_{n_t} = \sum_{i \in M} u^i_{n_t^-} + (1 - \delta_{n_t^-})x_{n_t^-}, \ n_t \in \mathbf{n}_0^{++}\setminus\{n_0\}, \ x_{n_0} = x_0, \tag{8.38}$$

where δ_{n_t} $(0 < \delta_{n_t} < 1)$ is a stochastic rate of pollution absorption by Mother Nature at node n_t. We suppose that δ_{n_t} can take two possible values, namely, $\delta_{n_t} \in \{\underline{\delta}, \overline{\delta}\}$, with $\underline{\delta} < \overline{\delta}$. The root node n_0 and nodes n^1_1, n^1_2, n^3_2 have the low rate $\underline{\delta}$ of pollution absorption, and nodes n^1_2, n^2_2, n^4_2 have the high level $\overline{\delta}$ of pollution absorption.

The environmental damage cost is given by $D_i(x_{n_t}) = \frac{1}{2}d_i x^2_{n_t}$, $i \in M$, where d_i is a strictly positive parameter. Denote by $\rho \in (0, 1)$ the discount factor. Assume that Player $i \in M$ maximizes the following objective functional:

$$W_i(x_0, \mathbf{u}) = \sum_{t=0}^{T-1} \rho^t \sum_{n_t \in \mathbf{n}_t} \pi_{n_t}\left(f_i\left(u^i_{n_t}\right) - D_i(x_{n_t})\right) + \rho^T \sum_{n_T \in \mathbf{n}_T} \pi_{n_T}\Phi_i(n_T, x_{n_T}),$$

subject to (8.38) and $u^i_{n_t} \geq 0$ for all $i \in M$ and any $n_t \in \mathbf{n}_t$, $t = 0, 1$. Suppose that the payoff function of Player i at a terminal node n_T is given by

$$\Phi_i(n_T, x_{n_T}) = -\gamma_i x_{n_T}.$$

Substituting for $\Phi_i(n_T, x_{n_T})$, $f_i\left(u^i_{n_t}\right)$ and $D_i(x_{n_t})$ by their values, we get

$$W_i(x_0, \mathbf{u}) = \sum_{t=0}^{T-1} \rho^t \sum_{n_t \in \mathbf{n}_t} \pi_{n_t}\left(\alpha_i u^i_{n_t} - \frac{1}{2}\beta_i\left(u^i_{n_t}\right)^2 - \frac{1}{2}d_i x^2_{n_t}\right) - \rho^T \sum_{n_T \in \mathbf{n}_T} \pi_{n_T}\gamma_i x_{n_T}.$$

We use the following parameter values in the numerical illustration:

$$x_0 = 10, \ \rho = 0.9, \ \underline{\delta} = 0.1, \ \overline{\delta} = 0.3.$$
$$\alpha_1 = 50, \ \alpha_2 = 40, \ \gamma_1 = 1.0, \ \gamma_2 = 0.9,$$
$$d_1 = 0.05, \ d_2 = 0.04,$$
$$\beta_1 = \beta_2 = 1.$$

Table 8.8 Cooperative control and state trajectories

Node	Controls u^{1*}	u^{2*}	State x^*
n_0	41.41	31.41	10.00
n_1^1	48.29	38.29	81.81
n_1^2	48.29	38.29	81.81
n_2^1, n_2^2			160.21
n_2^3, n_2^4			143.85

Now, we execute the different steps involved in the determination of an incentive equilibrium and the conditions for its credibility.

Step 1: Maximization of the joint payoffs. Applying Proposition 8.1, we get the optimal control and state values given in Table 8.8.

The cooperative players' payoffs in the game are 2060.51 and 1217.83, respectively.

Step 2: Determination of the penalty coefficients. Assuming that the incentive strategies are linear, as defined in (8.36), which take the form

$$u_{n_t}^i = \psi^i(u_{n_t}^{3-i}) = u_{n_t}^{i*} + p_{n_t}^i\left(u_{n_t}^{3-i} - u_{n_t}^{3-i*}\right),$$
$$u_{n_t}^{3-i} = \psi^{3-i}(u_{n_t}^i) = u_{n_t}^{3-i*} + p_{n_t}^{3-i}\left(u_{n_t}^i - u_{n_t}^{i*}\right),$$

then Player $i = 1, 2$ solves the following optimization problem, where a player assumes that the other is implementing her incentive strategy:

$$\max W_i(x_0, \mathbf{u}) \text{ subject to}$$
$$x_{n_t} = (1 - \delta_{n_t^-})x_{n_t^-} + u_{n_t^-}^i + u_{n_t^-}^{3-i*} + p_{n_t^-}^{3-i}\left(u_{n_t^-}^i - u_{n_t^-}^{i*}\right), \quad t = 1, 2, \ x_0 \text{ given},$$

where

$$W_i(x_0, \mathbf{u}) = \sum_{t=0}^{T-1} \rho^t \sum_{n_t \in \mathbf{n}_t} \pi_{n_t}\left(\alpha_i u_{n_t}^i - \frac{1}{2}\beta_i\left(u_{n_t}^i\right)^2 - \frac{1}{2}d_i x_{n_t}^2\right) - \rho^T \sum_{n_T \in \mathbf{n}_T} \pi_{n_T}\gamma_i x_{n_T}.$$

We can rewrite the state dynamics as follows:

$$x_{n_t} = (1 - \delta_{n_t^-})x_{n_t^-} + (1 + p_{n_t^-}^{3-i})u_{n_t^-}^i + u_{n_t^-}^{3-i*} - p_{n_t^-}^{3-i}u_{n_t^-}^{i*}.$$

In this problem, only Player i's control appears in the objective.

Player i's Hamiltonian is given by

$$
\begin{aligned}
H_i(n_t, \lambda^i_{\mathbf{n}^+_t}, x_{n_t}, u_{n_t}) = {} & \alpha_i u^i_{n_t} - \frac{1}{2}\beta_i\left(u^i_{n_t}\right)^2 - \frac{1}{2}d_i x^2_{n_t} \\
& + \rho \sum_{\nu \in \mathbf{n}^+_t} \pi^\nu_{n_t} \lambda^i_\nu \left((1-\delta_{n_t})x_{n_t} + (1+p^{3-i}_{n_t})u^i_{n_t} + u^{3-i*}_{n_t} - p^{3-i}_{n_t} u^{i*}_{n_t}\right),
\end{aligned}
$$

where $u^{i*}_{n_t}$, $i=1,2$, is a cooperative control represented in Table 8.8, and $\lambda^i_{n_t}$, $t = 0, \ldots, T$ is the costate variable appended by Player i to the state dynamics.

The costate variables satisfy the system

$$
\lambda^i_{n_t} = -d_i x_{n_t} + \rho(1-\delta_{n_t}) \sum_{\nu \in \mathbf{n}^+_t} \pi^\nu_{n_t} \lambda^i_\nu, \quad t = 0, 1,
$$

$$
\lambda^i_{n_T} = -\gamma_i, \quad T = 2.
$$

Minimizing the Hamiltonian function, we obtain the control

$$
u^i_{n_t} = \frac{1}{\beta_i}\left(\alpha_i + \rho(1+p^{3-i}_{n_t}) \sum_{\nu \in \mathbf{n}^+_t} \pi^\nu_{n_t} \lambda^i_\nu\right).
$$

Equating the control $u^i_{n_t}$ to the cooperative one $u^{i*}_{n_t}$, we find $p^{3-i}_{n_t}$:

$$
p^{3-i}_{n_t} = \frac{\beta_i u^{i*}_{n_t} - \alpha_i}{\rho \sum_{\nu \in \mathbf{n}^+_t} \pi^\nu_{n_t} \lambda^i_\nu} - 1.
$$

Substituting the cooperative controls and solving the dynamic system with respect to $\lambda^i_{n_t}$, $i=1,2$, we obtain

$$
p^{3-i}_{n_t} = \frac{\gamma_{3-i}}{\gamma_i} \quad \text{for any } n_t.
$$

For our numerical example we have

$$
p^1_{n_t} = p^1 = 1.11, \quad p^2_{n_t} = p^2 = 0.90, \text{ for all } n_t,
$$

and the following incentive strategies:

$$
\begin{aligned}
u^1_{n_t} &= u^{1*}_{n_t} + 1.11\left(u^2_{n_t} - u^{2*}_{n_t}\right), \\
u^2_{n_t} &= u^{2*}_{n_t} + 0.90\left(u^1_{n_t} - u^{1*}_{n_t}\right).
\end{aligned}
$$

In this example, the product of the penalty coefficients at each node is equal to one, i.e., $p_{n_t}^1 \times p_{n_t}^2 = 1$ for all n_t. These values ensure that each player's best reply to the other incentive strategy is to select the optimal control computed in the first step. Note that the result that the penalty terms do not vary across nodes is a special case due to the fact that the uncertainty only affects the decay rate in the dynamics.

Step 3: Determination of the credibility conditions. The incentive strategies are credible if the following two conditions hold:

$$
\begin{aligned}
W_1\big(x_0, (\psi^1(\mathbf{u}^2), \mathbf{u}^2)\big) &\geq W_1\big(x_0, (\mathbf{u}^{1*}, \mathbf{u}^2)\big), \quad \forall \mathbf{u}^2 \in \mathbf{U}^2, \\
W_2\big(x_0, (\mathbf{u}^1, \psi^2(\mathbf{u}^1))\big) &\geq W_2\big(x_0, (\mathbf{u}^1, \mathbf{u}^{2*})\big), \quad \forall \mathbf{u}^1 \in \mathbf{U}^1.
\end{aligned}
$$

That is, if a player is cheated, she would be better off implementing her incentive strategy than sticking to the agreement. The above quantities for Player 1 are as follows:

$$
\begin{aligned}
W_1\big(x_0, (\psi^1(\mathbf{u}^2), \mathbf{u}^2)\big) = \sum_{t=0}^{1} \rho^t \sum_{n_t \in \mathbf{n}_t} \pi_{n_t}\Big(&\alpha_1\big(u_{n_t}^{1*} + p^1(u_{n_t}^2 - u_{n_t}^{2*})\big) \\
&- \frac{1}{2}\beta_1\big(u_{n_t}^{1*} + p^1(u_{n_t}^2 - u_{n_t}^{2*})\big)^2 - \frac{1}{2}d_1\hat{x}_{n_t}^2\Big) - \rho^2 \gamma_1 \sum_{n_T \in \mathbf{n}_T} \pi_{n_T}\hat{x}_{n_T},
\end{aligned}
$$

$$
\hat{x}_{n_t} = (1 - \delta_{n_t^-})\hat{x}_{n_t^-} + u_{n_t^-}^{1*} + p^1(u_{n_t^-}^2 - u_{n_t^-}^{2*}) + u_{n_t^-}^2, \; \hat{x}_{n_0} = x_0,
$$

$$
\begin{aligned}
W_1\big(x_0, (\mathbf{u}^{1*}, \mathbf{u}^2)\big) = \sum_{t=0}^{1} \rho^t \sum_{n_t \in \mathbf{n}_t} \pi_{n_t}\Big(&\alpha_1 u_{n_t}^{1*} - \frac{1}{2}\beta_1(u_{n_t}^{1*})^2 - \frac{1}{2}d_1\tilde{x}_{n_t}^2\Big) \\
&- \rho^2 \gamma_1 \sum_{n_T \in \mathbf{n}_T} \pi_{n_T}\tilde{x}_{n_T},
\end{aligned}
$$

$$
\tilde{x}_{n_t} = (1 - \delta_{n_t^-})\tilde{x}_{n_t^-} + u_{n_t^-}^{1*} + u_{n_t^-}^2, \; \tilde{x}_{n_0} = x_0.
$$

The credibility condition is satisfied for Player 1 if

$$
\begin{aligned}
W_1\big(x_0, (\psi^1(\mathbf{u}^2), \mathbf{u}^2)\big) - W_1\big(x_0, (\mathbf{u}^{1*}, \mathbf{u}^2)\big) = \sum_{t=0}^{1} \rho^t \sum_{n_t \in \mathbf{n}_t} \pi_{n_t}\Big(&\alpha_1 p^1(u_{n_t}^2 - u_{n_t}^{2*}) \\
- \frac{1}{2}\beta_1 p^1(u_{n_t}^2 - u_{n_t}^{2*})(2u_{n_t}^{1*} + p^1(u_{n_t}^2 - u_{n_t}^{2*})) &- \frac{1}{2}d_1(\hat{x}_{n_t}^2 - \tilde{x}_{n_t}^2)\Big) \\
- \rho^2 \gamma_1 \sum_{n_T \in \mathbf{n}_T} \pi_{n_T}(\hat{x}_{n_T} - \tilde{x}_{n_T}) & \\
\geq 0, &
\end{aligned}
$$

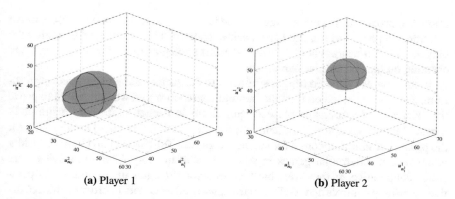

(a) Player 1   **(b)** Player 2

Fig. 8.6 Credibility conditions (in blue regions, incentive strategies are credible)

subject to state equations for $\hat{x}$ and $\tilde{x}$ given above. We can write the credibility condition for Player 2 in a similar way.

For Player 1, in the numerical example, the region where her incentive strategy is credible is

$$\Big\{ -1679.90 + 48.40u_{n_0}^2 - 0.70(u_{n_0}^2)^2 + 21.68u_{n_1}^2 - 0.28(u_{n_1}^2)^2 + 21.68u_{n_1^2}^2$$

$$- 0.28(u_{n_1^2}^2)^2 \ge 0, \ u_{n_0}^2 \in [0, 80], \ u_{n_1}^2 \in [0, 80], \ u_{n_1^2}^2 \in [0, 80] \Big\},$$

which is represented in Fig. 8.6a.

For Player 2, in the numerical example, the region where her incentive strategy is credible is

$$\Big\{ -1841.56 + 42.30u_{n_0}^1 - 0.46(u_{n_0}^1)^2 + 17.97u_{n_1}^1 - 0.18(u_{n_1}^1)^2 + 17.97u_{n_1^2}^1$$

$$- 0.18(u_{n_1^2}^1)^2 \ge 0, \ u_{n_0}^1 \in [0, 100], \ u_{n_1}^1 \in [0, 100], \ u_{n_1^2}^1 \in [0, 100] \Big\},$$

which is represented in Fig. 8.6b.

We should mention that the constraint on control variables $u_{n_t}^i$ is obtained from its nonnegativeness and the nonnegativeness of the revenues at any node ($u_{n_t}^i \le 2\alpha_i/\beta_i$).

8.4 Additional Readings

The literature in state-space dynamic games has dealt with the sustainability of the agreement over time along essentially two lines, namely, the design of time-consistent mechanisms (see Chaps. 4 and 7) and cooperative equilibria. In a cooperative equilibrium approach, as the name suggests, the idea is to make sure that the cooperative solution is an equilibrium, and hence, self-supported. This is achieved by letting the players use non-Markovian (or history-based) strategies that effectively deter any cheating on the cooperative agreement.

In this chapter we used trigger strategies to construct a subgame perfect equilibrium. This idea is based on the folk theorem approach which was suggested to prove the existence of a subgame perfect equilibrium in trigger strategies for infinitely repeated games (see, e.g., Aumann and Shapley 1994). See also Dutta (1995) for the proof of a similar theorem for stochastic games. Tolwinski et al. (1986) consider nonzero-sum differential games in strategies with memory. These strategies were called cooperative, as they were built as behavior strategies incorporating cooperative open-loop controls and feedback strategies used as threats in order to enforce the cooperative agreement. Chistyakov and Petrosyan (2013) examine the problem of strong strategic support for cooperative solutions in differential games, and Parilina and Tampieri (2018) state conditions for strategic support of cooperative solutions in stochastic games.

In Parilina and Zaccour (2015), the approximated equilibrium in DGPETs is constructed. In Parilina and Zaccour (2016), the approximated equilibrium is defined for DGPETs when players' payoffs are initially regularized by implementing the imputation distribution procedure. Linear-state dynamic games played over binary event trees are examined in Parilina et al. (2017), where the authors calculate the price of anarchy. In Wang et al. (2020), the authors examine conditions for strong strategic support of cooperative solutions in DGPETs.

Folk theorems are for infinite-horizon dynamic games. It is well known that enforcing cooperation in finite-horizon games is more difficult, not to say generally elusive. The reason is that, at the last stage, defection from the agreement is individually rational and this deviation cannot be punished. This clear-cut theoretical result has not always received empirical support, and in fact, experiments show that cooperation may be realized, at least partially, in finite-horizon games (see, e.g., Angelova et al. 2013). The literature has came out with different ways to cope with the difficulties in enforcing cooperation in finite-horizon dynamic games. Radner (1980) proposes the idea of ε-equilibrium and proves its existence for finitely repeated games. A third alternative can be used in the class of finitely repeated games when there exist more than one Nash equilibria in a one-shot game (Benoit and Krishna 1985).

Solan and Vieille (2003) consider a two-player game with perfect information that has no subgame perfect ε-equilibrium in pure strategies for a small ε, but has a subgame perfect ε-equilibrium in behavior strategies for any $\varepsilon > 0$. Flesch et al. (2014) also examine the existence of a subgame perfect ε-equilibrium in perfect information games with infinite horizon and Borel measurable payoffs. Flesch and Predtetchinski (2016) propose the concept of φ-tolerance equilibrium perfect-information games of infinite duration where φ is a function of history. A strategy profile is said to be a φ-tolerance equilibrium, if, for any history h, this strategy profile is a $\varphi(h)$-equilibrium in the subgame starting at h. This concept is close to the contemporaneous perfect

ε-equilibrium proposed in Mailath et al. (2005). In Flesch et al. (2019) it is proved that a game with individual upper semicontinuous payoff functions admits a subgame perfect ε-equilibrium for every $\varepsilon > 0$, in eventually pure-strategy profiles.

Another approach to endow the cooperative solution with an equilibrium property considered in this chapter is to use trigger strategies that credibly and effectively punish any player deviating from the agreement; see, e.g., Tolwinski et al. (1986), Haurie and Pohjola (1987), Dockner et al. (2000). The concept of incentive equilibrium assumes that trigger strategies may embody large discontinuities, i.e., a slight deviation from an agreed-upon path triggers harsh retaliation, generating a very different path from the agreed-upon one. An incentive equilibrium has the property that when both players implement their incentive strategies, the cooperative outcome is realized as an equilibrium. Therefore, no player should be tempted to deviate from the agreement during the course of the game, provided that the incentive strategies are credible. Ehtamo and Hämäläinen (1989, 1993) use linear incentive strategies in a dynamic resource game and demonstrate that such strategies are credible when deviations are not too large.

The concept of incentive strategies has of course been around for a long time in dynamic games (and economics), but it has often been often understood and used in a leader-follower (or principal-agent) sense. The idea is that the leader designs an incentive to induce the follower to reply in a certain way, which is often meant to be (only) in the leader's best interest, but may also be in the best collective interest (see the early contributions by Ho 1983; Başar 1984). In such a case, the incentive is one-sided. In this chapter, we focused on two-sided incentive strategies with the aim of implementing the joint optimization solution. Martín-Herrán and Zaccour (2005, 2009) characterize incentive strategies and their credibility for linear-state and linear-quadratic dynamic games (LQDG), but in a deterministic setting. The nonlinear incentive strategies to sustain an agreement over time are used in De Frutos and Martín-Herrán (2020). The incentive equilibrium in a dynamic game of pricing and advertising, modeling a distribution channel, consisting of a manufacturer and a retailer are examined in Jørgensen and Zaccour (2003). The incentive strategies in a dynamic closed-loop supply chain made up of one manufacturer and one retailer, with both players investing in a product recovery program are constructed in De Giovanni et al. (2016).

In this chapter we also considered an approach to compare cooperative and non-cooperative profits that is borrowed from algorithmic game theory (AGT), that is, the *price of anarchy* (PoA), which is a ratio of the players' reward in the worst equilibrium to that in the optimal solution (see Koutsoupias and Papadimitriou 1999). Depending on the context, the PoA can be computed in explicit form or not. In this last case, of particular interest is the determination of an upper bound that the PoA cannot exceed. For instance, it has been established that the PoA is less than or equal to $4/3$ in the problem of selfish routing with linear latency functions (Cole et al. 2006) and in a supply chain management problem with uniform demand (Perakis and Roels 2007).

Another measure of players' selfishness is proposed in Anshelevich et al. (2004) and is called the *price of stability* (PoS). In a profit maximization problem, the PoS is determined as a ratio of the players' reward in the optimal solution to that in the best equilibrium. When the optimal solution and the equilibrium are unique, which is the case here, the PoA and PoS coincide. Further, a typical assumption in AGT is that the players are symmetric. This assumption is methodologically interesting as it implies that the PoA only quantifies the loss in efficiency due to noncooperation (or the players' selfish behavior), and nothing else, e.g., the players' driving skills or differential travel costs.

The literature on PoA is vast, and we simply mention some references in those areas of operations research where the computation of the PoA has been an active topic. One of these is definitely transportation networks, where congestion may cause huge economic losses. In the famous Braess Paradox, congestion in networks is related to the selfish behavior of agents, and consequently, computing the PoA is the natural thing to do to assess the cost of lack of cooperation; see, e.g., Roughgarden and Tardos (2002), Roughgarden (2005), Cole et al. (2006).

In a one-manufacturer, one-retailer supply chain (called a dyad), it holds that individual optimization leads to a higher price to the consumer and a lower total profit than does joint optimization (see, e.g., the review in Ingene et al. 2012). The amount of losses in profits and the amount of losses in consumer surplus depend on the functional forms of the demand and cost functions. In supply chains exhibiting some competition at either the manufacturing or retailing level, the inefficiencies will depend on factors such as the degree of substitutability between products and between retail outlets. For examples of the determination of the PoA in different types of supply chains, see, e.g., Perakis and Roels (2007), Martínez-de Albéniz and Simchi-Levi (2009), Martínez-de Albéniz and Roels (2011), Perakis and Sun (2012). In a supply chain where the strategic interaction is vertical, the inefficiencies in the decentralized management case are due to double marginalization. In oligopolistic competition, the interaction is horizontal, with all firms competing in the same market, which leads to a lower price than under a monopoly. Also, the outcomes to producers depend on the type of competition, that is, in quantities à la Cournot or in prices à la Bertrand, and so is the PoA. See, e.g., Immorlica et al. (2010), Farahat and Perakis (2011a, b) for examples of the determination of the PoA in oligopolistic games.

In general, the PoA is computed in one-shot games. Başar and Zhu (2011) develop the idea of PoA in differential games, with an emphasis on the role of the information structure, that is, what the players know when they make their decisions. Zyskowski and Zhu (2013) consider the price (as well as the variance) of anarchy in a stochastic differential game.

8.5 Exercises

Exercise 8.1 In the example in Sect. 8.2.2, the damage cost is given by the quadratic function $D_i(x_{n_t}) = \frac{1}{2}d_i x_{n_t}^2$, $i \in M$, where d_i is a strictly positive parameter. Redo the computations in the simplified case where this cost is linear, that is, $D_i(x_{n_t}) = d_i x_{n_t}$.

Compute the price of anarchy and discuss the result.

Exercise 8.2 Consider a market with $m \geq 2$ firms that compete in quantities over a finite horizon. A firm $i \in M$ selects its quantity $q^i_{n_t} \in \mathbb{R}_+$ to be produced at node n_t. The market price is sticky.[3] At node n_t, the price $p_{n_t} \in \mathbb{R}_+$ is defined by the following state equation:

$$p_{n_t} = s_{n^-_i} p_{n^-_i} + \left(1 - s_{n^-_i}\right)\left(a - b \sum_{i=1}^n q^i_{n^-_i}\right), \qquad (8.39)$$

where $s_{n^-_i} \in [0, 1]$, and a and b are positive constants. The first term in the right-hand side of (8.39) represents the inertia in price, while the second term reflects the price adjustment. The initial market price is given by $p_{n_0} = p_0$. The profit of firm i at a nonterminal node n_t is given by

$$\phi_i(n_t, p_{n_t}, q_{n_t}) = p_{n_t} q^i_{n_t} - \frac{c_i}{2}\left(q^i_{n_t}\right)^2,$$

where parameter $c_i > 0$ represents firm i's unit cots. The profit at a terminal node n_T is zero for all players.

Let the game be played over a binary event tree depicted in Fig. 8.1. The conditional probability of realization of the left-hand (right-hand) node is 0.4 (0.6). Let $a = 100$, $b = 0.9$, and $p_0 = 10$. The variation of value s over the event tree is given in the table:

Node	n_0	n^1_1	n^2_1	n^1_2	n^2_2	n^3_2	n^4_2
s_{n_t}	0.30	0.20	0.50	0.30	0.15	0.24	0.78

1. Let $m = 3$ and the values of c_j be given by $c_1 = 1.8$, $c_2 = 4.8$, $c_3 = 3.4$. Find the Nash equilibrium, the cooperative strategy profile, and corresponding state trajectories. Construct the subgame perfect ε-equilibrium using the results of Theorem 8.1 in Sect. 8.2. Find the value of $\tilde{\varepsilon}$ defined by (8.10). Calculate the cost of cooperation in this game.
2. Let $m = 2$ and the values of c_j be given by $c_1 = 1.8$, $c_2 = 4.8$. Find the Nash equilibrium and the cooperative strategy profile and corresponding state trajectories. Define the incentive equilibrium and verify the credibility of the incentive strategies of the players in a linear form.
3. Let $m = 3$ and the values of c_j be given by $c_1 = 1.8$, $c_2 = 4.8$, $c_3 = 3.4$. Calculate the price of anarchy in this game. How does the price of anarchy change if parameter c_1 changes values from 1.8 to 3.8?

Exercise 8.3 Suppose that, in the example in Sect. 8.3.4, the decay parameter value $\delta_{n_t} = \delta$ for all $n_t \in \mathbf{n}_t$, $t \in \mathbb{T}$. Let the damage cost of Player i be given by $D_i(x_{n_t}) =$

---

[3] The model in this exercise is a stochastic and discrete-time version of the sticky-price continuous-time model in Fershtman and Kamien (1987). For a discrete-time version, see Parilina and Sedakov (2020).

Fig. 8.7 Event tree

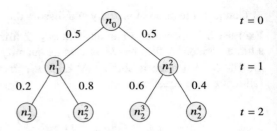

$\frac{1}{2} d^i_{n_t}(x_{n_t})^2$, where $d^i_{n_t} \in \{\underline{d}^i_{n_t}, \bar{d}^i_{n_t}\}$, where $\underline{d}^i_{n_t} = 0.3$ and $\bar{d}^i_{n_t} = 0.45$ for $i = 1, 2$ and $n_t \in \mathbf{n}_t, t \in \mathbb{T} \setminus \{T\}$.

Determine an incentive equilibrium with linear incentive strategies. Comment on the penalty terms at the different nodes.

Exercise 8.4   Consider the linear-quadratic dynamic game defined in (8.17)–(8.18). The event tree is shown in Fig. 8.7. The probabilities of passing through nodes are $\pi_{n_1^1} = \pi_{n_1^2} = 0.50$ in $t = 1$, and probabilities $\pi_{n_2^1} = 0.10$, $\pi_{n_2^2} = 0.40$, $\pi_{n_2^3} = 0.30$, $\pi_{n_2^4} = 0.20$ for $t = 2$.

The parameter values at the different nodes are given by

Node	r^1	r^2	c^1	c^2	d^1	d^2	g^1	g^2	k
n_0	5.00	6.50	0.70	1.00	0.30	0.50	1.00	1.50	−0.30
n_1^1	5.50	7.48	0.72	1.08	0.30	0.53	1.04	1.64	−0.33
n_1^2	4.50	5.53	0.70	0.92	0.30	0.48	0.96	1.37	−0.27
n_2^1	6.05	8.60	0.74	1.17	0.31	0.55	1.08	1.78	−0.36
n_2^2	4.95	6.35	0.70	0.99	0.30	0.50	0.99	1.49	−0.30
n_2^3	4.95	6.35	0.70	0.99	0.30	0.50	0.99	1.49	−0.30
n_2^4	4.05	4.70	0.66	0.85	0.29	0.45	0.99	1.24	−0.24

1. Compute an incentive equilibrium assuming that the players implement linear incentive strategies.
2. Determine the conditions under which this equilibrium is credible.

References

Angelova, V., Bruttel, L. V., Güth, W., and Kamcke, U. (2013). Can subgame perfect equilibrium threats foster cooperation? An experimental test of finite-horizon folk theorems. *Economic Inquiry*, 51(2):1345–1356.

Anshelevich, E., Dasgupta, A., Kleinberg, J., Tardos, E., Wexler, T., and Roughgarden, T. (2004). The price of stability for network design with fair cost allocation. In *45th Annual IEEE Symposium on Foundations of Computer Science*, pages 295–304.

Aumann, R. J. and Shapley, L. S. (1994). Long-term competition—a game-theoretic analysis. In Megiddo, N., editor, *Essays in Game Theory: In Honor of Michael Maschler*, pages 1–15, New York, NY. Springer New York.

Başar, T. and Zhu, Q. (2011). Prices of anarchy, information, and cooperation in differential games. *Dynamic Games and Applications*, 1:50–73.

Başar, T. (1974). A counterexample in linear-quadratic games: Existence of nonlinear Nash solutions. *Journal of Optimization Theory and Applications*, 14(4):425–430.

Başar, T. (1984). Affine incentive schemes for stochastic systems with dynamic information. *SIAM Journal on Control and Optimization*, 22(2):199–210.

Benoit, J.-P. and Krishna, V. (1985). Finitely repeated games. *Econometrica*, 53(4):905–22.

Chistyakov, S. and Petrosyan, L. (2013). Strong strategic support of cooperative solutions in differential games. *Annals of the International Society of Dynamic Games*, 12:99–107.

Cole, R., Dodis, Y., and Roughgarden, T. (2006). How much can taxes help selfish routing? *Journal of Computer and System Sciences*, 72(3):444–467. Network Algorithms 2005.

De Frutos, J. and Martín-Herrán, G. (2020). Non-linear incentive equilibrium strategies for a transboundary pollution differential game. In Pineau, P.-O., Sigué, S., and Taboubi, S., editors, *Games in Management Science: Essays in Honor of Georges Zaccour*, pages 187–204, Cham. Springer International Publishing.

De Giovanni, P., Reddy, P. V., and Zaccour, G. (2016). Incentive strategies for an optimal recovery program in a closed-loop supply chain. *European Journal of Operational Research*, 249(2):605–617.

Dockner, E. J., Jørgensen, S., Van Long, N., and Sorger, G. (2000). *Differential Games in Economics and Management Science*. Cambridge University Press.

Dockner, E. J. and Van Long, N. (1993). International pollution control: Cooperative versus noncooperative strategies. *Journal of Environmental Economics and Management*, 25(1):13–29.

Dutta, P. (1995). A folk theorem for stochastic games. *Journal of Economic Theory*, 66(1):1–32.

Ehtamo, H. and Hämäläinen, R. P. (1989). Incentive strategies and equilibria for dynamic games with delayed information. *Journal of Optimization Theory and Applications*, 63(3):355–369.

Ehtamo, H. and Hämäläinen, R. P. (1993). A cooperative incentive equilibrium for a resource management problem. *Journal of Economic Dynamics and Control*, 17:659–678.

Engwerda, J. (2005). *LQ dynamic optimization and differential games*. John Wiley & Sons.

Farahat, A. and Perakis, G. (2011a). On the efficiency of price competition. *Operations Research Letters*, 39(6):414–418.

Farahat, A. and Perakis, G. (2011b). Technical note–a comparison of Bertrand and Cournot profits in oligopolies with differentiated products. *Operations Research*, 59(2):507–513.

Fershtman, C. and Kamien, M. I. (1987). Dynamic duopolistic competition with sticky prices. *Econometrica*, 55(5):1151–1164.

Flesch, J., Herings, P. J.-J., Maes, J., and Predtetchinski, A. (2019). Individual upper semicontinuity and subgame perfect ε-equilibria in games with almost perfect information. *Economic Theory*.

Flesch, J., Kuipers, J., Mashiah-Yaakovi, A., Schoenmakers, G., Shmaya, E., Solan, E., and Vrieze, K. (2014). Non-existence of subgame-perfect ε-equilibrium in perfect information games with infinite horizon. *International Journal of Game Theory*, 43(4):945–951.

Flesch, J. and Predtetchinski, A. (2016). On refinements of subgame perfect ε-equilibrium. *International Journal of Game Theory*, 45(3):523–542.

Haurie, A. and Pohjola, M. (1987). Efficient equilibria in a differential game of capitalism. *Journal of Economic Dynamics and Control*, 11(1):65–78.

Ho, Y. C. (1983). On incentive problems. *Systems & Control Letters*, 3(2):63–68.

Immorlica, N., Markakis, E., and Piliouras, G. (2010). Coalition formation and price of anarchy in cournot oligopolies. In *Proceedings of the 6th International Conference on Internet and Network Economics*, WINE'10, page 270–281, Berlin, Heidelberg. Springer-Verlag.

Ingene, C. A., Taboubi, S., and Zaccour, G. (2012). Game-theoretic coordination mechanisms in distribution channels: Integration and extensions for models without competition. *Journal of Retailing*, 88(4):476–496.

Jørgensen, S., Martín-Herrán, G., and Zaccour, G. (2010). Dynamic games in the economics and management of pollution. *Environmental Modeling & Assessment*, 15:433–467.

Jørgensen, S. and Zaccour, G. (2003). Channel coordination over time: incentive equilibria and credibility. *Journal of Economic Dynamics and Control*, 27(5):801–822.

Kanani Kuchesfehani, E. and Zaccour, G. (2016). Incentive equilibrium strategies in dynamic games played over event trees. *Automatica*, 71:50–56.

Koutsoupias, E. and Papadimitriou, C. (1999). Worst-case equilibria. In Meinel, C. and Tison, S., editors, *STACS 99*, pages 404–413, Berlin, Heidelberg. Springer Berlin Heidelberg.

Mailath, G., Postlewaite, A., and Samuelson, L. (2005). Contemporaneous perfect epsilon-equilibria. *Games and Economic Behavior*, 53(1):126–140.

Martín-Herrán, G. and Zaccour, G. (2005). Credibility of incentive equilibrium strategies in linear-state differential games. *Journal of Optimization Theory and Applications*, 126(2):367–389.

Martín-Herrán, G. and Zaccour, G. (2009). Credible linear-incentive equilibrium strategies in linear-quadratic differential games. In Pourtallier, O., Gaitsgory, V., and Bernhard, P., editors, *Advances in Dynamic Games and Their Applications: Analytical and Numerical Developments*, pages 1–31, Boston. Birkhäuser Boston.

Martínez-de Albéniz, V. and Roels, G. (2011). Competing for shelf space. *Production and Operations Management*, 20(1):32–46.

Martínez-de Albéniz, V. and Simchi-Levi, D. (2009). Competition in the supply option market. *Operations Research*, 57(5):1082–1097.

Parilina, E. M. and Sedakov, A. (2020). Stable Coalition Structures in Dynamic Competitive Environment. In Pineau, P.-O., Sigué, S., and Taboubi, S., editors, *Games in Management Science: Essays in Honor of Georges Zaccour*, pages 381–396, Cham. Springer International Publishing.

Parilina, E. M., Sedakov, A., and Zaccour, G. (2017). Price of anarchy in a linear-state stochastic dynamic game. *European Journal of Operational Research*, 258(2):790–800.

Parilina, E. M. and Tampieri, A. (2018). Stability and cooperative solution in stochastic games. *Theory and Decision*, 84(4):601–625.

Parilina, E. M. and Zaccour, G. (2015). Approximated cooperative equilibria for games played over event trees. *Operations Research Letters*, 43(5):507–513.

Parilina, E. M. and Zaccour, G. (2016). Strategic support of node-consistent cooperative outcomes in dynamic games played over event trees. *International Game Theory Review*, 18(2).

Perakis, G. and Roels, G. (2007). The price of anarchy in supply chains: Quantifying the efficiency of price-only contracts. *Management Science*, 53(8):1249–1268.

Perakis, G. and Sun, W. (2012). Price of anarchy for supply chains with partial positive externalities. *Operations Research Letters*, 40(2):78–83.

Radner, R. (1980). Collusive behavior in noncooperative epsilon-equilibria of oligopolies with long but finite lives. *Journal of Economic Theory*, 22(2):136–154.

Roughgarden, T. (2005). *Selfish Routing and the Price of Anarchy*. MIT Press.

Roughgarden, T. and Tardos, E. (2002). How bad is selfish routing? *Journal of the ACM*, 49(2):236–259.

Solan, E. and Vieille, N. (2003). Deterministic multi-player Dynkin games. *Journal of Mathematical Economics*, 39(8):911–929.

Tolwinski, B., Haurie, A., and Leitmann, G. (1986). Cooperative equilibria in differential games. *Journal of Mathematical Analysis and Applications*, 119(1):182–202.

Wang, L., Liu, C., Gao, H., and Lin, C. (2020). Strongly strategic support of cooperative solutions for games over event trees. *Operations Research Letters*, 48(1):61–66.

Zyskowski, M. and Zhu, Q. (2013). Price and variance of anarchy in mean-variance cost density-shaping stochastic differential games. In *52nd IEEE Conference on Decision and Control*, pages 1720–1725.

Index

Printed in the United States
by Baker & Taylor Publisher Services